CISM International Centre for Mechanical Sciences

Courses and Lectures

Volume 577

The series presents lecture notes, monographs, edited works and proceedings in the field of Mechanics, Engineering, Computer Science and Applied Mathematics. Purpose of the series is to make known in the international scientific and technical community results obtained in some of the activities organized by CISM, the International Centre for Mechanical Sciences.

More information about this series at http://www.springer.com/series/76

David J. Steigmann
Editor

The Role of Mechanics in the Study of Lipid Bilayers

Springer

Editor
David J. Steigmann
Faculty of Mechanical Engineering
University of California, Berkeley
Berkeley, CA
USA

ISSN 0254-1971 ISSN 2309-3706 (electronic)
CISM International Centre for Mechanical Sciences
Courses and Lectures
ISBN 978-3-319-56347-3 ISBN 978-3-319-56348-0 (eBook)
DOI 10.1007/978-3-319-56348-0

Library of Congress Control Number: 2017937266

Printed on acid-free paper

This Springer imprint is published by Springer Nature
The registered company is Springer International Publishing AG
The registered company address is: Gewerbestrasse 11, 6330 Cham, Switzerland

Preface

This volume consists of Lecture Notes based on lectures delivered at the Advanced Summer School entitled "The Role of Mechanics in the Study of Lipid Bilayers" held at the International Centre for Mechanical Sciences (CISM) in Udine, Italy, during the period July 11–15, 2016. The course was presented by six lecturers, from Germany, Mexico, Spain, the UK and the USA (2).

The purpose of the six chapters comprising the volume is to provide a state-of-the-art account of the continuum theory underpinning the mechanics and physics of lipid bilayers and its applications.

Chapter "Mechanics and Physics of Lipid Bilayers" outlines an approach to the theory of lipid bilayers through an appeal to three-dimensional liquid-crystal theory. This provides an over-arching framework that encompasses the classical theory while facilitating its extension to embrace nonstandard effects associated with lipid tilt and distension, dissipative processes involving flow and diffusion, and electromagnetic interactions.

Chapter "Elasticity and Hereditariness" is devoted to the study of the energetics of lipid membranes, the nature and origin of the line tension accompanying phase transitions, and the role played by viscoelastic effects.

Chapter "Lipid Membranes: From Self-Assembly to Elasticity" delves into the physical basis of lipid bilayer arrangements, their self-assembly and associated thermodynamics, their elastic moduli, and the physical origins of lipid tilt.

In Chapter "The Geometry of Fluid Membranes: Variational Principles, Symmetries and Conservation Laws" the theory of lipid membranes is cast in a variational and differential-geometric setting, facilitating a development of the notion of membrane stress and its role in the associated mathematics. Included here are the consequences of reparametrization invariance and Euclidean invariance and the systematic treatment of constraints.

Chapter "On the Computational Modeling of Lipid Bilayers Using Thin-shell Theory" is devoted to the numerical analysis of lipid membrane behavior by advanced finite element methods. This provides an opportunity, through several examples, to exhibit the potential of the theory of lipid bilayers to predict of the emergence of various geometric features such as filaments and buds.

Chapter "Onsager's Variational Principle in Soft Matter: Introduction and Application to the Dynamics of Adsorption of Proteins onto Fluid Membranes" describes a far-ranging investigation into Onsager's variational principle with applications to chemo-mechanical problems in soft matter and the dynamics of protein adsorption.

These chapters combine to provide a unique perspective on this important branch of bio-physics from the vantage point of mechanics and applied mathematics.

It is a pleasure to acknowledge the efforts of my colleagues, Profs. Arroyo, Deseri, Deserno, Guven and Sauer, for presenting their lectures and for preparing the chapters of this volume, and the students for attending the lectures and contributing to the discussions.

I particularly thank the Rector, officers and staff at CISM for their encouragement, enthusiasm, assistance, and warm hospitality, which were essential to the success of the School. I am also grateful to Prof. Paolo Serafini, Executive Editor of CISM, for his guidance and encouragement in the preparation of these lecture notes.

Berkeley, USA David J. Steigmann

Contents

Mechanics and Physics of Lipid Bilayers

David J. Steigmann

Abstract In this chapter we review recent work by the writer and coworkers on various aspects of the mechanics and physics of lipid bilayers. A framework for lipid bilayer surface, based on a dimension reduction procedure applied to three-dimensional liquid crystal theory, is reviewed in Sect. 1. This accommodates the non-standard effects of lipid distension and tilt. A special case of the general model in which tilt is suppressed but distension, and accompanying surface dilation, are permitted, is also derived. This is further specialized, in Sect. 2, to obtain a model of the classical type, due to Canham and Helfrich. Our approach facilitates understanding of the place of the classical theory, and its logical extensions, in a larger context. Section 3 provides a further development of the theory with surface dilation—reported here for the first time—to accommodate dissipative effects, including intra-membrane viscous flow and the diffusion of trans-membrane embedded proteins. This may be viewed as a theory of generalized capillarity, accounting for various higher order gradient effects of the Cahn–Hilliard type in the constitutive equations. A simpler variant of this model is described in Sect. 4, in which non-standard gradient effects are suppressed. This furnishes the simplest thermodynamically consistent extension of the classical theory to cover diffusion and viscosity. Finally, Sect. 5 is devoted to the electromechanical theory. This is limited to the simplest extension of the classical model to accommodate surface flexo-electricity and the coupling of surface shape with a polarization field. Restrictions on the latter, consistent with the three-dimensional electromechanical theory for liquid crystals, yield a relatively simple generalization of the classical theory appropriate for analyzing membrane response to a remote applied electric field.

D.J. Steigmann (✉)
Department of Mechanical Engineering, University of California,
6133 Etcheverry Hall, Mailstop 1740, Berkeley, CA 94720-1740, USA
e-mail: dsteigmann@berkeley.edu

D.J. Steigmann (ed.), *The Role of Mechanics in the Study of Lipid Bilayers*,
CISM International Centre for Mechanical Sciences 577,
DOI 10.1007/978-3-319-56348-0_1

1 A Model for Lipid Membranes with Tilt and Distension, Derived from Three-Dimensional Liquid Crystal Theory

1.1 Introduction

Research on the mechanics and physics of lipid bilayer membranes emphasizes the classical Canham/Helfrich theory (Canham 1970; Helfrich 1973) and its variants, in which the lipid molecules are constrained to remain fixed in length and orthogonal to the membrane surface as it deforms. In this framework the mechanical response is determined solely by the geometry of the membrane surface. The subject has therefore benefited from a close connection to the differential geometry of surfaces, as exemplified by the work of Guven (Capovilla and Guven 2002; Deserno et al. 2007; Müller et al. 2005a, b, 2007), described elsewhere in this book. Canham/Helfrich-type models are appropriate when the density of the lipids on the surface is sufficiently high as to promote alignment of the lipids with the surface normal. At lower values of lipid density it is possible for the lipids to *tilt* relative to the surface normal.

Models of the Canham/Helfrich type may be viewed in the framework of the nonlinear Kirchhoff–Love theory of elastic shells, whereas models that accommodate tilt belong to the Cosserat theory of shells (Naghdi 1972). Naturally, the latter models may be shown to subsume the former upon the introduction of appropriate kinematic constraints (Steigmann 1999a). Here we extract the relevant two-dimensional theory from the mechanics of three-dimensional liquid crystals, regarded as Cosserat fluids.

In this section we present a concise outline of three-dimensional liquid crystal theory (Ericksen 1961, 1962, 1976; Virga 1994). This is followed by a description of the dimension reduction procedure, which entails the identification of the leading-order term in the thickness-wise expansion of the energy of a thin liquid-crystal film. We show that the resulting expression may be refined by imposing a condition on the thickness-wise derivative of the director field that describes the lipid trajectories. This condition ensures that the two-dimensional model is energetically optimal. The resulting model, thus optimized, is effectively a Cosserat shell theory for fluid films. The model allows for lipid distension, in contrast to the more common theory of liquid crystals in which distension is suppressed and an associated Lagrange multiplier is incorporated. This is motivated by our interest in certain unconventional effects, associated with the gradient of surface dilation, that accompany lipid distension. Relevant theory is developed in Sect. 2.

Molecular dynamics simulations in progress (Mandadapu 2016) indicate, in contrast to conventional models, that lipid tilt is relevant near interfaces with local embedded structures such as trans-membrane proteins, where hydrophobic lipid–protein interactions require either localized distension or tilt to prevent the hydrophobic surfaces of these structures from coming into contact with the surrounding aqueous solution. Various plausible modes of interaction involving distension or tilt are depicted in Fig. 1. Initial efforts to model such interactions on the basis of continuum theory have yielded remarkably good quantitative predictions vis a vis MD simulations down to extremely small length scales (Rangamani et al. 2014). This

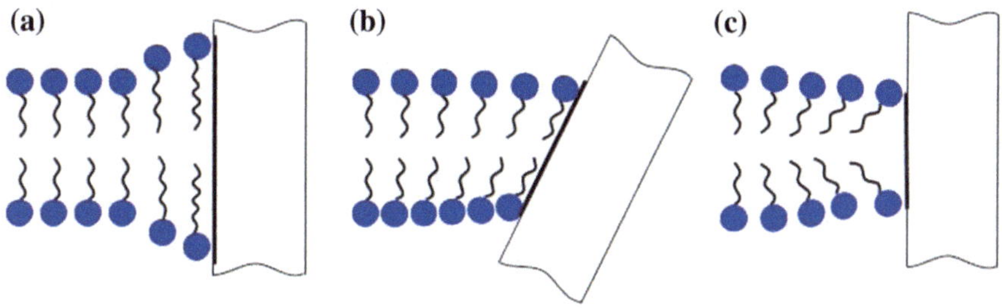

Fig. 1 Distension and tilt modes

fact provides impetus for the present emphasis on continued development of the continuum approach.

1.2 Liquid Crystal Theory

In this section we present a brief overview of the static theory of liquid crystals in three dimensions, to set the stage for the derivation of a two-dimensional model that follows.

The basic kinematic variables in the static theory are a deformation map, $\chi(\boldsymbol{x})$, taking the material point with position $\boldsymbol{x}$ in a reference configuration κ_r to its position $\boldsymbol{y} = \chi(\boldsymbol{x})$ in a current configuration κ_c; and a director field $\boldsymbol{d}$ associated with the oriented molecules of the liquid crystal. Each point $\boldsymbol{x} \in \kappa_r$ is associated with such a molecule, so that $\boldsymbol{d}$ may be regarded as field defined on κ_r. Roughly, the latter furnishes the direction field of the trajectories of these molecules and deforms with them. We express this either as a function of $\boldsymbol{x}$ or $\boldsymbol{y}$, the two alternatives being equivalent because of the presumed invertibility of $\chi(\cdot)$.

The deformation and director fields are assumed to be smooth. Let the gradient of the former be denoted by $\boldsymbol{F}$. Two gradients of $\boldsymbol{d}$ are needed: that with respect to $\boldsymbol{y}$ and that with respect to $\boldsymbol{x}$. These are denoted by $\boldsymbol{D}$ and $\boldsymbol{G}$, respectively, and connected by

$$\boldsymbol{G} = \boldsymbol{D}\boldsymbol{F}. \tag{1.1}$$

The aforementioned invertibility property implies that $\boldsymbol{F}$ is non-singular. If κ_r is a configuration which could in principle be occupied by the material, then we have the conventional restriction $\det \boldsymbol{F} > 0$. Accordingly, $\boldsymbol{D} = \boldsymbol{G}\boldsymbol{F}^{-1}$.

The director and deformation fields are kinematically independent; that is, the former is not to be confused with a conventional material vector that is convected by the deformation map, although—like a principal axis of strain—it may be regarded as a material vector in any *one* deformation. This point is essential in the theory as developed by Ericksen (1961), which requires the fields χ and $\boldsymbol{d}$ to be independently variable. Thus, suppose $\dot{\chi}$ and $\dot{d}$ are variations of these fields at a particular material point. Let $x(\varsigma)$ be the parametric representation of a given trajectory of oriented molecules of the liquid crystal in κ_r. These molecules occupy the trajectory $y(\varsigma) = \chi(x(\varsigma))$ in κ_c and we assume ς to be scaled such that $d = y'$, the derivative with respect to ς. The chain rule then yields $d = Fd_r$, where $d_r = x'$ is the direction field in κ_r. Kinematic independence amounts to the stipulation that $\dot{\chi}$ and $\dot{d}$ can be assigned independently. For example, if $\dot{\chi}$ vanishes then $\dot{d} = F\dot{d}_r$, which imposes no restriction on $\dot{d}$. However, if $\dot{d}$ vanishes then $\dot{d}_r = -F^{-1}\dot{F}F^{-1}d$, with no restriction on $\dot{\chi}$. Thus it is clear that kinematic independence entails the fact that d_r is allowed to vary, and hence that it is not a material vector with respect to a variable deformation map;. i.e., a map that can vary with a parameter (e.g., time). A similar situation exists with respect to principal axes of strain. These behave like material vectors in any one deformation, but are generally non-material when the deformation is allowed to evolve.

Following the originators of the subject, in the case of incompressibility we assume the existence of a function $U(d, D)$, representing the strain energy per unit volume of κ_c. This is presumed to be a quadratic function of D; i.e., (Virga 1994; Eq. (3.7))

$$U = l(d) + L(d) \cdot D + \tfrac{1}{2} D \cdot \mathcal{L}(d)[D], \tag{1.2}$$

in which l, L and $\mathcal{L}$ are scalar, second-order tensor and fourth-order tensor-valued functions, respectively, with $\mathcal{L} = \mathcal{L}^t$. Quadratic energies are justified by the fact that the characteristic length scale for variation of the director field is typically large compared to molecular dimensions. The gradient of the director field, non-dimensionalized by the only local scale available—the molecular length—is then sufficiently small that a quadratic approximation proves to be sufficient. Whether quadratic or not, the function U is subject to the restriction

$$U(d, D) = U(Qd, QDQ^t) \quad \text{for all rotations} \quad Q, \tag{1.3}$$

associated with the frame invariance of the constitutive response. A representation theorem for quadratic functions meeting this requirement, with d restricted to be a unit vector in accordance with the conventional theory, is given in (Virga 1994). This is the classical Frank formula for the energy of a liquid crystal.

Following Dafermos (1970), we assume that $U(d, \cdot)$ is convex. In the quadratic case this in turn is equivalent to the positivity of the tensor $\mathcal{L}(d)$. In the general case it implies that

$$A \cdot U_{DD}[A] > 0 \quad \text{for all} \quad A \neq 0, \tag{1.4}$$

where bold subscripts identify derivatives of scalar-valued functions with respect to tensors.

It is well known that convexity of the strain energy (with respect to the deformation gradient) is incompatible with frame invariance in conventional elasticity theory (Ciarlet 1993). Here there is no such conflict. In particular convexity is satisfied by imposing appropriate restrictions on the coefficients arising in the representation theorem for functions of the form (1.2) that satisfy (1.4) (Virga 1994).

Regarding balance laws, Ericksen's approach (1961; 1962; 1976) is based on a virtual-work expression derived from the energy

$$\mathcal{E} = \int_{\kappa_c} U dv \tag{1.5}$$

associated with the region κ_c occupied by the liquid crystal. His approach differs from others (e.g., (Virga 1994)) that account solely for variations in the director field, the deformation field remaining fixed. Instead, Ericksen fixes the material points and allows variations in the deformation and director fields. These variations may be subject to local constraints such as incompressibility ($\det \boldsymbol{F} = 1$) or director inextensibility ($|\boldsymbol{d}| = const.$), but are otherwise independent.

Ericksen's approach is facilitated by using the expression

$$\mathcal{E} = \int_{\kappa_r} U^* dV, \tag{1.6}$$

where

$$U^*(\boldsymbol{F}, \boldsymbol{d}, \boldsymbol{G}) = \theta U(\boldsymbol{d}, \boldsymbol{G}\boldsymbol{F}^{-1}) \quad \text{and} \quad \theta = \det \boldsymbol{F}, \tag{1.7}$$

is the strain energy per unit reference volume. For, the domain κ_r remains fixed under variations of the type considered.

It follows trivially from (1.7) that $U^*(\boldsymbol{F}, \boldsymbol{d}, \boldsymbol{G}) = U^*(\boldsymbol{F}\boldsymbol{H}, \boldsymbol{d}, \boldsymbol{G}\boldsymbol{H})$ for all $\boldsymbol{H}$ with $\det \boldsymbol{H} = 1$, implying that the material is fluid in the sense of Noll (Truesdell 1991). Thus the strain-energy function is invariant under the action of the proper unimodular group. In this work we will assume bulk incompressibility of the liquid crystal and thus impose the isochoricity constraint $\theta = 1$.

Further adjustment is needed if the director length is constrained to remain fixed, as in the conventional theory, although Ericksen (1961, 1962, 1976) does not confine attention to this case. In this work we allow for lipid distension and thus do not impose inextensibility.

1.3 Dimension Reduction

The three-dimensional theory is used to develop a two-dimensional model of lipid membranes in terms of an areal strain-energy function defined on a surface ω occupied by a layer of lipids. This layer is only two molecules thick, and the molecules are densely distributed on ω. While it is permissible to regard the membrane as a two-dimensional continuum; it is not a continuum in the usual three-dimensional sense. For this reason it is appropriate to propose a two-dimensional model at the outset, as advocated in (Ericksen 1979). However, appeal to three-dimensional theory furnishes a useful guide as to the variables that may be expected to figure in a two-dimensional theory of lipid membranes.

Let t be the total arclength of a fixed number of lipid molecules occupying a trajectory piercing the pre-image of a regular orientable material surface $\omega \in \kappa_c$ in a reference configuration κ_r. To extract a two-dimensional theory, we proceed to estimate the three-dimensional energy $\mathcal{E}$ in terms of t, and seek an areal surface energy W, independent of t, such that

$$\mathcal{E}/t = E + o(t)/t, \quad \text{as} \quad t \to 0, \tag{1.8}$$

where

$$E = \int_\omega W da. \tag{1.9}$$

The areal energy density thus furnishes the leading-order energy in the limit as t tends to zero. Our objective is an expression for the explicit form of W. To qualify as a useful approximation to the three-dimensional energy, the areal energy density should be well posed in the sense that (1.9) possesses a minimizer. This follows immediately from (1.8); any minimizer of the three-dimensional energy $\mathcal{E}$ necessarily minimizes E, *at leading order*. Later, we discuss local necessary conditions for this requirement to be satisfied.

Proceeding, we first parametrize the surface ω using a system of coordinates $\{\theta^\alpha\}$. Position on the surface is thus given by a function $\boldsymbol{r}(\theta^\alpha)$. This in turn induces the vectors $\boldsymbol{a}_\alpha = \boldsymbol{r}_{,\alpha}$ at the point on ω with coordinates θ^α. If these are not collinear, then they constitute a natural basis for the tangent plane $T_{\omega(\theta^\alpha)}$ at the considered point. Using these we define a positive definite metric $a_{\alpha\beta} = \boldsymbol{a}_\alpha \cdot \boldsymbol{a}_\beta$ and a dual metric $a^{\alpha\beta}$, given by the inverse of the matrix of metric components. These furnish a dual basis $\boldsymbol{a}^\alpha = a^{\alpha\beta}\boldsymbol{a}_\beta$ for $T_{\omega(\theta^\alpha)}$ and a surface orientation $\boldsymbol{n} = \frac{1}{2}\varepsilon^{\alpha\beta}\boldsymbol{a}_\alpha \times \boldsymbol{a}_\beta$, where $\varepsilon^{\alpha\beta} = e^{\alpha\beta}/\sqrt{a}$, $a = \det(a_{\alpha\beta})$ and $e^{\alpha\beta}$ is the usual permutation symbol ($e^{12} = -e^{21} = 1$, $e^{11} = e^{22} = 0$). The area measure on ω is $da = \sqrt{a}d\theta^1 d\theta^2$.

We regard the lipids as being permanently attached to ω and thus view the latter as a material surface in the sense of being convected by the deformation map $\boldsymbol{\chi}(\cdot)$. Thus we may regard $\{\theta^\alpha\}$ as a convected-coordinate system; that is, the coordinates may be regarded as maintaining fixed values while computing variational derivatives in the sense of Ericksen. Details on the convected-coordinate formalism are developed in Sect. 3.6 below.

To convert the volume integral in (1.5) to an expression involving integration over ω is it necessary to compute a liquid-crystal volume measure using a suitable parametrization of three-space. To this end we use a three-dimensional coordinate system $\{\theta^i\} = \{\theta^\alpha, \varsigma\}$, defined in an open neighborhood of a point on the surface, where ς parametrizes the trajectory of liquid crystal molecules that intersects ω at the point (θ^α). The surface ω is identified with the coordinate surface $\varsigma = 0$. Position of a point in this neighborhood is thus given by

$$\boldsymbol{y}(\theta^i) = \boldsymbol{r}(\theta^\alpha) + \boldsymbol{p}(\theta^\alpha, \varsigma); \quad \boldsymbol{p}(\theta^\alpha, 0) = \boldsymbol{0}, \tag{1.10}$$

where $\boldsymbol{p}$ is the position of this point relative to the surface point (θ^α). As before this induces the natural basis vectors $\boldsymbol{g}_i = \boldsymbol{y}_{,i}$; the (positive definite) metric $g_{ij} = \boldsymbol{g}_i \cdot \boldsymbol{g}_j$; the dual metric g^{ij}, whose matrix is the inverse of the metric components; and the dual basis $\boldsymbol{g}^i = g^{ij}\boldsymbol{g}_j$. The associated volume measure is

$$dv = \sqrt{g} d\theta^1 d\theta^2 d\varsigma = \mu d\varsigma da, \quad \text{where} \quad \mu = \sqrt{g/a} \tag{1.11}$$

and $g = \det(g_{ij}) = \left|\boldsymbol{g}_1 \times \boldsymbol{g}_2 \cdot \boldsymbol{g}_3\right|^2$ (Sokolnikoff 1964).

The parameter ς is scaled such that

$$\boldsymbol{d} = \boldsymbol{p}', \tag{1.12}$$

where $\boldsymbol{d}$ is the director field of the three-dimensional theory and $(\cdot)' = \partial(\cdot)/\partial\varsigma$. Thus the function $\boldsymbol{p}(\theta^\alpha, \cdot)$ generates the trajectory of liquid crystal molecules that intersect ω transversely at the point (θ^α). We assume the coordinate ς to be convected with the lipid trajectories. Accordingly, if ς measures arclength along these trajectories in the reference configuration κ_r, then $\varsigma \in (-t/2, t/2)$, where t is the total arclength of a lipid trajectory piercing the associated material surface in that configuration.

The energy may then be written as

$$\mathcal{E} = \int_\Omega \left(\int_{-t/2}^{t/2} \mu_r U^* d\varsigma \right) dA, \tag{1.13}$$

where Ω is the image of ω in the reference configuration κ_r and μ_r is the associated value of μ, given by $\mu_r = \sqrt{G/A}$ in which G and A are the values of g and a in κ_r and on Ω, respectively; and

$$U^* = \theta U, \quad \text{where} \quad \theta = \sqrt{g/G}, \tag{1.14}$$

is the strain energy per unit reference volume. Here we have used the easily derived result $\det \boldsymbol{F} = \sqrt{g/G}$. Of course $U^* = U$ in the case of bulk incompressibility, but it is useful to maintain a conceptual distinction between these energies.

For a given director distribution the interior integral in (1.13) may be regarded as a function of t. This is estimated for small t by combining a Taylor expansion with Leibniz' rule, yielding

$$\mathcal{E} = t \int_\Omega W^* dA + o(t), \tag{1.15}$$

where

$$W^* = (\mu_r U^*)_{|\Omega}, \tag{1.16}$$

the restriction of the integrand to the surface Ω. This is the leading-order areal energy density on Ω. This in turn yields (1.9) with

$$JW = W^*, \tag{1.17}$$

where

$$J = \sqrt{a/A} \tag{1.18}$$

is the surface dilation. Combining these results yields $W = (U^*\sqrt{G/a})_{|\Omega} = (\mu U^*\sqrt{G/g})_{|\Omega}$ and therefore

$$W = (\mu U)_{|\omega} \tag{1.19}$$

in the case of bulk incompressibility.

We make this explicit by using (1.10) to obtain

$$\boldsymbol{g}_\alpha = \boldsymbol{a}_\alpha + \boldsymbol{p}_{,\alpha}, \quad \boldsymbol{g}_3 = \boldsymbol{d} \quad \text{and} \quad \sqrt{g} = \left|(\boldsymbol{a}_1 + \boldsymbol{p}_{,1}) \times (\boldsymbol{a}_2 + \boldsymbol{p}_{,2}) \cdot \boldsymbol{d}\right|. \tag{1.20}$$

Sufficient regularity is assumed in the trajectory field of the liquid-crystal molecules to support the thickness-wise Taylor expansion

$$\boldsymbol{p} = \varsigma\boldsymbol{\delta}(\theta^\alpha) + \tfrac{1}{2}\varsigma^2\boldsymbol{\eta}(\theta^\alpha) + ..., \tag{1.21}$$

where the *independent* vector fields $\boldsymbol{\delta}$ and $\boldsymbol{\eta}$, respectively, are the restrictions of $\boldsymbol{d}$ and $\boldsymbol{d}'$ to the surface ω. Using (1.11) and (1.20) we then obtain

$$W = |\boldsymbol{n} \cdot \boldsymbol{\delta}|\, U_{|\omega}. \tag{1.22}$$

To make this explicit we require the restriction to ω of the three-dimensional director gradient $\boldsymbol{D}$. Expressed in the $\{\theta^i\}$ system, this is

$$\boldsymbol{D} = \boldsymbol{d}_{,i} \otimes \boldsymbol{g}^i. \tag{1.23}$$

This in turn requires the restrictions $\boldsymbol{g}^i_{|\omega}$. To construct these we note, from (1.20) and (1.21), that $\{\boldsymbol{g}_\alpha\}_{|\omega} = \{\boldsymbol{a}_\alpha\}$. Combining this with $\boldsymbol{g}^i \cdot \boldsymbol{g}_j = \delta^i_j$ (the Kronecker delta) and $\boldsymbol{g}^3 = grad\varsigma = \alpha\boldsymbol{n}$ for some $\alpha \in \mathbb{R}$ (because ω is a surface on which $\varsigma = const.$), we conclude that $\boldsymbol{g}^3_{|\omega} = (\boldsymbol{n} \cdot \boldsymbol{\delta})^{-1}\boldsymbol{n}$. Then, in the case of bulk incompressibility we have

$$W = |\boldsymbol{n} \cdot \boldsymbol{\delta}|\, U(\boldsymbol{\delta}, D_{|\omega}), \tag{1.24}$$

with

$$\boldsymbol{D}_{|\omega} = \boldsymbol{\delta}_{,\alpha} \otimes \boldsymbol{g}^{\alpha} + (\boldsymbol{n} \cdot \boldsymbol{\delta})^{-1} \eta \otimes \boldsymbol{n}, \tag{1.25}$$

wherein all terms are evaluated on ω.

To reduce this we resolve $\boldsymbol{g}^{\alpha}$ in the basis $\{\boldsymbol{a}^{\beta}, \boldsymbol{n}\}$ and use $\{\boldsymbol{g}_{\alpha}\}_{|\omega} = \{\boldsymbol{a}_{\alpha}\}$ with $\delta^{\alpha}_{\beta} = \boldsymbol{g}^{\alpha} \cdot \boldsymbol{g}_{\beta} = \boldsymbol{g}^{\alpha} \cdot \boldsymbol{a}_{\beta}$; thus,

$$\boldsymbol{g}^{\alpha} = \boldsymbol{a}^{\alpha} + (\boldsymbol{g}^{\alpha} \cdot \boldsymbol{n})\boldsymbol{n}. \tag{1.26}$$

Next, we resolve $\boldsymbol{n}$ in the basis $\{\boldsymbol{g}_i\}$. Recalling that $\boldsymbol{g}_3 = \boldsymbol{\delta}$ and $\boldsymbol{n} \cdot \boldsymbol{g}^3 = (\boldsymbol{n} \cdot \boldsymbol{\delta})^{-1}$, we derive

$$\boldsymbol{n} = (\boldsymbol{n} \cdot \boldsymbol{g}^i)\boldsymbol{g}_i = (\boldsymbol{n} \cdot \boldsymbol{g}^{\alpha})\boldsymbol{a}_{\alpha} + (\boldsymbol{n} \cdot \boldsymbol{\delta})^{-1}\boldsymbol{\delta}. \tag{1.27}$$

Accordingly,

$$\boldsymbol{n} \cdot \boldsymbol{g}^{\beta} = -(\boldsymbol{n} \cdot \boldsymbol{\delta})^{-1} \delta^{\beta}, \tag{1.28}$$

where

$$\delta^{\beta} = \boldsymbol{a}^{\beta} \cdot \boldsymbol{\delta}. \tag{1.29}$$

Substitution into the second expression above then yields

$$\boldsymbol{g}^{\alpha} = \boldsymbol{a}^{\alpha} - (\boldsymbol{n} \cdot \boldsymbol{\delta})^{-1} \delta^{\alpha} \boldsymbol{n}. \tag{1.30}$$

The expression for $\boldsymbol{D}_{|\omega}$ is then given by substituting into the first expression above:

$$\boldsymbol{D}_{|\omega} = \nabla \boldsymbol{\delta} + (\boldsymbol{n} \cdot \boldsymbol{\delta})^{-1} (\eta - \delta^{\alpha} \boldsymbol{\delta}_{,\alpha}) \otimes \boldsymbol{n}, \tag{1.31}$$

where

$$\nabla \boldsymbol{\delta} = \boldsymbol{\delta}_{,\alpha} \otimes \boldsymbol{a}^{\alpha} \tag{1.32}$$

is the surface gradient of the director field.

This expression for the limit energy is meaningful provided that $\boldsymbol{n} \cdot \boldsymbol{\delta} \neq 0$, but not if $\boldsymbol{n} \cdot \boldsymbol{\delta}$ vanishes. However, in the latter state the lipids lie in the tangent plane $T_{\omega(\theta^{\alpha})}$ and their hydrophobic tail groups are exposed to the surrounding aqueous solution; this is a highly energetic and unstable condition. To avoid this we introduce the additional ad hoc requirement that the local two-dimensional energy should grow without bound as $\boldsymbol{n} \cdot \boldsymbol{\delta} \to 0$.

1.4 The Optimal Two-Dimensional Model

The areal energy density is a function of the independent fields $\boldsymbol{r}$ (via $\boldsymbol{a}^{\alpha}$ and $\boldsymbol{n}$), $\boldsymbol{\delta}$ and $\boldsymbol{\eta}$. Among these, dependence on the latter is purely algebraic, and the Euler equation associated with it is

$$U_{\eta} = \mathbf{0}. \tag{1.33}$$

This of course is an equilibrium condition, associated with states that render the total energy stationary. It is also necessary if the field $\boldsymbol{\eta}$ is to minimize the energy, all other fields remaining fixed. This suggests a strategy whereby the energy is minimized with respect to $\boldsymbol{\eta}$ *a priori*, to derive an expression for the energy that depends on the fields $\boldsymbol{r}$ and $\boldsymbol{\delta}$. Here we show that this is indeed feasible, granted the hypotheses made thus far. The resulting model emerges as a special case of the theory of elastic Cosserat surfaces (Naghdi 1972).

To this end we fix all variables other than $\boldsymbol{\eta}$ and consider the function defined by $G(\boldsymbol{\eta}) = U(\boldsymbol{\delta}, \boldsymbol{D}_{|\omega})$. Let $\sigma(u) = G(\boldsymbol{\eta}(u))$, where $\boldsymbol{\eta}(u)$ is a one-parameter family of vectors. The chain rule then furnishes

$$\dot{\sigma} = (\boldsymbol{n}\cdot\boldsymbol{\delta})^{-1}(U_{\boldsymbol{D}})\boldsymbol{n}\cdot\dot{\boldsymbol{\eta}} \quad \text{and} \quad \ddot{\sigma} = (\boldsymbol{n}\cdot\boldsymbol{\delta})^{-1}(U_{\boldsymbol{D}})\boldsymbol{n}\cdot\ddot{\boldsymbol{\eta}} + (\boldsymbol{n}\cdot\boldsymbol{\delta})^{-2}\dot{\boldsymbol{\eta}}\otimes\boldsymbol{n}\cdot U_{\boldsymbol{DD}}[\dot{\boldsymbol{\eta}}\otimes\boldsymbol{n}], \tag{1.34}$$

which imply that

$$G_{\eta} = (\boldsymbol{n}\cdot\boldsymbol{\delta})^{-1}(U_{\boldsymbol{D}})\boldsymbol{n} \quad \text{and} \quad \dot{\boldsymbol{\eta}}\cdot(G_{\eta\eta})\dot{\boldsymbol{\eta}} = (\boldsymbol{n}\cdot\boldsymbol{\delta})^{-2}\dot{\boldsymbol{\eta}}\otimes\boldsymbol{n}\cdot U_{\boldsymbol{DD}}[\dot{\boldsymbol{\eta}}\otimes\boldsymbol{n}]. \tag{1.35}$$

Accordingly, $G_{\eta\eta}$ is positive definite by virtue of (1.4), and the stationarity condition (1.33) (equivalent to the vanishing of G_{η}) is equivalent to the restriction

$$(U_{\boldsymbol{D}})\boldsymbol{n} = \mathbf{0} \quad \text{on} \quad \omega. \tag{1.36}$$

That (1.33) is solvable uniquely for $\boldsymbol{\eta}$ is an immediate consequence of the implicit function theorem. Moreover, the solution minimizes the energy automatically. For energies of the form (1.2), this is obtained by solving

$$(\mathcal{L}[\boldsymbol{D}])\boldsymbol{n} = -\boldsymbol{L}\boldsymbol{n} \tag{1.37}$$

in which $\boldsymbol{D}$ is given by (1.31). Explicitly,

$$\boldsymbol{\eta} = \delta^{\alpha}\boldsymbol{\delta}_{,\alpha} - (\boldsymbol{n}\cdot\boldsymbol{\delta})\boldsymbol{A}^{-1}\{\boldsymbol{L}\boldsymbol{n} + (\mathcal{L}[\nabla\boldsymbol{\delta}])\boldsymbol{n}\} \tag{1.38}$$

in which $\boldsymbol{L}$ and $\mathcal{L}$ are evaluated at $\boldsymbol{d} = \boldsymbol{\delta}$ and $\boldsymbol{A}(\boldsymbol{\delta})$ is the positive-definite tensor-valued function defined, for any vector $\boldsymbol{v}$, by

$$\boldsymbol{A}\boldsymbol{v} = (\mathcal{L}[\boldsymbol{v}\otimes\boldsymbol{n}])\boldsymbol{n}. \tag{1.39}$$

We have established the existence of a unique solution $\eta = \bar{\eta}(\boldsymbol{n}, \boldsymbol{\delta}, \nabla\boldsymbol{\delta})$ to (1.33). This solution delivers the areal energy density (cf. (1.24))

$$\bar{W}(\boldsymbol{n}, \boldsymbol{\delta}, \nabla\boldsymbol{\delta}) = |\boldsymbol{n} \cdot \boldsymbol{\delta}|\, U[\nabla\boldsymbol{\delta} + (\boldsymbol{n} \cdot \boldsymbol{\delta})^{-1}(\bar{\eta} - \delta^{\alpha}\boldsymbol{\delta}_{,\alpha}) \otimes \boldsymbol{n}], \tag{1.40}$$

which furnishes the optimal value of the areal energy density W with respect to $\boldsymbol{\eta}$. For quadratic three-dimensional energies this procedure furnishes areal energy densities that are quadratic in $\nabla\boldsymbol{\delta}$. The desired functional is then given by

$$E = \int_{\omega} \bar{W}(\boldsymbol{n}, \boldsymbol{\delta}, \nabla\boldsymbol{\delta})da. \tag{1.41}$$

In view of (1.8) this expression furnishes the rigorous leading-order energy for a thin liquid crystal film. Moreover, the presumed convexity of the three-dimensional energy U with respect to $\boldsymbol{D}$ implies that $\bar{W}$ is a convex function of $\nabla\boldsymbol{\delta}$. This is important for the well-posedness of the two-dimensional model.

1.5 Euler Equations and Boundary Conditions

The Euler equations for the lipid membrane follow from the stationarity, in the sense of Ericksen, of the energy functional. To facilitate analysis, we first convert this energy to an integral over a fixed reference surface Ω, obtaining

$$E = \int_{\Omega} W^* dA, \quad \text{where} \quad W^*(\boldsymbol{a}_{\alpha}, \boldsymbol{\delta}, \boldsymbol{\delta}_{,\alpha}; \theta^{\alpha}) = J\bar{W}. \tag{1.42}$$

Here we have used the fact that the argument $\boldsymbol{n}$ of the function $\bar{W}$ is determined by $\{\boldsymbol{a}_{\alpha}\}$. Defining

$$J\boldsymbol{N}^{\alpha} = \partial W^*/\partial \boldsymbol{a}_{\alpha}, \quad J\boldsymbol{M}^{\alpha} - \partial W^*/\partial \boldsymbol{\delta}_{,\alpha} \quad \text{and} \quad J\boldsymbol{m} = \partial W^*/\partial \boldsymbol{\delta}, \tag{1.43}$$

we conclude that the Euler equations are

$$(J\boldsymbol{N}^{\alpha})_{|\alpha} = \boldsymbol{0} \quad \text{and} \quad (J\boldsymbol{M}^{\alpha})_{|\alpha} - J\boldsymbol{m} = \boldsymbol{0}, \tag{1.44}$$

where the $(\cdot)_{|\alpha}$ is the covariant derivative on Ω. To accommodate lateral pressures $P^{\pm}$, say, at the major surfaces, we assume that these scale as $P^{\pm} = tp^{\pm} + o(t)$, where $p^{\pm}$ are independent of t. Following (Steigmann 2010), this is found to yield

$$\int_{\partial\kappa_{c(n)}} P\boldsymbol{n} \cdot \dot{\boldsymbol{\chi}} da = -t \int_{\omega} p\boldsymbol{n} \cdot \dot{\boldsymbol{r}} da + o(t), \tag{1.45}$$

where, on the right-hand side, $\boldsymbol{n}$ is the unit normal to ω and $p = p^+ - p^-$ is the leading-order pressure difference across ω in the direction of the surface normal. The leading-order stationarity condition is then given by

$$\dot{E} = \int_\omega p\boldsymbol{n} \cdot \dot{\boldsymbol{r}} da. \tag{1.46}$$

The identity (Naghdi 1972) $(J\boldsymbol{N}^\alpha)_{|\alpha} = J\boldsymbol{N}^\alpha_{;\alpha}$, where $(\cdot)_{;\alpha}$ is the covariant derivative on the *current surface* ω, then furnishes

$$\boldsymbol{N}^\alpha_{;\alpha} + p\boldsymbol{n} = \boldsymbol{0} \quad \text{and} \quad \boldsymbol{M}^\alpha_{;\alpha} - \boldsymbol{m} = \boldsymbol{0}. \tag{1.47}$$

Standard boundary conditions entail the specification of $\boldsymbol{r}$ and $\boldsymbol{N}^\alpha \nu_\alpha$, and of $\boldsymbol{\delta}$ and $\boldsymbol{M}^\alpha \nu_\alpha$, on (possibly different) complementary parts of the edge $\partial\omega$ with exterior unit normal $\boldsymbol{\nu} = \nu_\alpha \boldsymbol{a}^\alpha$. These coincide with the equilibrium equations and boundary conditions for a Cosserat shell (Naghdi 1972; Steigmann 1999a).

It is interesting to observe that the assumption of bulk incompressibility imposes no constraint on the foregoing variational problem. To see this we recall that the local dilation is $\theta = \sqrt{g/G}$, with $\sqrt{g} = \mu\sqrt{a}$ and $\sqrt{G} = \mu_r\sqrt{A}$, where $\mu = |\boldsymbol{n} \cdot \boldsymbol{\delta}|$ and $\mu_r = |\boldsymbol{N} \cdot \boldsymbol{\delta}_r|$. Here the subscript $_r$ is used to identify quantities associated with the reference surface Ω, with unit normal $\boldsymbol{N}$. It follows that $\theta = J\mu/\mu_r$, with $J = \sqrt{a/A}$. The variational derivative of the constraint $\theta \equiv 1$ then yields $(J\mu)^\cdot = J\mu\dot{\mu}_r/\mu_r$. Because $\boldsymbol{\delta}_r$ is not a material vector with respect to the deformation map, this expression serves merely to impose a restriction on $\dot{\boldsymbol{\delta}}_r$ in terms of the primary variables involved in the statement of the variational problem. Thus, because $\boldsymbol{\delta}_r$ is not included in that list of variables, it follows that bulk incompressibility does not impose any constraint among the primary variables.

Exceptionally, if lipid tilt is suppressed; i.e., if $\boldsymbol{\delta} = \lambda\boldsymbol{n}$ for some scalar field $\lambda > 0$, then with $\boldsymbol{\delta}_r = \boldsymbol{N}$ ($\mu = \lambda$ and $\mu_r = 1$) we derive $\theta = \lambda J$. If, in addition, lipid distension is suppressed; i.e., if $\lambda \equiv 1$, then bulk incompressibility implies areal incompressibility; i.e., $J \equiv 1$. Because the determinant A of the metric is fixed on Ω, this in turn imposes a restriction on the determinant a of the metric on the evolving surface ω, and therefore constitutes a *bona fide* constraint on the variational problem. This is the more commonly studied case, and is discussed fully in Sect. 2. The case of distension without tilt, the theory for which is quite different, is discussed in the next subsection.

However, before proceeding, we pause to solve a problem in which lipid tilt figures prominently. The problem addressed is inspired by an unusual feature predicted by recent molecular dynamics simulations (Mandadapu 2016). This entails tilt in the immediate vicinity of an embedded hydrophobic cylinder representing a transmembrane protein. The lipids tilt locally so as to cover the hydrophobic surface of the cylinder (Fig. 1), exposing a vacuum gap adjoining the lipid tails. Molecular dynamics simulations indicate that this unusual configuration of lipids is energetically favorable when the height of the hydrophobic surface of the cylinder is less than that of the adjacent lipids.

We model this situation by adapting the foregoing tilt theory to one of the leaves of the bilayer. The actual configuration is then obtained by invoking reflection symmetry with respect to the underlying surface ω, which is assumed to remain flat as the lipids tilt non-uniformly. This restriction means that $\boldsymbol{n} = \boldsymbol{k}$, the fixed unit normal to the plane ω. Accordingly, the energy reduces to a function of $\boldsymbol{\delta}$ and $\nabla\boldsymbol{\delta}$, in which $\nabla(\cdot)$ is the gradient on the plane ω. We decompose the director field in the form

$$\boldsymbol{\delta} = \boldsymbol{\phi} + \lambda\boldsymbol{k}, \tag{1.48}$$

where $\boldsymbol{\phi}$, the tilt field, lies tangential to ω. In terms of these components,

$$\nabla\boldsymbol{\delta} = \nabla\boldsymbol{\phi} + \boldsymbol{k} \otimes \nabla\lambda. \tag{1.49}$$

To obtain a simple explicit model for purposes of illustration, we adopt the quadratic frame-invariant energy (Rangamani and Steigmann 2014)

$$W = \tfrac{1}{2}k\,|\nabla\boldsymbol{\delta}|^2 + G(\xi, \lambda); \quad \xi = |\nabla\boldsymbol{\phi}|\,, \tag{1.50}$$

where k is a positive constant and with G chosen to penalize lipid collapse ($G \to \infty$ as $\lambda \to 0$). Treating the lipids as springs with stiffness C, we might propose that

$$G(\xi, \lambda) = \tfrac{1}{2}C(\sqrt{\xi^2 + \lambda^2} - \lambda_0)^2 + H(\lambda) \tag{1.51}$$

where λ_0 is the relaxed lipid length, with $H(\lambda)$ constructed such that $H(\lambda_0) = 0$, $H'(\lambda_0) = 0$ and $H \to \infty$ as $\lambda \to 0$.

The Euler equations are easily derived and given by

$$div\left(\frac{\partial W}{\partial \nabla\boldsymbol{\phi}}\right) = \frac{\partial W}{\partial \boldsymbol{\phi}} \quad \text{and} \quad div\left(\frac{\partial W}{\partial \nabla\lambda}\right) = \frac{\partial W}{\partial \lambda}, \tag{1.52}$$

where $div(\cdot)$ is the two-dimensional divergence on ω. Standard boundary conditions entail the specification of $\boldsymbol{\phi}$ or $(\partial W/\partial\nabla\boldsymbol{\phi})\boldsymbol{\nu}$, and λ or $\boldsymbol{\nu} \cdot \partial W/\partial\nabla\lambda$, on appropriate parts of the edge $\partial\omega$ with exterior unit normal $\boldsymbol{\nu}$.

For energies of the form (1.50) it is a simple matter to obtain

$$\frac{\partial W}{\partial \nabla\boldsymbol{\phi}} = k\nabla\boldsymbol{\phi}, \quad \frac{\partial W}{\partial \nabla\lambda} = k\nabla\lambda \quad \text{and} \quad \frac{\partial W}{\partial \lambda} = G_\lambda. \tag{1.53}$$

To compute $\partial W/\partial\boldsymbol{\phi}$ we use the chain rule in the form

$$\frac{\partial W}{\partial \boldsymbol{\phi}} \cdot \dot{\boldsymbol{\phi}} = G_\xi \dot{\xi} = \xi^{-1} G_\xi \boldsymbol{\phi} \cdot \dot{\boldsymbol{\phi}}. \tag{1.54}$$

Accordingly,

$$\frac{\partial W}{\partial \boldsymbol{\phi}} = \xi^{-1} G_\xi \boldsymbol{\phi} \tag{1.55}$$

and the Euler equations reduce to the nonlinear system

$$k\Delta \boldsymbol{\phi} = \xi^{-1} G_\xi \boldsymbol{\phi} \quad \text{and} \quad k\Delta\lambda = G_\lambda, \tag{1.56}$$

where $\Delta(\cdot)$ is the Laplacian on ω. This system is coupled through the function G.

Explicit solutions are available for the linearized problem. Using superposed dots to denote small changes (not to be confused with the use of the same notation for variations) the linearized system is found to be

$$k\Delta\dot{\boldsymbol{\phi}} = (\xi^{-1}G_\xi)^{\cdot}\boldsymbol{\phi}_0 + (\xi^{-1}G_\xi)_0\dot{\boldsymbol{\phi}} \quad \text{and} \quad k\Delta\dot{\lambda} = (G_{\lambda\lambda})_0\dot{\lambda} + (G_{\lambda\xi})_0\dot{\xi}, \tag{1.57}$$

where the subscript $(\cdot)_0$ identifies quantities associated with a base state relative to which the system is linearized. Here we choose the base state $(\boldsymbol{\phi}, \lambda)_0 = (\mathbf{0}, \lambda_0)$, corresponding to a relaxed state with no tilt. For energies of the form (1.50), with (1.51), this state is easily seen to furnish a solution to the nonlinear problem (1.56). Moreover, in these circumstances the linearized problem reduces to the simple decoupled system

$$\Delta\dot{\boldsymbol{\phi}} = \mathbf{0} \quad \text{and} \quad \Delta\dot{\lambda} = \kappa\dot{\lambda}, \tag{1.58}$$

where $\kappa = (C + H''(\lambda_0))/k$ is a nominally positive constant.

The solution to the second equation, exterior to a circle and decaying in the far field ($\dot{\lambda} \to 0$, corresponding to $\lambda \to \lambda_0$), is given, in polar coordinates (r, θ), by

$$\dot{\lambda} = \sum_{m=0}^{\infty} K_m(\sqrt{\kappa}r)(C_m \cos m\theta + D_m \sin m\theta), \tag{1.59}$$

where K_m is the modified Bessel function of the second kind, of order m, and the Fourier coefficients C_m and D_m are determined by the assigned function $\dot{\lambda}_0 (= \lambda - \lambda_0)$ on the circle.

1.6 Distension Without Tilt

There is experimental evidence (Deseri et al. 2008) that lipid membranes may exhibit thickness distension without tilt. Theory for this is subsumed under the foregoing model by requiring $\boldsymbol{\delta}$ to be aligned with the surface normal; i.e.,

$$\boldsymbol{\delta} = \lambda\boldsymbol{n}, \tag{1.60}$$

for some scalar field $\lambda > 0$. We find that the areal energy density reduces to a function of the list $\{\lambda, \nabla\lambda, \boldsymbol{n}, \nabla\boldsymbol{n}\}$ in which $\nabla(\cdot)$ is the surface gradient on the current surface ω and $\nabla\boldsymbol{n} = -\boldsymbol{b}$, where

$$\boldsymbol{b} = b_{\alpha\beta}\boldsymbol{a}^{\alpha} \otimes \boldsymbol{a}^{\beta}, \quad \text{with} \quad b_{\alpha\beta} = \boldsymbol{n} \cdot \boldsymbol{r}_{,\alpha\beta}, \tag{1.61}$$

is the curvature tensor of ω. The energy is therefore expressible in the form

$$W = \hat{W}(\lambda, \nabla\lambda, \boldsymbol{n}, \boldsymbol{b}), \tag{1.62}$$

where

$$\hat{W}(\lambda, \nabla\lambda, \boldsymbol{n}, \boldsymbol{b}) = \lambda U(\boldsymbol{\delta}, \nabla\boldsymbol{\delta} + \lambda^{-1}\bar{\boldsymbol{\eta}} \otimes \boldsymbol{n}), \tag{1.63}$$

in which (1.60) is incorporated; i.e.,

$$\nabla\boldsymbol{\delta} = \boldsymbol{n} \otimes \nabla\lambda - \lambda\boldsymbol{b}. \tag{1.64}$$

This is subject to the frame-invariance requirement

$$\hat{W}(\lambda, \nabla\lambda, \boldsymbol{n}, \boldsymbol{b}) = \hat{W}(\lambda, \boldsymbol{Q}\nabla\lambda, \boldsymbol{Qn}, \boldsymbol{QbQ}^t) \quad \text{for all rotations} \quad \boldsymbol{Q}. \tag{1.65}$$

To examine the consequences of this restriction we choose rotations having axis $\boldsymbol{n}$ (i.e., $\boldsymbol{Qn} = \boldsymbol{n}$), yielding

$$\hat{W}(\lambda, \nabla\lambda, \boldsymbol{n}, \boldsymbol{b}) = \hat{W}(\lambda, \boldsymbol{Q}\nabla\lambda, \boldsymbol{n}, \boldsymbol{QbQ}^t) \tag{1.66}$$

for all such $\boldsymbol{Q}$. Here it is only the action of $\boldsymbol{Q}$ on T_ω that is relevant, and so we may regard this as a restriction that must hold for all *two-dimensional* rotations that map T_ω to itself. Then, the function

$$F(\nabla\lambda, \boldsymbol{b}) = \hat{W}(\lambda, \nabla\lambda, \boldsymbol{n}, \boldsymbol{b}), \tag{1.67}$$

obtained by fixing λ and $\boldsymbol{n}$, satisfies

$$F(\nabla\lambda, \boldsymbol{b}) = F(\boldsymbol{Q}\nabla\lambda, \boldsymbol{QbQ}^t) \tag{1.68}$$

for all two-dimensional rotations $\boldsymbol{Q}$. It is well known (Zheng 1993; Table 2) that F then depends on its arguments through the list $\{|\nabla\lambda|, \nabla\lambda \cdot \boldsymbol{b}(\nabla\lambda), H, K\}$, where

$$H = \tfrac{1}{2}tr\boldsymbol{b} \quad \text{and} \quad K = \det \boldsymbol{b} \tag{1.69}$$

are the mean and Gaussian curvatures of ω. The energy is thus expressible in the form

$$\hat{W}(\lambda, \nabla\lambda, \boldsymbol{n}, \boldsymbol{b}) = \hat{F}(\lambda, |\nabla\lambda|, \nabla\lambda \cdot \boldsymbol{b}(\nabla\lambda), H, K, \boldsymbol{n}), \tag{1.70}$$

in which $\hat{F}$ is invariant under replacement of $\boldsymbol{n}$ by $\boldsymbol{Qn}$ for *all* rotations $\boldsymbol{Q}$. Dependence on the latter thus occurs through $|\boldsymbol{n}|$ $(= 1)$; that is, the energy is independent of $\boldsymbol{n}$. Combining these facts, we find that the areal energy density has the canonical representation

$$\hat{W}(\lambda, \nabla\lambda, \boldsymbol{n}, \boldsymbol{b}) = \tilde{W}(\lambda, |\nabla\lambda|, \nabla\lambda \cdot \boldsymbol{b}(\nabla\lambda), H, K), \tag{1.71}$$

which satisfies (1.66) automatically, for *all* rotations.

We have remarked that if the three-dimensional energy U is quadratic in the director gradient $\boldsymbol{D}$ (cf. (1.2)), then the procedure given in Sects. 1.3 and 1.4 to derive the areal energy density delivers a function that is quadratic in $\nabla\boldsymbol{\delta}$. This implies, via (1.64), that the energy density is then quadratic in $\nabla\lambda$ and $\boldsymbol{b}$ jointly. It also implies that the invariant $\nabla\lambda \cdot \boldsymbol{b}(\nabla\lambda)$, which is of third order in $\nabla\boldsymbol{\delta}$, should be suppressed in the list of arguments of the energy function in (1.71).

1.7 Surface Dilation and Its Gradient

As noted previously, the constraint of bulk incompressibility, in the absence of director tilt, implies that

$$\lambda J = 1. \tag{1.72}$$

We may then use $\nabla\lambda = -J^{-2}\nabla J$ to conclude that the areal energy density (1.71) is expressible in the form

$$\tilde{W}(\lambda, |\nabla\lambda|, \nabla\lambda \cdot \boldsymbol{b}(\nabla\lambda), H, K) = W(J, |\nabla J|, \nabla J \cdot \boldsymbol{b}(\nabla J), H, K) \tag{1.73}$$

for some function W.

We observe that the constitutive dependence of the energy on J and its gradient is equivalent, by virtue of the conservation of mass, to a constitutive dependence on current areal mass density and its gradient, together with a parametric dependence on the (convected) coordinates via the referential density. In this context the notion of uniformity of material response presumes the existence of a reference surface on which the mass density is uniform. Similar remarks apply to the three-dimensional theory, and serve to connect the present model to more common theories of gradient effects in fluids.

In the quadratic case the energy is expressible in the form

$$W = \alpha(J) + k(J)H^2 + \bar{k}(J)K + \pi(J)\,|\nabla J| + \omega(J)\,|\nabla J|^2, \tag{1.74}$$

where we have imposed bilayer symmetry—the requirement that the energy be an even function of H—merely for the sake of discussion. This generalizes the conventional Helfrich theory of lipid membranes by including the effects of surface dilation and dilation gradient. Suppression of the curvature dependence yields a theory of generalized capillarity that accounts for the dilation and dilation gradient. We note that both the dilation *and* its gradient figure in the leading-order two-dimensional model derived from three-dimensional liquid crystal theory. From this point of view it is therefore not appropriate to regard the energy as a function of dilation unless its gradient is also included, contrary to the view advanced in (Baesu et al. 2004).

Further, using

$$\nabla\boldsymbol{\delta} = -J^{-1}(J^{-1}\boldsymbol{n}\otimes\nabla J + \boldsymbol{b}) \tag{1.75}$$

we infer that ∇J is linear in $\nabla\boldsymbol{\delta}$. Then, $|\nabla J|^2$ is a positive, homogeneous quadratic (hence convex) function of $\nabla\boldsymbol{\delta}$, and it follows from the convexity of the energy with respect to the director gradient that $\omega > 0$. This has significant implications for the well-posedness of energy minimization problems. Moreover, because H^2 is a positive, homogeneous quadratic function of $\boldsymbol{b}$, which in turn is a linear function of $\nabla\boldsymbol{\delta}$, it follows from convexity of the energy that $k > 0$.

1.8 Equilibrium Equations and Edge Conditions

The suppression of director tilt yields equilibrium equations and edge conditions that differ in their basic structure from those without constraints. To obtain them we proceed from the virtual-work principle. Here the energy per unit area of Ω is $W^* = JW$, and it variation is

$$\dot{E} = \int_\omega (\dot{W} + W\dot{J}/J)da, \tag{1.76}$$

where $\dot{J}/J = \boldsymbol{a}^\alpha\cdot\dot{\boldsymbol{r}}_{,\alpha}$. We seek vector fields $\boldsymbol{N}^\alpha$ and $\boldsymbol{M}^{\alpha\beta}$ such that

$$\dot{W} + W\dot{J}/J = \boldsymbol{N}^\alpha\cdot\boldsymbol{u}_{,\alpha} + \boldsymbol{M}^{\alpha\beta}\cdot\boldsymbol{u}_{;\alpha\beta}, \tag{1.77}$$

where $\boldsymbol{u} = \dot{\boldsymbol{r}}$ and $\boldsymbol{u}_{;\alpha\beta} = \boldsymbol{u}_{,\alpha\beta} - \Gamma^\lambda_{\alpha\beta}\boldsymbol{u}_{,\lambda}$ is the second covariant derivative of the virtual displacement $\boldsymbol{u}$; here, $\Gamma^\lambda_{\alpha\beta} = \boldsymbol{a}^\lambda\cdot\boldsymbol{a}_{\alpha,\beta}$ are the Christoffel symbols on ω. We then proceed as in (Steigmann 2013) to write

$$\dot{W} + W\dot{J}/J = \varphi^\alpha_{;\alpha} - \boldsymbol{u}\cdot\boldsymbol{T}^\alpha_{;\alpha}, \tag{1.78}$$

where

$$\boldsymbol{T}^\alpha = \boldsymbol{N}^\alpha - \boldsymbol{M}^{\alpha\beta}_{;\beta}, \tag{1.79}$$

with

$$\boldsymbol{M}^{\beta\alpha}_{;\beta} = \boldsymbol{M}^{\beta\alpha}_{,\beta} + \boldsymbol{M}^{\beta\alpha}\Gamma^{\lambda}_{\lambda\beta} + \boldsymbol{M}^{\beta\lambda}\Gamma^{\alpha}_{\lambda\beta}, \tag{1.80}$$

and

$$\varphi^{\alpha} = \boldsymbol{T}^{\alpha} \cdot \boldsymbol{u} + \boldsymbol{M}^{\alpha\beta} \cdot \boldsymbol{u}_{,\beta}. \tag{1.81}$$

The integral of the first term in (1.78) may be transformed to $\partial\omega$ using Stokes' theorem, and the remaining integral over ω is then leads to the Euler equation

$$\boldsymbol{T}^{\alpha}_{;\alpha} + p\boldsymbol{n} = \boldsymbol{0}, \tag{1.82}$$

where allowance has been made for pressure loading.

It is customary to decompose the equilibrium equation into tangential and normal parts, and to this end we write

$$\boldsymbol{N}^{\alpha} = N^{\beta\alpha}\boldsymbol{a}_{\beta} + N^{\alpha}\boldsymbol{n}, \quad \boldsymbol{M}^{\alpha\beta} = M^{\lambda\alpha\beta}\boldsymbol{a}_{\lambda} + M^{\alpha\beta}\boldsymbol{n}, \tag{1.83}$$

substitute into (1.79) and invoke the Gauss and Weingarten equations of surface theory (Sokolnikoff 1964), obtaining

$$\boldsymbol{T}^{\alpha} = (N^{\lambda\alpha} + M^{\beta\alpha}b^{\lambda}_{\beta} - M^{\lambda\beta\alpha}_{;\beta})\boldsymbol{a}_{\lambda} + (N^{\alpha} - M^{\beta\alpha}_{;\beta} - M^{\lambda\beta\alpha}b_{\lambda\beta})\boldsymbol{n}, \tag{1.84}$$

where

$$M^{\lambda\beta\alpha}_{;\beta} = M^{\lambda\beta\alpha}_{,\beta} + M^{\lambda\beta\alpha}\Gamma^{\mu}_{\mu\beta} + M^{\lambda\beta\mu}\Gamma^{\alpha}_{\mu\beta} + M^{\mu\beta\alpha}\Gamma^{\lambda}_{\mu\beta}. \tag{1.85}$$

Projection of (1.82) onto $\boldsymbol{a}^{\mu}$ and $\boldsymbol{n}$ then yields

$$(N^{\mu\alpha} + M^{\beta\alpha}b^{\mu}_{\beta} - M^{\mu\beta\alpha}_{;\beta})_{;\alpha} + (M^{\beta\alpha}_{;\beta} + M^{\lambda\beta\alpha}b_{\lambda\beta} - N^{\alpha})b^{\mu}_{\alpha} = 0 \tag{1.86}$$

and

$$(N^{\alpha} - M^{\beta\alpha}_{;\beta} - M^{\lambda\beta\alpha}b_{\lambda\beta})_{;\alpha} + (N^{\beta\alpha} + M^{\lambda\alpha}b^{\beta}_{\lambda} - M^{\beta\mu\alpha}_{;\mu})b_{\beta\alpha} + p = 0. \tag{1.87}$$

The natural boundary conditions are

$$\boldsymbol{T}^{\alpha}\nu_{\alpha} - (\boldsymbol{M}^{\alpha\beta}\nu_{\alpha}\tau_{\beta})' = \boldsymbol{f} \quad \text{and} \quad \boldsymbol{M}^{\alpha\beta}\nu_{\alpha}\nu_{\beta} = \boldsymbol{c} \quad \text{on} \quad \partial\omega_{n}, \tag{1.88}$$

where $\boldsymbol{\tau}$ is the unit tangent to $\partial\omega$ in the sense of Stokes' theorem, $\boldsymbol{\nu} = \boldsymbol{\tau} \times \boldsymbol{n}$ is the unit normal to $\partial\omega$ lying in the tangent plane, $(\cdot)'$ is the arclength derivative on $\partial\omega$; and $\boldsymbol{f}$ and $\boldsymbol{c}$ are the assigned force and double-force densities on the part $\partial\omega_{n}$ of the edge where natural boundary conditions are imposed. These follow from the boundary integral, $\int_{\omega} \varphi^{\alpha}\nu_{\alpha}ds$, referred to previously, by writing $\boldsymbol{u}_{,\beta} = \tau_{\beta}\boldsymbol{u}' + \nu_{\beta}\boldsymbol{u}_{\nu}$, where $\boldsymbol{u}'$ and $\boldsymbol{u}_\nu$ are the tangential and normal derivatives of $\boldsymbol{u}$ on $\partial\omega$, and integrating the terms involving tangential derivatives by parts, assuming $\partial\omega$ to be smooth and thus $\boldsymbol{\tau}$ (hence $\boldsymbol{\nu}$) to be continuous; if the boundary is merely piecewise smooth then this procedure leads to additional corner forces similar to those found in the classical Kirchhoff–Love theory of shells. Equation (1.88) then follow provided that the virtual work of the edge loads may be written in the form $\int_{\partial\omega_n}(\boldsymbol{f}\cdot\boldsymbol{u}+\boldsymbol{c}\cdot\boldsymbol{u}_\nu)ds$. Details of the procedure are provided in (Steigmann 2013) in the setting of shell theory.

To make these equations explicit we use the energy density (1.74) to express $\dot{W}$ as a linear form in $\boldsymbol{u}_{,\alpha}$ and $\boldsymbol{u}_{;\alpha\beta}$. We then use (1.77) to read off the relevant expressions for $\boldsymbol{N}^\alpha$ and $\boldsymbol{M}^{\alpha\beta}$. To this end we write the energy as a function of the list $\{J, H, K, G\}$, where

$$G = |\nabla J|, \tag{1.89}$$

and use

$$\dot{W} = W_J\dot{J} + W_H\dot{H} + W_K\dot{K} + W_G\dot{G}, \tag{1.90}$$

where (Agrawal and Steigmann 2009)

$$2\dot{H} = a^{\alpha\beta}\boldsymbol{n}\cdot\boldsymbol{u}_{;\alpha\beta} - 2b^{\alpha\beta}\boldsymbol{a}_\beta\cdot\boldsymbol{u}_{,\alpha} \quad \text{and} \quad \dot{K} = \tilde{b}^{\alpha\beta}\boldsymbol{n}\cdot\boldsymbol{u}_{;\alpha\beta} - 2K\boldsymbol{a}^\alpha\cdot\boldsymbol{u}_{,\alpha}, \tag{1.91}$$

and where

$$\tilde{b}^{\alpha\beta} = 2Ha^{\alpha\beta} - b^{\alpha\beta} \tag{1.92}$$

is the cofactor of the curvature tensor. To compute the variation of G we begin by observing that $\nabla J = J_{,\alpha}\boldsymbol{a}^\alpha$, with

$$J_{,\alpha} = (\sqrt{a/A})_{,\alpha} = JS^\lambda_{\lambda\alpha}, \tag{1.93}$$

where

$$S^\lambda_{\alpha\beta} = \Gamma^\lambda_{\alpha\beta} - \bar{\Gamma}^\lambda_{\alpha\beta}, \tag{1.94}$$

$\bar{\Gamma}^\lambda_{\alpha\beta}$ are the Christoffel symbols induced by the convected coordinates θ^α on the reference surface Ω, and use has been made of the identities $(\sqrt{a})_{,\alpha}/\sqrt{a} = \Gamma^\lambda_{\lambda\alpha}$ and $(\sqrt{A})_{,\alpha}/\sqrt{A} = \bar{\Gamma}^\lambda_{\lambda\alpha}$ (Sokolnikoff 1964). We then have

$$(\nabla J)^{\cdot} = \dot{J}_{,\alpha}\boldsymbol{a}^\alpha + J_{,\alpha}\dot{\boldsymbol{a}}^\alpha, \tag{1.95}$$

where $\dot{\boldsymbol{a}}^\alpha$ is given by (1.120) below and

$$\dot{J}_{,\alpha} = \dot{J}S^\lambda_{\lambda\alpha} + J\dot{S}^\lambda_{\lambda\alpha}, \tag{1.96}$$

in which (see the Appendix to this section; Eq. (1.124))

$$\dot{S}^{\beta}_{\beta\alpha} = \boldsymbol{a}^{\beta} \cdot \boldsymbol{u}_{;\alpha\beta} + b^{\beta}_{\alpha}\boldsymbol{n} \cdot \boldsymbol{u}_{,\beta}. \tag{1.97}$$

After some algebra we derive

$$\begin{aligned} \dot{G} &= G^{-1}\nabla J \cdot (\nabla J)^{\cdot} \\ &= JG^{-1}J_{,\lambda}a^{\alpha\lambda}\boldsymbol{a}^{\beta} \cdot \boldsymbol{u}_{;\alpha\beta} \\ &\quad + \{G\boldsymbol{a}^{\alpha} + JG^{-1}J_{,\lambda}a^{\lambda\alpha}\nabla J + JG^{-1}J_{,\lambda}b^{\lambda\alpha}\boldsymbol{n}\} \cdot \boldsymbol{u}_{,\alpha}. \end{aligned} \tag{1.98}$$

Comparing (1.78) and (1.90) and taking account of the symmetries of $\boldsymbol{u}_{;\alpha\beta}$ and $\boldsymbol{M}^{\alpha\beta}$, we conclude that

$$\boldsymbol{M}^{\alpha\beta} = \tfrac{1}{2}JG^{-1}W_G J_{,\mu}(a^{\alpha\mu}\boldsymbol{a}^{\beta} + a^{\beta\mu}\boldsymbol{a}^{\alpha}) + (\tfrac{1}{2}W_H a^{\alpha\beta} + W_K\tilde{b}^{\alpha\beta})\boldsymbol{n} \tag{1.99}$$

and

$$\begin{aligned} \boldsymbol{N}^{\alpha} &= \{[W + JW_J - 2(HW_H + KW_K)]a^{\alpha\beta} + W_H\tilde{b}^{\alpha\beta}\}\boldsymbol{a}_{\beta} \\ &\quad + W_G(G\boldsymbol{a}^{\alpha} + JG^{-1}J_{,\lambda}a^{\lambda\alpha}\nabla J + JG^{-1}J_{,\lambda}b^{\lambda\alpha}\boldsymbol{n}). \end{aligned} \tag{1.100}$$

It follows immediately that

$$\begin{aligned} M^{\alpha\beta} &= \tfrac{1}{2}W_H a^{\alpha\beta} + W_K\tilde{b}^{\alpha\beta}, \\ M^{\lambda\alpha\beta} &= \tfrac{1}{2}JG^{-1}W_G J_{,\mu}(a^{\alpha\mu}a^{\beta\lambda} + a^{\beta\mu}a^{\alpha\lambda}) \end{aligned} \tag{1.101}$$

and

$$\begin{aligned} N^{\alpha} &= JG^{-1}W_G J_{,\lambda}(2Ha^{\lambda\alpha} - \tilde{b}^{\lambda\alpha}), \\ N^{\alpha\beta} &= [W + JW_J - 2(HW_H + KW_K) + GW_G]a^{\alpha\beta} \\ &\quad + JG^{-1}W_G J_{,\lambda}J_{,\mu}a^{\lambda\alpha}a^{\mu\beta} + W_H\tilde{b}^{\alpha\beta}. \end{aligned} \tag{1.102}$$

We observe that the response functions $M^{\alpha\beta}$ and N^{α} vanish whenever the surface curvature vanishes. If the strain energy of the underlying three-dimensional liquid crystal is purely quadratic in the director gradient, then (1.75) implies that π vanishes in (1.74), yielding

$$W = \alpha(J) + k(J)H^2 + \bar{k}(J)K + \omega(J)G^2. \tag{1.103}$$

We show in Sect. 2 that (1.87) reduces to the conventional *shape equation* for lipid films (Jenkins 1977; Steigmann 1999b) in the presence of the areal incompressibility constraint $J = 1$, which is typically imposed in the classical Canham/Helfrich model.

1.9 Legendre–Hadamard Necessary Condition for Energy Minimizers

If equilibria are energy minimizers, then they satisfy the operative Legendre–Hadamard (or Weierstrass–Graves) necessary condition pointwise on ω. In the present context, this is the inequality (Hilgers and Pipkin 1993)

$$\mathcal{W}(\boldsymbol{r}_{,\alpha};\, \boldsymbol{r}_{,\alpha\beta} + \boldsymbol{a} b_\alpha b_\beta) - \mathcal{W}(\boldsymbol{r}_{,\alpha};\, \boldsymbol{r}_{,\alpha\beta}) \geq \boldsymbol{a} \cdot \partial \mathcal{W} / \partial \boldsymbol{r}_{,\alpha\beta} b_\alpha b_\beta \tag{1.104}$$

for all three-vectors $\boldsymbol{a}$ and for all b_α, where

$$\mathcal{W}(\boldsymbol{r}_{,\alpha};\, \boldsymbol{r}_{,\alpha\beta}) = W(J, H, K, G). \tag{1.105}$$

We have used the fact that the present energy is determined by the first and second derivatives of the position field $\boldsymbol{r}(\theta^\alpha)$ on ω. In (1.104) this position field is assumed to correspond to an energy minimizer, and the derivatives on the right-hand side are evaluated at the minimizing configuration.

To compute these derivatives, we proceed from the chain rule, obtaining

$$\partial \mathcal{W} / \partial \boldsymbol{r}_{,\alpha\beta} \cdot \dot{\boldsymbol{r}}_{,\alpha\beta} = \dot{W} = W_H \dot{H} + W_K \dot{K} + W_G \dot{G}, \tag{1.106}$$

where superposed dots refer to variations in which the $\boldsymbol{r}_{,\alpha}$, and hence J, are held fixed. In this case (1.91) and (1.98) reduce to

$$\dot{H} = \tfrac{1}{2} a^{\alpha\beta} \boldsymbol{n} \cdot \dot{\boldsymbol{r}}_{,\alpha\beta}, \quad \dot{K} = \tilde{b}^{\alpha\beta} \boldsymbol{n} \cdot \dot{\boldsymbol{r}}_{,\alpha\beta} \quad \text{and} \quad \dot{G} = J G^{-1} J_{,\lambda} a^{\alpha\lambda} \boldsymbol{a}^\beta \cdot \dot{\boldsymbol{r}}_{,\alpha\beta}. \tag{1.107}$$

Substituting into (1.106) and comparing with (1.77), we conclude that

$$\partial \mathcal{W} / \partial \boldsymbol{r}_{,\alpha\beta} = \boldsymbol{M}^{\alpha\beta}, \tag{1.108}$$

and hence that

$$\boldsymbol{a} \cdot \partial \mathcal{W} / \partial \boldsymbol{r}_{,\alpha\beta} b_\alpha b_\beta = M^{\alpha\beta} b_\alpha b_\beta \boldsymbol{a} \cdot \boldsymbol{n} + M^{\gamma\alpha\beta} b_\alpha b_\beta \boldsymbol{a} \cdot \boldsymbol{a}_\gamma, \tag{1.109}$$

where $M^{\alpha\beta}$ and $M^{\gamma\alpha\beta}$ are given by (1.101). In particular,

$$M^{\alpha\beta} b_\alpha b_\beta = \tfrac{1}{2} W_H + \varsigma W_K, \tag{1.110}$$

where

$$\varsigma = \tilde{b}^{\alpha\beta} b_\alpha b_\beta \tag{1.111}$$

in which the normalization condition $a^{\alpha\beta} b_\alpha b_\beta = 1$ has been imposed without loss of generality; and

$$M^{\gamma\alpha\beta}b_\alpha b_\beta = JG^{-1}W_G J_{,\lambda}b^\lambda b^\gamma, \quad \text{where} \quad b^\lambda = a^{\beta\lambda}b_\beta. \tag{1.112}$$

To reduce (1.104) we first evaluate the finite perturbations of H and K induced by $\boldsymbol{r}_{,\alpha\beta} \to \boldsymbol{r}_{,\alpha\beta} + \boldsymbol{a}b_\alpha b_\beta$ with $\boldsymbol{r}_{,\alpha}$ fixed. To this end we observe that $b_{\alpha\beta} \to b_{\alpha\beta} + \boldsymbol{n}\cdot\boldsymbol{a}b_\alpha b_\beta$. Consequently, $H \to H + \Delta H$ and $K \to K + \Delta K$, where $\Delta H = 1/2\boldsymbol{n}\cdot\boldsymbol{a}$ and $\Delta K = \varsigma\boldsymbol{n}\cdot\boldsymbol{a} = 2\varsigma\Delta H$. Further, using (1.94) we then infer that replacement of $\boldsymbol{r}_{,\alpha\beta}$ by $\boldsymbol{r}_{,\alpha\beta} + \boldsymbol{a}b_\alpha b_\beta$ is tantamount to the replacement of $S^\lambda_{\lambda\alpha}$ by $S^\lambda_{\lambda\alpha} + a^\lambda b_\lambda b_\alpha$, where $a^\lambda = \boldsymbol{a}\cdot\boldsymbol{a}^\lambda$; and hence to the replacement of ∇J by $\nabla J + J(\boldsymbol{a}\cdot\boldsymbol{b})\boldsymbol{b}$, where $\boldsymbol{a} = a^\lambda \boldsymbol{a}_\lambda$ and $\boldsymbol{b} = b_\lambda \boldsymbol{a}^\lambda$. Using (1.89) and (1.93), we then find that (1.104) reduces to

$$\begin{aligned} &W(J, H + \Delta H,\ K + \Delta K,\ |\nabla J + J(\boldsymbol{a}\cdot\boldsymbol{b})\boldsymbol{b}|) - W(J, H, K, G) \\ &\geq (\Delta H)W_H(J, H, K, G) + (\Delta K)W_K(J, H, K, G) \\ &\quad + G^{-1}J(\boldsymbol{a}\cdot\boldsymbol{b})(\boldsymbol{b}\cdot\nabla J)W_G(J, H, K, G). \end{aligned} \tag{1.113}$$

This is the version of the Weierstrass–Graves inequality (Graves 1939) appropriate to the present model.

To derive a necessary condition for (1.113) we select $\boldsymbol{a} = (\boldsymbol{n}\cdot\boldsymbol{a})\boldsymbol{n}$ (i.e., $\boldsymbol{a}\cdot\boldsymbol{b} = 0$) and linearize with respect to $\theta = \boldsymbol{n}\cdot\boldsymbol{a}$. To this end we fix J, H, K, G, and ς and set $F(\theta) = W(J, H + \theta/2,\ K + \varsigma\theta, G)$. Then, $F'(\theta) = 1/2W_H + \varsigma W_K$ and (1.113) is seen to be equivalent to the convexity of $F(\theta)$ at $\theta = 0$; i.e., $F(\theta) \geq F(0) + \theta F'(0)$. This in turn implies that $\theta^2[F''(0) + o(\theta^2)/\theta^2] \geq 0$. Dividing by θ^2 and passing to the limit, we obtain the Legendre–Hadamard condition $F''(0) \geq 0$, which is equivalent to

$$\tfrac{1}{4}W_{HH} + \varsigma W_{HK} + \varsigma^2 W_{KK} \geq 0, \tag{1.114}$$

in which the derivatives are evaluated at H and K. Here, ς is bounded between the largest and smallest eigenvalues of the cofactor of the curvature of the equilibrium surface at the (arbitrary) point in question.

Inequality (1.114) is easily seen to be equivalent to the restriction

$$k(J) \geq 0 \tag{1.115}$$

for energies of the form (1.103). We have already seen that this is implied by the convexity of the energy as a function of the director gradient. It is interesting that there is no corresponding restriction on the coefficient $\bar{k}$. To explore the further consequences of (1.113) in the present theory we consider the case $\boldsymbol{a}\cdot\boldsymbol{n} = 0$. This implies that $G^2 = a^{\alpha\beta}J_{,\alpha}J_{,\beta}$ is replaced by

$$G^2 + J^2(a^\lambda b_\lambda)^2 a^{\alpha\beta}b_\alpha b_\beta + 2J(a^\lambda b_\lambda)J_{,\alpha}b^\alpha. \tag{1.116}$$

For the energy function (1.103), we thus reduce (1.113), in the case $\boldsymbol{a}\cdot\boldsymbol{n} = 0$, to the inequality

$$\omega(J)[G^2 + J^2(a^\lambda b_\lambda)^2 a^{\alpha\beta} b_\alpha b_\beta + 2J(a^\lambda b_\lambda)J_{,\alpha} b^\alpha] - \omega(J)G^2 \geq 2J\omega(J)(a^\gamma b_\gamma)J_{,\lambda} b^\lambda, \tag{1.117}$$

which is equivalent to $J^2\omega(J)(\boldsymbol{a} \cdot \boldsymbol{b})^2 |\boldsymbol{b}|^2 \geq 0$ for all a^λ and b_λ; this in turn is equivalent to

$$\omega(J) \geq 0. \tag{1.118}$$

We have seen that this inequality is assured by the convexity condition (1.4).

Naturally the notion of energy minimization is meaningful only when the applied loads are conservative, this putting a restriction on the nature of the lateral pressure, for example. A catalog of admissible pressure loadings is discussed in (Steigmann 1991). Inequality (1.104) remains necessary without amendment in the presence of such pressure loads and also in the presence of conservative edge loads.

Appendix

We use $\dot{\boldsymbol{a}}^\alpha = (a^{\alpha\beta}\boldsymbol{a}_\beta)^\cdot$ with $\dot{a}^{\alpha\beta} = -a^{\alpha\nu}a^{\beta\mu}\dot{a}_{\nu\mu}$, which follows from $a^{\alpha\gamma}a_{\gamma\beta} = \delta^\alpha_\beta$, the Kronecker delta. Using $a_{\nu\mu} = \boldsymbol{a}_\nu \cdot \boldsymbol{a}_\mu$ we then obtain

$$\dot{a}^{\alpha\beta} = -a^{\beta\mu}\boldsymbol{a}^\alpha \cdot \dot{\boldsymbol{a}}_\mu - a^{\alpha\mu}\boldsymbol{a}^\beta \cdot \dot{\boldsymbol{a}}_\mu \tag{1.119}$$

and

$$\dot{\boldsymbol{a}}^\alpha = a^{\alpha\mu}\dot{\boldsymbol{a}}_\mu - [\boldsymbol{a}^\beta \cdot (a^{\alpha\mu}\dot{\boldsymbol{a}}_\mu)]\boldsymbol{a}_\beta - \boldsymbol{a}^\mu(\boldsymbol{a}^\alpha \cdot \dot{\boldsymbol{a}}_\mu). \tag{1.120}$$

Combine the first two terms as

$$a^{\alpha\mu}\dot{\boldsymbol{a}}_\mu - [\boldsymbol{a}^\beta \cdot (a^{\alpha\mu}\dot{\boldsymbol{a}}_\mu)]\boldsymbol{a}_\beta = (\boldsymbol{I} - \boldsymbol{a}_\beta \otimes \boldsymbol{a}^\beta)(a^{\alpha\mu}\dot{\boldsymbol{a}}_\mu) = \boldsymbol{n} \cdot (a^{\alpha\mu}\dot{\boldsymbol{a}}_\mu)\boldsymbol{n}, \tag{1.121}$$

where $\boldsymbol{I}$ is the identity for three-space, and use $(\boldsymbol{I} - \boldsymbol{a}_\beta \otimes \boldsymbol{a}^\beta) = \boldsymbol{n} \otimes \boldsymbol{n}$.

Next, we use

$$J\boldsymbol{n} = \tfrac{1}{2}\mu^{\alpha\beta}\boldsymbol{a}_\alpha \times \boldsymbol{a}_\beta, \tag{1.122}$$

where $\mu^{\alpha\beta} = J\varepsilon^{\alpha\beta} = e^{\alpha\beta}/\sqrt{A}$, with $e^{12} = -e^{21} = 1$, $e^{11} = e^{22} = 0$, obtaining

$$\dot{J}\boldsymbol{n} + J\dot{\boldsymbol{n}} = \mu^{\beta\alpha}\boldsymbol{a}_\beta \times \dot{\boldsymbol{a}}_\alpha. \tag{1.123}$$

Finally, (1.97) follows by using (1.94) to derive $\dot{S}^\lambda_{\alpha\beta} = \dot{\Gamma}^\lambda_{\alpha\beta} = (\boldsymbol{a}^\lambda \cdot \boldsymbol{a}_{\alpha,\beta})^\cdot$. Using (1.120) together with the Gauss equation $\boldsymbol{a}_{\alpha,\beta} = \Gamma^\varphi_{\alpha\beta}\boldsymbol{a}_\varphi + b_{\alpha\beta}\boldsymbol{n}$ and $\dot{\boldsymbol{r}}_{;\alpha\beta} = \dot{\boldsymbol{r}}_{,\alpha\beta} - \Gamma^\varphi_{\alpha\beta}\dot{\boldsymbol{r}}_{,\varphi}$, we find that

$$\dot{S}^\lambda_{\alpha\beta} = \boldsymbol{a}^\lambda \cdot \dot{\boldsymbol{r}}_{;\alpha\beta} + a^{\lambda\mu}b_{\alpha\beta}\boldsymbol{n} \cdot \dot{\boldsymbol{r}}_{,\mu}, \tag{1.124}$$

which yields (1.97) upon recalling that $\boldsymbol{u} = \dot{\boldsymbol{r}}$.

2 The Classical Canham/Helfrich Model

The classical theory, due to Canham (1970) and Helfrich (1973), is long established and constitutes the basis of the bulk of the literature on the continuum theory of lipid bilayers. It is a special case of the foregoing in which both tilt and distension are suppressed. As we have seen, the suppression of distension in the absence of tilt is appropriate in the presence of bulk incompressibility provided that areal incompressibility also obtains. This constraint has interesting mathematical consequences which alter the basic character of the theory. Accordingly, we develop the classical model anew in a variational setting.

2.1 *Surface Geometry and the Energy Functional*

With reference to (1.71) and in the absence of lipid distension, the response of the membrane is seen to be embodied in an areal free-energy density function $W(H, K; \theta^\alpha)$, where H is the mean curvature of the membrane surface and K is the Gaussian curvature. These are defined by

$$H = \tfrac{1}{2}a^{\alpha\beta}b_{\alpha\beta} \quad \text{and} \quad K = \tfrac{1}{2}\varepsilon^{\alpha\beta}\varepsilon^{\lambda\mu}b_{\alpha\lambda}b_{\beta\mu}, \tag{2.1}$$

where $(a^{\alpha\beta})$ is the matrix of dual metric components, the inverse of the metric $(a_{\alpha\beta})$; $\varepsilon^{\alpha\beta} = e^{\alpha\beta}/\sqrt{a}$ is the permutation tensor density with $a = \det(a_{\alpha\beta})$; $e^{12} = -e^{21} = 1$, $e^{11} = e^{22} = 0$; and $b_{\alpha\beta}$ are the coefficients of the second fundamental form. The latter are the covariant components of the surface curvature tensor. The contravariant *cofactor* of the curvature is given by

$$\tilde{b}^{\alpha\beta} = \varepsilon^{\alpha\lambda}\varepsilon^{\beta\gamma}b_{\lambda\gamma}, \tag{2.2}$$

and satisfies

$$b^\beta_\mu \tilde{b}^{\mu\alpha} = Ka^{\beta\alpha}, \tag{2.3}$$

where b^β_μ are the mixed components of the curvature. These components figure in the Gauss and Weingarten equations

$$\boldsymbol{a}_{\alpha;\beta} = b_{\alpha\beta}\boldsymbol{n} \quad \text{and} \quad \boldsymbol{n}_{,\alpha} = -b^\beta_\alpha \boldsymbol{a}_\beta, \tag{2.4}$$

respectively. Here, $\boldsymbol{a}_\alpha = \boldsymbol{r}_{,\alpha}$ are the tangent vectors to ω induced by the parametrization $\boldsymbol{r}(\theta^\alpha)$, the position in 3-space of a point on the surface with coordinates θ^α, and the unit-vector field $\boldsymbol{n}(\theta^\alpha) = \frac{1}{2}\varepsilon^{\alpha\beta}\boldsymbol{a}_\alpha \times \boldsymbol{a}_\beta$ is the local surface orientation. The $\boldsymbol{a}_\alpha$ are related to the metric by $a_{\alpha\beta} = \boldsymbol{a}_\alpha \cdot \boldsymbol{a}_\beta$, where the dot refers to the conventional Euclidean inner product. Further, semi-colons are used to denote surface covariant differentiation. Thus, for example,

$$\boldsymbol{a}_{\alpha;\beta} = \boldsymbol{a}_{\alpha,\beta} - \Gamma^{\lambda}_{\alpha\beta}\boldsymbol{a}_{\lambda}, \tag{2.5}$$

where $\Gamma^{\lambda}_{\alpha\beta}$ are the Christoffel symbols induced by the coordinates on ω.

Equilibria are those configurations that render stationary the potential energy defined by

$$E = \int_{\omega} W(H, K; \theta^{\alpha})da. \tag{2.6}$$

To accommodate the present constraint on surface area, we consider the augmented energy functional

$$E = \int_{\Omega} [JW(H, K; \theta^{\alpha}) + \lambda(\theta^{\alpha})(J - 1)]dA, \tag{2.7}$$

where $\lambda(\theta^{\alpha})$ is a Lagrange-multiplier field,

$$J = \sqrt{a/A} \tag{2.8}$$

is the local areal stretch induced by the map from a fixed reference surface Ω to the actual surface ω, and A is the value of a on Ω.

2.2 *The Shape Equation*

To compute the variation of the energy it is necessary to have explicit formulas for the variational derivatives of J, H, and K induced by the virtual displacement $\boldsymbol{u}(\theta^{\alpha}) = \dot{\boldsymbol{r}}$ of the equilibrium position field $\boldsymbol{r}(\theta^{\alpha})$. The induced variation of the energy is

$$\dot{E} = \int_{\omega} [\dot{W} + (W + \lambda)\dot{J}/J]da, \tag{2.9}$$

where

$$\dot{W} = W_H\dot{H} + W_K\dot{K}. \tag{2.10}$$

Here and henceforth the subscripts H and K refer to partial derivatives. We note that, while the augmented functional may be considered to depend on both the position and Lagrange multiplier, regarded as independent fields, variation with respect to the latter merely returns the constraint and so need not be made explicit.

For example, under *tangential* variations we have (Steigmann et al. 2003)

$$\dot{J}/J = u^{\alpha}_{;\alpha}, \quad \dot{H} = u^{\alpha}H_{,\alpha} \quad \text{and} \quad \dot{K} = u^{\alpha}K_{,\alpha}, \tag{2.11}$$

wherein J, H and K are equilibrium fields. These furnish

$$\dot{W} = u^\alpha(W_H H_{,\alpha} + W_K K_{,\alpha}) \quad \text{and}$$
$$(W+\lambda)\dot{J}/J = [(W+\lambda)u^\alpha]_{;\alpha} - u^\alpha(W+\lambda)_{,\alpha}, \tag{2.12}$$

which may be combined with (2.9) and Stokes' theorem to obtain the associated variation

$$\dot{E} = \int_\omega u^\alpha(W_H H_{,\alpha} + W_K K_{,\alpha} - W_{,\alpha} - \lambda_{,\alpha})da + \int_{\partial\omega}(W+\lambda)u^\alpha \nu_\alpha ds, \tag{2.13}$$

where ν_α are the covariant components of the exterior unit normal to the edge $\partial\omega$, lying in the tangent plane of ω as it is traversed in the direction of increasing arclength s. The associated Euler equation is equivalent to the vanishing of the parenthetical term in the first integral. To reduce this, we use

$$W_{,\alpha} = W_H H_{,\alpha} + W_K K_{,\alpha} + \partial W/\partial\theta^\alpha, \tag{2.14}$$

where the partial derivative on the right is due to the explicit coordinate dependence in the function W. This arises from the possible non-uniformity of the film properties in the present context. Accordingly, the relevant Euler equation is expressible in the form

$$\lambda_{,\alpha} = -\partial W/\partial\theta^\alpha. \tag{2.15}$$

This yields $\lambda = const.$ in the special case of a film with properties that are uniform in the sense that W does not depend explicitly on the coordinates θ^α. A class of problems pertaining to non-uniform films is discussed in Sect. 4.

For *normal* variations we have (Steigmann et al. 2003)

$$\dot{J}/J = -2Hu, \quad 2\dot{H} = \Delta u + u(4H^2 - 2K) \quad \text{and} \quad \dot{K} = 2KHu + (\tilde{b}^{\alpha\beta}u_{,\alpha})_{;\beta}, \tag{2.16}$$

where $\Delta(\cdot) = a^{\alpha\beta}(\cdot)_{;\alpha\beta}$ is the surface Laplacian, also known as the Beltrami operator. These lead to

$$(W+\lambda)\dot{J}/J = -2Hu(W+\lambda) \tag{2.17}$$

and, with some effort (Steigmann et al. 2003), to

$$\begin{aligned}\dot{W} &= u[\Delta(\tfrac{1}{2}W_H) + (W_K)_{;\beta\alpha}\tilde{b}^{\beta\alpha} + W_H(2H^2 - K) + 2KHW_K] \\ &\quad + \tfrac{1}{2}W_H(a^{\alpha\beta}u_{,\alpha})_{;\beta} - \tfrac{1}{2}[(W_H)_{,\beta}a^{\alpha\beta}u]_{;\alpha} + (W_K\tilde{b}^{\alpha\beta}u_{,\alpha})_{;\beta} - [(W_K)_{,\beta}\tilde{b}^{\alpha\beta}u]_{;\alpha}.\end{aligned} \tag{2.18}$$

Thus normal variations induce the energy variation

$$\dot{E} = \int_{\omega} u[\Delta(\tfrac{1}{2}W_H) + (W_K)_{;\beta\alpha}\tilde{b}^{\beta\alpha} + W_H(2H^2 - K) + 2KHW_K - 2H(W + \lambda)]da$$
$$+ \int_{\partial\omega} [\tfrac{1}{2}W_H\nu^{\alpha}u_{,\alpha} - \tfrac{1}{2}(W_H)_{,\alpha}\nu^{\alpha}u + W_K\tilde{b}^{\alpha\beta}\nu_{\beta}u_{,\alpha} - (W_K)_{,\alpha}\tilde{b}^{\alpha\beta}\nu_{\beta}u]ds. \quad (2.19)$$

Suppose the membrane bounds a volume of incompressible liquid. We assume the membrane to be impermeable and regard the membrane and liquid as a closed system. The bulk incompressibility of the liquid is taken into account through the replacement

$$E \to E - \int_B p(X)[\det(Grad\chi) - 1]dV, \quad (2.20)$$

where $\chi(X)$ is a one-to-one deformation that maps a fixed configuration B of the bulk fluid to its current configuration, $Grad$ is the gradient with respect to position $X \in B$, $\boldsymbol{y} = \chi(X; \epsilon)$ is the position of a material point in the current configuration, and $p(X)$ is a Lagrange multiplier field. Incompressibility is associated with the constraint $\det(Grad\chi) = 1$. This leads to the substitution

$$\dot{E} \to \dot{E} + \int_R \dot{\chi} \cdot gradp dv - \int_{\omega} upda, \quad (2.21)$$

where R is the volume occupied by the liquid in its equilibrium configuration ($\omega \subset \partial R$) and $grad$ is the gradient with respect to $\boldsymbol{y} \in R$. Accordingly, the Euler equations are $gradp = \boldsymbol{0}$, implying that p is uniform in R, and

$$\Delta(\tfrac{1}{2}W_H) + (W_K)_{;\beta\alpha}\tilde{b}^{\beta\alpha} + W_H(2H^2 - K) + 2KHW_K - 2H(W + \lambda) = p \quad \text{on} \quad \omega. \quad (2.22)$$

It follows that p is mechanically equivalent to a net lateral pressure exerted on the membrane in the direction of its orientation $\boldsymbol{n}$. This may be compared to (1.87).

From (1.114) we deduce that a further necessary condition for an equilibrium state to be energy *minimizing*, in addition to (2.15) and (2.22), is

$$\tfrac{1}{4}W_{HH} + \varsigma W_{HK} + \varsigma^2 W_{KK} > 0 \quad (2.23)$$

at all points of the film, where ς is defined in (1.114).

For bilayers that have no natural orientation the energy function W satisfies the symmetry relation $W(H, K; \theta^{\alpha}) = W(-H, K; \theta^{\alpha})$ (Steigmann 1999b). A well-known example—a special case of (1.103)—is

$$W = kH^2 + \bar{k}K \quad (2.24)$$

wherein k and $\bar{k}$ are empirical constants, and pertains to films with uniform properties. This is compatible with (2.23) if and only if $k > 0$, while $\bar{k}$ is unrestricted, and the shape equation reduces to

$$k[\Delta H + 2H(H^2 - K)] - 2\lambda H = p. \tag{2.25}$$

In general, bilayer symmetry implies that $p = 0$ if the membrane is flat anywhere. For, the odd-order derivatives of W with respect to H vanish at zero curvature and (2.22) reduces to $p = 0$. The uniformity of p then implies that it vanishes everywhere.

2.3 Edge Conditions

With (2.15) and (2.22) satisfied the variation of the energy reduces to $\dot{E} = \dot{E}_B$, where

$$\dot{E}_B = B_t + B_n, \tag{2.26}$$

and

$$B_t = \int_{\partial\omega} (W + \lambda)u^\alpha \nu_\alpha ds, \tag{2.27}$$

$$B_n = \int_{\partial\omega} [\tfrac{1}{2} W_H \nu^\alpha u_{,\alpha} - \tfrac{1}{2}(W_H)_{,\alpha}\nu^\alpha u + W_K \tilde{b}^{\alpha\beta}\nu_\beta u_{,\alpha} - (W_K)_{,\alpha}\tilde{b}^{\alpha\beta}\nu_\beta u]ds, \tag{2.28}$$

respectively, are the contributions to the boundary working arising from tangential and normal variations. To reduce the latter to a usable form it is necessary to express the derivatives $u_{,\alpha}$ of a normal variation in terms of its independent arclength and normal derivatives $u'(s)$ and $u_{,\nu}$ on $\partial\omega$. These are the restrictions to $\partial\omega$ of $\tau^\alpha u_{,\alpha}$ and $\nu^\alpha u_{,\alpha}$, respectively, where $\boldsymbol{\tau}$ is the unit tangent to $\partial\omega$, pointing in the direction of increasing arclength. Let $\theta^\alpha(s)$ be the parametrization of $\partial\omega$. Then,

$$\boldsymbol{\nu} = \nu_\alpha \boldsymbol{a}^\alpha = \boldsymbol{\tau} \times \boldsymbol{n}, \tag{2.29}$$

where

$$\boldsymbol{\tau} = \tfrac{d}{ds}\boldsymbol{r}(\theta^\alpha(s)) = \tau^\alpha \boldsymbol{a}_\alpha \quad \text{and} \quad \tau^\alpha = d\theta^\alpha/ds. \tag{2.30}$$

The required expression follows from the orthonormality of $\{\boldsymbol{\nu}, \boldsymbol{\tau}\}$. Thus (Naghdi 1972),

$$u_{,\alpha} = \tau_\alpha u' + \nu_\alpha u_{,\nu}. \tag{2.31}$$

Combining this with (Steigmann 1999b)

$$\tilde{b}^{\alpha\beta} = 2Ha^{\alpha\beta} - b^{\alpha\beta} \tag{2.32}$$

and

$$a^{\alpha\beta} = \tau^\alpha\tau^\beta + \nu^\alpha\nu^\beta \tag{2.33}$$

results in

$$a^{\alpha\beta}u_{,\alpha} = \tau^\beta u' + \nu^\beta u_{,\nu} \tag{2.34}$$

and

$$W_K\tilde{b}^{\alpha\beta}\nu_\beta u_{,\alpha} = u(\tau W_K)' - (\tau W_K u)' + (2H - \kappa_\nu)W_K u_{,\nu}. \tag{2.35}$$

Here,

$$\tau = b^{\alpha\beta}\tau_\alpha\nu_\beta \tag{2.36}$$

is the *twist* of the surface ω on the $\boldsymbol{\nu}, \boldsymbol{\tau}$ - axes, and

$$\kappa_\nu = b^{\alpha\beta}\nu_\alpha\nu_\beta \tag{2.37}$$

is its *normal* curvature in the direction of $\boldsymbol{\nu}$. Thus, for $\partial\omega$ piecewise smooth in the sense that $\boldsymbol{a}_\alpha$ is continuous there, while $\bar{\boldsymbol{r}}(s) = \boldsymbol{r}(\theta^\alpha(s))$ is piecewise differentiable (so that $\boldsymbol{\tau}$ can have a finite number of jumps), we obtain

$$\begin{aligned} B_n = &\int_{\partial\omega}[(\tau W_K)' - \tfrac{1}{2}\nu^\beta(W_H)_{,\beta} - (W_K)_{,\beta}\tilde{b}^{\alpha\beta}\nu_\alpha]u\,ds \\ &+ \int_{\partial\omega}(\tfrac{1}{2}W_H + \kappa_\tau W_K)u_{,\nu}ds + \sum uW_K\lfloor\tau\rfloor, \end{aligned} \tag{2.38}$$

where the square bracket in the summand identifies the forward jump of the enclosed quantity at a corner of $\partial\omega$ (a point where $\boldsymbol{\tau}$ is discontinuous), and

$$\kappa_\tau = b_{\alpha\beta}\tau^\alpha\tau^\beta = 2H - \kappa_\nu \tag{2.39}$$

is the normal curvature of ω in the direction of $\boldsymbol{\tau}$. The twist and normal curvatures are simply the components of the curvature tensor $\boldsymbol{b} = b_{\alpha\beta}\boldsymbol{a}^\alpha \otimes \boldsymbol{a}^\beta$ on the $\boldsymbol{\nu}$, $\boldsymbol{\tau}$ - axes; i.e.,

$$\boldsymbol{b} = \kappa_\nu\boldsymbol{\nu}\otimes\boldsymbol{\nu} + \kappa_\tau\boldsymbol{\tau}\otimes\boldsymbol{\tau} + \tau(\boldsymbol{\nu}\otimes\boldsymbol{\tau} + \boldsymbol{\tau}\otimes\boldsymbol{\nu}). \tag{2.40}$$

Some conceptual clarity is gained by expressing u and $u_{,\nu}$ in terms of the virtual displacement $\boldsymbol{u}$ and its normal derivative $\boldsymbol{u}_{,\nu} = \nu^{\alpha}\boldsymbol{u}_{,\alpha}$. To this end we use $u = \boldsymbol{u} \cdot \boldsymbol{n}$ to obtain

$$u_{,\nu} = \nu^{\alpha}(\boldsymbol{u} \cdot \boldsymbol{n})_{,\alpha} = \boldsymbol{n} \cdot \boldsymbol{u}_{,\nu} + \boldsymbol{u} \cdot \boldsymbol{n}_{,\nu}, \tag{2.41}$$

where

$$\boldsymbol{n}_{,\nu} = \nu^{\alpha}\boldsymbol{n}_{,\alpha} = -\nu^{\alpha}b_{\alpha}^{\beta}\boldsymbol{a}_{\beta}. \tag{2.42}$$

The latter may be simplified by substituting

$$\boldsymbol{a}_{\beta} = \tau_{\beta}\boldsymbol{\tau} + \nu_{\beta}\boldsymbol{\nu}, \tag{2.43}$$

yielding

$$\boldsymbol{n}_{,\nu} = -\tau\boldsymbol{\tau} - \kappa_{\nu}\boldsymbol{\nu}. \tag{2.44}$$

Further, it is possible to show that (Steigmann 1999a)

$$\dot{\boldsymbol{n}} = -(\boldsymbol{n} \cdot \boldsymbol{u}_{,\nu})\boldsymbol{\nu} - (\boldsymbol{n} \cdot \boldsymbol{u}')\boldsymbol{\tau}, \tag{2.45}$$

where $\dot{\boldsymbol{n}}$ is the variation of $\boldsymbol{n}$. Because the latter is a unit vector in all configurations of the surface, it follows that $\dot{\boldsymbol{n}} = \boldsymbol{\omega} \times \boldsymbol{n}$ for some vector $\boldsymbol{\omega}$. Thus,

$$-\boldsymbol{n} \cdot \boldsymbol{u}_{,\nu} = \boldsymbol{\tau} \cdot \boldsymbol{\omega} \tag{2.46}$$

and hence

$$u_{,\nu} = -\boldsymbol{\tau} \cdot \boldsymbol{\omega} - b\boldsymbol{\nu} \cdot \boldsymbol{u}. \tag{2.47}$$

We also note the relations

$$\boldsymbol{n} \cdot \boldsymbol{u}' = \boldsymbol{\nu} \cdot \boldsymbol{\omega} \quad \text{and} \quad u' = \boldsymbol{\nu} \cdot \boldsymbol{\omega} - b\boldsymbol{\tau} \cdot \boldsymbol{u}, \tag{2.48}$$

which are derived similarly.

The foregoing may be used to reduce (2.26) to the compact form

$$\dot{E}_B = \int_{\partial\omega} (F_{\nu}\boldsymbol{\nu} + F_{\tau}\boldsymbol{\tau} + F_n\boldsymbol{n}) \cdot \boldsymbol{u} ds - \int_{\partial\omega} M\boldsymbol{\tau} \cdot \boldsymbol{\omega} ds + \sum \boldsymbol{f}_i \cdot \boldsymbol{u}_i, \tag{2.49}$$

where

$$M = \tfrac{1}{2}W_H + \kappa_{\tau}W_K \tag{2.50}$$

is the bending couple applied to ω per unit length of $\partial\omega$,

$$F_\nu = W + \lambda - \kappa_\nu M, \quad F_\tau = -\tau M \quad \text{and}$$
$$F_n = (\tau W_K)' - (\tfrac{1}{2} W_H)_{,\nu} - (W_K)_{,\beta} \tilde{b}^{\alpha\beta} \nu_\alpha, \tag{2.51}$$

respectively, are the $\boldsymbol{\nu}$ -,$\boldsymbol{\tau}$ - and $\boldsymbol{n}$ - components of the force per unit length applied to $\partial\omega$, and

$$\boldsymbol{f}_i = W_K[\tau]_i \boldsymbol{n} \tag{2.52}$$

is the force applied to the film at the ith corner of $\partial\omega$.

3 Dissipative Effects: Diffusion and Viscous Flow

This section provides theory for viscous flow and protein transport in the bilayer surface. The basic aim is to further advance recent efforts to establish models for viscous flow, and the interaction of different species, on lipid bilayers. These processes are known to play an essential role in regulating a wide range of cellular functions. The focus here is on the case of lipid distension without tilt, which is appropriate in the presence of a sufficiently high surface concentration of lipids.

3.1 Effect of a Second Chemical Species

Lipid distension λ and areal surface dilation J are connected via bulk incompressibility; thus, $\lambda J = 1$. Variable distension is needed to accommodate hydrophobic mismatch in the absence of tilt at the boundaries between lipids and transmembrane proteins, regarded as a second species on the lipid bilayer. Because of the bulk incompressibility of the liquid crystal, this is tantamount to a spatial variation of the dilation J. In the present setting proteins are accommodated by using a continuum model based on an areal density of proteins rather than interactions with discrete proteins. This is appropriate if the number of proteins per unit area of the membrane surface is sufficiently high. Locally, the distension adjusts near a discrete protein. Accordingly, surface dilation adjusts to accommodate the proteins.

We have seen that in a dimension reduction approach based on 3D liquid crystal theory, and in the absence of lipid tilt, the free-energy density, W, is found to depend on dilation J, the norm of its gradient, $G = |\nabla J|$, and the mean and Gaussian curvatures, H and K, respectively. Here we allow a further dependence on the areal protein density, σ. Following the work of Cahn and Hilliard (1958), and in line with other recent work on multi-component membranes (Embar et al. 2013), we also assume a constitutive sensitivity to the gradient of concentration $\nabla\sigma$. This provides an energetic penalty to the formation of sharp interfaces between different species on

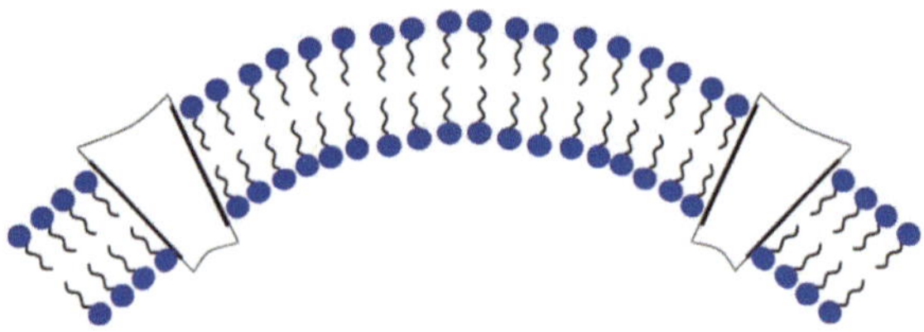

Fig. 2 Transmembrane proteins generate spontaneous curvature

the bilayer. Invariance requirements imply that this enters the constitutive response through the norm $L = |\nabla\sigma|$. Thus the areal energy density is a function W of the list

$$\{J, G, H, K, \sigma, L\}. \tag{3.1}$$

A specific example is afforded by an expression of the form (cf. (1.103))

$$W(J, G, H, K, \sigma, L) = \alpha(J,\sigma) + k(J,\sigma)[H - C(J,\sigma)]^2 \\ + \bar{k}(J,\sigma)K + \omega(J,\sigma)G^2 + \varpi(J,\sigma)L^2, \tag{3.2}$$

Here, α is a chemical interaction term accounting for the contribution of proteins to the energy; and C, the well-known *spontaneous curvature*, serves to couple local membrane shape to local protein concentration. Hydrophobic interactions between the lipids and embedded proteins tend to promote local membrane curvature as required to adjust to protein geometry, as shown in Fig. 2. This effect is taken into account in the function C.

The term involving G penalizes abrupt transitions in J and thus promotes smooth transitions from regions having large and small dilations, or small and large distensions. The functions α and C are phenomenological in nature and plausible forms for them must be specified. The quadratic dependencies on the curvature and gradient terms may be justified by scaling arguments of the kind adopted in conventional liquid crystal theory: The length scales for the spatial variations of the associated functions are assumed to substantially exceed the intrinsic local scale, the lipid length, which is on the order of molecular dimensions. The leading-order gradient terms are then quadratic. For this reason we suppress an otherwise permissible constitutive dependence on the variables $\nabla J \cdot \boldsymbol{b}(\nabla J)$, $\nabla\sigma \cdot \boldsymbol{b}(\nabla\sigma)$ and $\nabla J \cdot \boldsymbol{b}(\nabla\sigma)$, where $\boldsymbol{b}$ is the surface curvature tensor, on the grounds that these are cubic in the gradients and hence of higher order than those retained in (3.2).

3.2 *Equilibrium Theory*

When a potential energy exists for a given system it is conventional to regard equilibria as energy minimizers. This affords a straightforward and convenient framework for extracting the equilibrium equations via a variational procedure. Of course this presumes that any loading that may be present admits a load potential, so that an overall potential energy can be defined. When this is not the case the virtual-work theorem preserves the utility of the variational approach.

Suppose, for the moment, that there is no external loading applied to the bilayer. Then, the net free energy is precisely the potential energy and the problem is to minimize the expression

$$E = \int_{\omega} W da, \tag{3.3}$$

where ω is the surface currently occupied by the bilayer. To this end we re-write the energy in the form

$$E = \int_{\Omega} JW dA, \tag{3.4}$$

where Ω is a fixed reference surface and J is the surface dilation.

Equilibria render the energy stationary and thus satisfy $\dot{E} = 0$, where the superposed dot is the variational derivative. A direct derivation is pursued on the basis of (3.1), rather than by enforcing the constraint $\lambda J = 1$ via Lagrange multipliers. This procedure affords a more direct comparison with conventional models based on the Canham/Helfrich theory. Accordingly,

$$0 = \int_{\Omega} (\dot{J}W + J\dot{W}) dA = \int_{\omega} (\dot{W} + W\dot{J}/J) da, \tag{3.5}$$

where

$$\dot{W} = W_J \dot{J} + W_H \dot{H} + W_K \dot{K} + W_G \dot{G} + W_\sigma \dot{\sigma} + W_L \dot{L} \tag{3.6}$$

is the variation of the energy density. Here subscripts refer to partial derivatives with respect to the indicated variables. Using formulas developed in (Steigmann 2013) for the variational derivatives of the associated variables, it is then possible, following a lengthy and detailed procedure, to reduce the integrand in (3.5) to the form

$$\dot{W} + W\dot{J}/J = \varphi^{\alpha}_{;\alpha} - \boldsymbol{u} \cdot \boldsymbol{T}^{\alpha}_{;\alpha} + \mu\dot{\sigma} + (L^{-1} W_L a^{\alpha\beta} \sigma_{,\beta} \dot{\sigma})_{;\alpha}, \tag{3.7}$$

where $\boldsymbol{u}$ is the virtual displacement of the lipids (the variational derivative of the membrane position field); the Greek subscripts preceded by commas indicate partial derivatives with respect to the surface coordinates the subscripted semi-colons refer to surface covariant derivatives;

$$\mu = W_\sigma - (L^{-1} W_L a^{\alpha\beta} \sigma_{,\beta})_{;\alpha} \tag{3.8}$$

is the chemical potential; and,

$$\varphi^{\alpha} = \boldsymbol{T}^{\alpha} \cdot \boldsymbol{u} + \boldsymbol{M}^{\alpha\beta} \cdot \boldsymbol{u}_{,\beta} \tag{3.9}$$

with

$$\boldsymbol{T}^{\alpha} = \boldsymbol{N}^{\alpha} - \boldsymbol{M}^{\alpha\beta}_{;\beta}, \tag{3.10}$$

where

$$\boldsymbol{N}^{\alpha} = N^{\beta\alpha}\boldsymbol{a}_{\beta} + N^{\alpha}\boldsymbol{n} \quad \text{and} \quad \boldsymbol{M}^{\alpha\beta} = M^{\lambda\alpha\beta}\boldsymbol{a}_{\lambda} + M^{\alpha\beta}\boldsymbol{n}, \tag{3.11}$$

in which the coefficients are the constitutive functions. In the present context these are (Steigmann 2013)

$$\begin{aligned} N^{\beta\alpha} &= [W + JW_J - 2(HW_H + KW_K) + GW_G]a^{\alpha\beta} \\ &\quad + (JG^{-1}W_G J_{,\lambda} J_{,\mu} - L^{-1}W_L \sigma_{,\lambda}\sigma_{,\mu})a^{\alpha\lambda}a^{\beta\mu} + W_H \tilde{b}^{\alpha\beta}, \end{aligned} \tag{3.12}$$

$$N^{\alpha} = JG^{-1}W_G J_{,\lambda}(2Ha^{\lambda\alpha} - \tilde{b}^{\lambda\alpha}), \tag{3.13}$$

$$M^{\alpha\beta} = \tfrac{1}{2}W_H a^{\alpha\beta} + W_K \tilde{b}^{\alpha\beta} \tag{3.14}$$

and

$$M^{\lambda\alpha\beta} = \tfrac{1}{2}JG^{-1}W_G J_{,\mu}(a^{\alpha\mu}a^{\beta\lambda} + a^{\beta\mu}a^{\alpha\lambda}), \tag{3.15}$$

in which the effect of concentration gradient is reflected in (3.12). Here $a^{\alpha\beta}$ is the contravariant surface metric and $\tilde{b}^{\alpha\beta}$ is the cofactor of the surface curvature tensor. Because of the identities $\tilde{b}^{\alpha\beta}_{;\beta} = 0$, which are equivalent to the Mainardi–Codazzi compatibility conditions of surface theory (Steigmann 1999b), it is convenient to express the constitutive equations in terms of the cofactor of the curvature rather than the curvature itself, to facilitate simplification of the balance laws.

Loads admitted by the present model may include a distributed load over the surface ω, such as a normal pressure, as well as edge tractions and couple tractions, $\boldsymbol{f}$ and $\boldsymbol{c}$, representing forces and double forces per unit length. Additional terms arise from the effects of concentration. Here these are confined to a flux of power, ξ, due to variations in concentration at the edge of the bilayer. When these are operative, the stationary-energy statement $\dot{E} = 0$ is replaced by the virtual-work statement

$$\dot{E} = \int_{\omega} p\boldsymbol{n}\cdot\boldsymbol{u}da + \int_{\partial\omega}(\boldsymbol{f}\cdot\boldsymbol{u} + \boldsymbol{c}\cdot\boldsymbol{u}_{,\nu} + \xi\dot{\sigma})da, \tag{3.16}$$

where p is the net lateral pressure in the direction of the surface normal $\boldsymbol{n}$, due, for example, to osmotic pressure, and $\boldsymbol{u}_{,\nu}$ is the derivative of the virtual displacement in the direction of the normal to the edge $\partial\omega$, lying in the surface tangent plane. This statement is required to hold for arbitrary variations $\dot{\sigma}$ of the concentration and arbitrary kinematically admissible virtual displacements $\boldsymbol{u}$.

The necessary and sufficient conditions for this are the Euler–Lagrange equations (Steigmann 2013)

$$\boldsymbol{T}^{\alpha}_{;\alpha} + p\boldsymbol{n} = \boldsymbol{0} \quad \text{and} \quad \mu = 0 \quad \text{in} \quad \omega, \tag{3.17}$$

and the natural boundary conditions

$$\xi = L^{-1} W_L \boldsymbol{\nu} \cdot \nabla\sigma, \quad \boldsymbol{f} = \boldsymbol{T}^{\alpha}\nu_{\alpha} - (\boldsymbol{M}^{\alpha\beta}\nu_{\alpha}\tau_{\beta})' \quad \text{and} \quad \boldsymbol{c} = \boldsymbol{M}^{\alpha\beta}\nu_{\alpha}\nu_{\beta}, \tag{3.18}$$

where $\boldsymbol{\tau} = \boldsymbol{n} \times \boldsymbol{\nu}$ is the unit tangent to the edge and $(\cdot)'$ is the arclength derivative in the direction of $\boldsymbol{\tau}$. Alternative boundary conditions entail the specification of the concentration, position and surface orientation at an edge of the bilayer, in the case of open bilayers.

These equations, which have not appeared in the literature at this level of generality, are the necessary and sufficient conditions for the mechanical and chemical equilibrium of the lipid–protein bilayer.

The literature on lipid bilayers emphasizes the well-known shape equation. This is the projection of the equation of mechanical equilibrium onto the direction of the surface normal. The version of this equation in the present theory is somewhat more complicated than its classical counterpart. In the present context it is given by (cf. (1.87))

$$(N^{\alpha} - M^{\beta\alpha}_{;\beta} - M^{\lambda\beta\alpha}b_{\lambda\beta})_{;\alpha} + (N^{\beta\alpha} + M^{\lambda\alpha}b^{\beta}_{\lambda} - M^{\beta\mu\alpha}_{;\mu})b_{\beta\alpha} + p = 0, \tag{3.19}$$

where p is the net lateral pressure and $b_{\beta\alpha}$ are the components of the surface curvature. The tangential components of the equilibrium equation are (cf. (1.86))

$$(N^{\mu\alpha} + M^{\beta\alpha}b^{\mu}_{\beta} - M^{\mu\beta\alpha}_{;\beta})_{;\alpha} + (M^{\beta\alpha}_{;\beta} + M^{\lambda\beta\alpha}b_{\lambda\beta} - N^{\alpha})b^{\mu}_{\alpha} = 0. \tag{3.20}$$

In contrast to classical diffusion, in which the species concentration is uniformly distributed in thermodynamic equilibrium (granted compatible boundary conditions), in the present setting the coupling with bilayer shape means that in equilibrium the species concentration may be non-uniformly distributed. The present model thus has the potential to predict complex geometries that characterize structures such as the rough endoplasmic reticulum, where membrane shape is known to be correlated with non-uniform distributions of trans-membrane proteins.

3.3 *Diffusion*

To describe the dynamic evolution of the lipid–protein mixture, a model is proposed that accommodates protein diffusion and intra-membrane lipid viscosity, and their interplay with the evolving shape dynamics of the bilayer. This model is constructed so as to reduce to the foregoing static theory in the equilibrium limit, while ensuring that the modeling of transient response is inherently dissipative. The latter condition serves as surrogate for the second law of thermodynamics in the present setting. Elastic, diffusive and viscous effects are combined into a single model, in the spirit of similar work on coupled bulk-fluid/membrane dynamics described in (Arroyo and DeSimone 2009), on the understanding that not all effects can be expected to play equal roles in a given problem. In this way different regimes of response can be explored within a single over-arching framework.

In view of the small length scales involved in typical applications of the model, and the disparate time-scales associated with inertial and dissipative phenomena, the effects of mechanical inertia are here suppressed, so that the bilayer is deemed to remain in mechanical equilibrium during the course of diffusion and viscous flow; that is, the dynamics are regarded as being quasi-static from the purely mechanical point of view. This implies that Eqs. (3.17a) and (3.18b, c) remain in effect during the course of a dynamical process. The only additional balance law required is that associated with diffusive transport of the proteins over the evolving bilayer surface. In the absence of bulk protein sources supplied by the surrounding aqueous solution, this reduces to

$$\frac{d}{dt}\int_{\pi} \sigma da = -\int_{\partial\pi} \boldsymbol{m}\cdot\boldsymbol{\nu} ds, \tag{3.21}$$

where t is the time, $\boldsymbol{m}$ is the diffusive flux vector and π is an arbitrary subsurface of the bilayer surface ω. Converting the left-hand integral to a fixed surface, as in (3.4), carrying out the derivative, and invoking Stokes' theorem leads to the local form

$$\dot{\sigma} + \sigma\dot{J}/J + m^{\alpha}_{;\alpha} = 0, \tag{3.22}$$

where the third term involving the surface covariant derivative is the divergence of the flux, m^{α} are the contravariant components of the flux vector, and $\dot{\sigma}$ is the *material* time derivative (not to be confused with the variational derivative of the previous subsection); i.e., the time derivative following a lipid molecule. This is given by

$$\dot{\sigma} = \sigma_t + v^{\alpha}\sigma_{,\alpha}, \tag{3.23}$$

where σ_t is the time derivative at a fixed surface point and v^{α} are the components of the tangential velocity field.

The evolution of surface dilation is described by the kinematic rule $\dot{J}/J = v^{\alpha}_{;\alpha} - 2Hw$, where w is the velocity in the direction of the surface normal. Here and elsewhere, the velocity field is that of the lipids. This yields the evolution equation

$$(\ln J)_t + v^{\alpha}(\ln J)_{,\alpha} = v^{\alpha}_{;\alpha} - 2Hw \tag{3.24}$$

for the surface dilation. Using these relations, Eq. (3.22) may be reduced to the diffusive balance law

$$\sigma_t + v^{\alpha}\sigma_{,\alpha} + \sigma(v^{\alpha}_{;\alpha} - 2Hw) + m^{\alpha}_{;\alpha} = 0. \tag{3.25}$$

3.4 Dissipative Dynamic Evolution

Under isothermal conditions, and for negligible inertia, the energy balance is:

$$D = P - dE/dt, \tag{3.26}$$

where E is the energy content of the lipid–protein system, P is the power supplied to the system by mechanical means and by protein flux, and D is the dissipation, assumed to satisfy $D \geq 0$ in every process. This requirement serves as a surrogate for the second law of thermodynamics. The rate of change of the energy is

$$dE/dt = \int_{\pi} (\dot{W} + W\dot{J}/J)da, \tag{3.27}$$

as in (3.5), except that here the superposed dots refer to material time derivatives (time derivatives following fixed lipid molecules) rather than variational derivatives; π is an arbitrary subsurface of the bilayer surface ω. Further, (3.16) remains valid, with the same interpretation. Following the earlier variational development, this may be cast in the form

$$dE/dt = \int_{\partial\pi} \varphi^{\alpha}\nu_{\alpha}ds - \int_{\pi} \boldsymbol{u} \cdot \boldsymbol{T}^{\alpha}_{;\alpha}da + \int_{\pi} \mu\dot{\sigma}da + \int_{\partial\pi} \dot{\sigma}(L^{-1}W_L\boldsymbol{\nu} \cdot \nabla\sigma)ds, \tag{3.28}$$

in which the dots now refer to material time derivatives, while the power supply is given by

$$P = \int_{\omega} p\boldsymbol{n} \cdot \boldsymbol{u}da + \int_{\partial\omega} (\boldsymbol{f} \cdot \boldsymbol{u} + \boldsymbol{c} \cdot \boldsymbol{u}_{,\nu} + \xi\dot{\sigma} - \boldsymbol{q} \cdot \boldsymbol{\nu})da, \tag{3.29}$$

where $\boldsymbol{u}$ is now the material (lipid) velocity and $\boldsymbol{q}$ is the flux of energy transported by the proteins through the edge $\partial\omega$. This is given by

$$\boldsymbol{q} = \mu\boldsymbol{m}. \tag{3.30}$$

Substituting the diffusive balance law (3.22), which holds in ω, reduces (3.28) to

$$\begin{aligned} dE/dt = &\int_{\partial\pi} \varphi^{\alpha}\nu_{\alpha}ds - \int_{\pi} \boldsymbol{u} \cdot \boldsymbol{T}^{\alpha}_{;\alpha}da - \int_{\pi} \mu\sigma\dot{J}/Jda \\ &+ \int_{\pi} \boldsymbol{m} \cdot \nabla\mu da + \int_{\partial\pi} (\dot{\sigma}L^{-1}W_L\nabla\sigma - \mu\boldsymbol{m}) \cdot \boldsymbol{\nu}ds, \end{aligned} \tag{3.31}$$

where Stokes' theorem has been used to treat a term involving the divergence of the diffusive flux. Because $\dot{J}$ is linear in the velocity gradient, the associated term may be absorbed into the stress; that is, the stress vectors $\boldsymbol{T}^{\alpha}$ defined in (3.10)–(3.15) are amended by the addition of the terms $\mu\sigma\boldsymbol{a}^{\alpha}$, arising from $\dot{J}/J = \boldsymbol{a}^{\alpha} \cdot \boldsymbol{u}_{,\alpha}$, yielding a non-equilibrium contribution to the stress arising from diffusion. This has the effect of adding the expression $\mu\sigma a^{\beta\alpha}$ to the right-hand side of (3.12). The effects of intra-membrane lipid viscosity are taken into account by further appending a viscous stress $\pi^{\beta\alpha}$ to the right-hand side of the same equation, yielding the full expression for the non-equilibrium stress. This is given by (3.10)–(3.15), with (3.12) replaced by

$$N^{\alpha\beta} = [W + JW_J - 2(HW_H + KW_K) + GW_G]a^{\alpha\beta} + (JG^{-1}W_G J_{,\lambda}J_{,\mu} - L^{-1}W_L\sigma_{,\lambda}\sigma_{,\mu})a^{\alpha\lambda}a^{\beta\mu} + W_H\tilde{b}^{\alpha\beta} + \pi^{\alpha\beta} + \mu\sigma a^{\alpha\beta}, \tag{3.32}$$

which clearly reduces to (3.12) in (thermodynamic) equilibrium. Explicit constitutive equations for the dissipative terms are discussed below. With these adjustments to the stress, (3.31) reduces to

$$dE/dt = \int_{\partial\pi} \varphi^\alpha \nu_\alpha ds - \int_\pi \boldsymbol{u} \cdot \boldsymbol{T}^\alpha_{;\alpha} da - \int_\pi \boldsymbol{m} \cdot \nabla\mu da + \int_{\partial\pi} (\dot{\sigma} L^{-1} W_L \nabla\sigma - \mu\boldsymbol{m}) \cdot \boldsymbol{\nu} ds. \tag{3.33}$$

Imposing the mechanical equilibrium equation (3.17a) and the boundary conditions (3.18a, b) on the *non-equilibrium* stress, as appropriate for quasi-static evolution problems, Eqs. (3.29)–(3.31) are then used to reduce the expression for the dissipation, Eq. (3.26), to

$$D = \int_\pi (\tfrac{1}{2}\pi^{\alpha\beta}\dot{a}_{\alpha\beta} - \boldsymbol{m} \cdot \nabla\mu) da. \tag{3.34}$$

The arbitrariness of $\pi \subset \omega$ combines with the requirement $D \geq 0$ to imply that the integrand is pointwise non-negative on ω. These impose restrictions on the constitutive equations for the viscous stress $\pi^{\alpha\beta}$ and the diffusive flux $\boldsymbol{m}$. In conventional models in which these effects operate separately, this necessarily entails the restrictions

$$\pi^{\alpha\beta}\dot{a}_{\alpha\beta} \geq 0 \quad \text{and} \quad \boldsymbol{m} \cdot \nabla\mu \leq 0, \tag{3.35}$$

which are also sufficient for the dynamics to be dissipative in the general case. These ensure that the dynamical theory reduces to the equilibrium theory described in the previous section in the absence of dissipation; i.e., when the surface straining $\dot{a}_{\alpha\beta}$ and the diffusive flux $\boldsymbol{m}$ both vanish pointwise on the bilayer surface.

3.5 *Constitutive Equations for the Dissipative Variables*

Following Scriven (1960) and Aris (1989) the viscous stress is assumed to be given by the surface analog of the Navier–Stokes theory. Thus,

$$\pi^{\alpha\beta} = (\nu a^{\alpha\eta} a^{\beta\mu} + \epsilon E^{\alpha\eta\beta\mu})\dot{a}_{\eta\mu}, \tag{3.36}$$

where ν and ϵ are the positive intra-membrane viscosities, and

$$E^{\alpha\eta\beta\mu} = a^{\alpha\beta} a^{\eta\mu} + a^{\alpha\mu} a^{\eta\beta} - a^{\alpha\eta} a^{\beta\mu}. \tag{3.37}$$

In practice the surface straining is expressed in the form (Rangamani et al. 2013)

$$\dot{a}_{\alpha\beta} = v_{\alpha;\beta} + v_{\beta;\alpha} - 2wb_{\alpha\beta}, \tag{3.38}$$

in terms of the tangential velocities v_α and the normal velocity w.

Regarding the diffusive flux, for illustrative purposes we may adopt the simplest constitutive assumption; i.e.,

$$\boldsymbol{m} = -c\nabla\mu, \tag{3.39}$$

where c, another positive constant, is the diffusivity.

These relations satisfy (3.35) and hence ensure dissipative membrane dynamics. More general constitutive equations coupling viscosity and diffusion could also be considered in this framework.

3.6 *Convected Coordinates versus Surface-Fixed Coordinates*

In the foregoing we have implicitly made use of convected coordinates in the formulation of balance laws. We pause to elaborate on this formalism in the present subsection. We identify the underlying material manifold with a convected-coordinate system ξ^α. This may be identified with the system θ^α at a fixed instant t_0, say. The associated surface Ω, with parametric representation $\boldsymbol{x}(\xi^\alpha) = \boldsymbol{r}(\xi^\alpha, t_0)$, is fixed and may serve as a reference surface in a Lagrangian or referential description of the motion. That is, we regard these coordinates as being convected in the sense that they identify, via a map $\boldsymbol{r} = \hat{\boldsymbol{r}}(\xi^\alpha, t)$, the current position at time t of a material point that was located at $\boldsymbol{x}(\xi^\alpha) \in \Omega$ at time t_0. The notion may be generalized by regarding Ω as a surface that is in one-to-one correspondence with that occupied at time t_0, so that Ω need not actually be occupied in the course of the motion. The connection with the θ^α - parametrization of ω is provided by

$$\hat{\boldsymbol{r}}(\xi^\alpha, t) = \boldsymbol{r}(\theta^\alpha(\xi^\beta, t), t). \tag{3.40}$$

Thus, we specify the fixed surface coordinates θ^α as functions of ξ^α and t subject to $\theta^\alpha(\xi^\beta, t_0) = \xi^\alpha$. We assume the relations giving θ^α in terms of ξ^α to be invertible, to reflect the notion that at fixed t, the coordinates θ^α can be associated with a unique material point (identified by fixed values of ξ^α). Any function, $f(\theta^\alpha, t)$, say, may then be expressed in terms of convected coordinates as $\hat{f}(\xi^\alpha, t)$, where

$$\hat{f}(\xi^\alpha, t) = f(\theta^\alpha(\xi^\beta, t), t). \tag{3.41}$$

The material derivative of f is its partial time derivative in the convected-coordinate representation; i.e., $\dot{f} = \partial \hat{f}(\xi^\alpha, t)/\partial t$, whereas its local time derivative in the surface-fixed coordinate parametrization is $f_t = \partial f(\theta^\alpha, t)/\partial t$. By the chain rule, the two are related by $\dot{f} = f_t + (\theta^\alpha)^{\cdot} f_{,\alpha}$.

The velocity of a material point on Ω that has been convected by the motion to ω is $\boldsymbol{u} = \dot{\boldsymbol{r}} = \partial \hat{\boldsymbol{r}}/\partial t$. We may write this in terms of components on the natural basis induced by the surface-fixed (θ^α) parametrization. Thus,

$$\boldsymbol{u} = v^\alpha \boldsymbol{a}_\alpha + w\boldsymbol{n}. \tag{3.42}$$

This is not the same as the time derivative $\boldsymbol{r}_t$. However, the two are related by

$$\boldsymbol{u} = (\theta^\alpha)^{\cdot} \boldsymbol{a}_\alpha + \boldsymbol{r}_t. \tag{3.43}$$

Following Scriven (1960) we adopt the surface-fixed parametrization defined by

$$\tfrac{d}{dt}\theta^\alpha = v^\alpha(\theta^\beta, t), \quad \theta^\alpha_{|t_0} = \xi^\alpha, \tag{3.44}$$

where the derivative is evaluated at a fixed value of the doublet $\{\xi^\alpha\}$ and is therefore equal to $(\theta^\alpha)^{\cdot}$. Accordingly, the normal velocity in (3.42) is given by

$$w\boldsymbol{n} = \boldsymbol{r}_t, \tag{3.45}$$

and the convective and surface-fixed time derivatives satisfy the connection

$$\dot{f} = f_t + v^\alpha f_{,\alpha}. \tag{3.46}$$

We require an expression for the material derivative $\dot{a}_{\alpha\beta}$ in terms of the surface-fixed θ^α - parametrization. To this end we adopt convected coordinates ξ^α whose values coincide with the instantaneous values of θ^α. The two sets of coordinate systems will of course differ at different instants due to the fact that material is convecting with respect to the θ^α - system. Said differently, the material point instantaneously located at the place with surface-fixed coordinates θ^α will have different locations at different instants and hence be associated with different values of θ^α, whereas the values of ξ^α remain invariant. Accordingly, while it is always permissible to identify ξ^α with θ^α at any particular instant t_0, say, it is not possible to do so over an interval of time. However, for our purposes this limitation is not restrictive. Using $\dot{a}_{\lambda\mu} = \dot{\boldsymbol{a}}_\lambda \cdot \boldsymbol{a}_\mu + \boldsymbol{a}_\lambda \cdot \dot{\boldsymbol{a}}_\mu$ and

$$\dot{\boldsymbol{a}}_\lambda = (\partial \boldsymbol{r}/\partial\theta^\lambda)^{\cdot} = [\partial \boldsymbol{r}/\partial\xi^\mu(\partial\xi^\mu/\partial\theta^\lambda)]^{\cdot}$$
$$= \partial \boldsymbol{u}/\partial\xi^\mu(\partial\xi^\mu/\partial\theta^\lambda) + \partial \boldsymbol{r}/\partial\xi^\mu(\partial^2\xi^\mu/\partial\theta^\lambda\partial\theta^\alpha)v^\alpha, \tag{3.47}$$

together with $\partial\xi^\mu/\partial\theta^\lambda = \delta^\mu_\lambda$ (the Kronecker delta) and hence $\partial^2\xi^\mu/\partial\theta^\lambda\partial\theta^\alpha = 0$ at time t_0, we derive $\dot{\boldsymbol{a}}_\alpha = \partial\boldsymbol{u}/\partial\xi^\alpha$ and

$$\dot{a}_{\lambda\mu} = \boldsymbol{u}_{,\lambda}\cdot\boldsymbol{a}_\mu + \boldsymbol{a}_\lambda\cdot\boldsymbol{u}_{,\mu}, \tag{3.48}$$

where $\boldsymbol{u}_{,\lambda} = \partial\boldsymbol{u}/\partial\theta^\lambda$ at the considered instant.

Combining (3.42) with the Gauss and Weingarten equations yields

$$\boldsymbol{u}_{,\lambda} = (v_{\alpha;\lambda} - wb_{\alpha\lambda})\boldsymbol{a}^\alpha + (v^\alpha b_{\alpha\lambda} + w_{,\lambda})\boldsymbol{n}, \tag{3.49}$$

where $\boldsymbol{a}^\alpha = a^{\alpha\beta}\boldsymbol{a}_\beta$ and $v_{\alpha;\lambda}$ is the covariant derivative defined by

$$v_{\alpha;\lambda} = v_{\alpha,\lambda} - v_\beta\Gamma^\beta_{\alpha\lambda}, \tag{3.50}$$

in which $\Gamma^\beta_{\alpha\lambda}$ are the Christoffel symbols on ω computed using the θ^α - system. This delivers the expression (3.38); i.e.:

$$\dot{a}_{\lambda\mu} = v_{\mu;\lambda} + v_{\lambda;\mu} - 2wb_{\lambda\mu}. \tag{3.51}$$

The corresponding result in Aris' book (1989; Eqs. 10.21.3,4) is given, in our notation, by

$$\dot{a}_{\lambda\mu} = v_{\mu;\lambda} + v_{\lambda;\mu} + (a_{\lambda\mu})_t, \tag{3.52}$$

where $(\cdot)_t$ is computed at fixed θ^α. The latter is $(a_{\lambda\mu})_t = (\boldsymbol{a}_\lambda)_t\cdot\boldsymbol{a}_\mu + \boldsymbol{a}_\lambda\cdot(\boldsymbol{a}_\mu)_t$, where $(\boldsymbol{a}_\lambda)_t = (\boldsymbol{r}_{,\lambda})_t = (\boldsymbol{r}_t)_{,\lambda}$, and (cf. (3.45))

$$(\boldsymbol{r}_t)_{,\lambda} = (w\boldsymbol{n})_{,\lambda} = w_{,\lambda}\boldsymbol{n} - wb_{\lambda\alpha}\boldsymbol{a}^\alpha, \tag{3.53}$$

yielding $(a_{\lambda\mu})_t = -2wb_{\lambda\mu}$, in agreement with (3.51).

The foregoing relationships facilitate the derivation of balance laws. Thus, if f is the areal density of a particular quantity on ω, then the rate of change of the total quantity in a part π of ω is

$$\frac{d}{dt}\int_\pi f da = \frac{d}{dt}\int_\Pi fJdA = \int_\pi(\dot{f} + f\dot{J}/J)da, \tag{3.54}$$

where Π is the part of the fixed surface Ω that is convected to π and J is the local areal dilation of the surface; i.e.,

$$\int_\pi da = \int_\Pi JdA \quad \text{for all} \quad \Pi\subset\Omega. \tag{3.55}$$

To express the right-hand side of (3.54) in terms of the surface-fixed parametrization, we combine $\dot{J}/J = \frac{1}{2}a^{\alpha\beta}\dot{a}_{\alpha\beta}$ (Steigmann et al. 2003) with (3.51), obtaining

$$\dot{J}/J = v^{\alpha}_{;\alpha} - 2Hw. \tag{3.56}$$

For example, mass conservation is expressed by

$$\tfrac{d}{dt}\int_{\pi}\rho da = 0. \tag{3.57}$$

Using (3.54) with $f = \rho$ and invoking the arbitrariness of π then yields the local conservation law

$$0 = \dot{\rho} + \rho\dot{J}/J = \rho_t + v^{\alpha}\rho_{,\alpha} + \rho(v^{\alpha}_{;\alpha} - 2Hw). \tag{3.58}$$

It is well known that lipid membranes are relatively stiff against areal dilation in the absence of lipid distension, in comparison to bending or shearing in the tangent plane (Evans and Skalak 1980). To model this we can impose $J = 1$, as in the next section, as a local constraint at material points. In this case $\dot{J}$ vanishes and (3.58) simplifies to

$$0 = \dot{\rho} = \rho_t + v^{\alpha}\rho_{,\alpha}, \tag{3.59}$$

the first of which implies that ρ is independent of t in the convected-coordinate description; that is, ρ is independent of t when expressed as a function of ξ^{μ} and t. Accordingly, it's value at a particular material point is invariant in time and thus given by the density at that point in the fixed configuration associated with Ω. The constraint on J is thus seen to be equivalent to the constraint that ρ be invariant at any material point.

Suppose the membrane is such that the mass density is uniformly distributed on the fixed surface Ω used in the definition of convected coordinates. The presumed existence of such a configuration, even if it is never actually occupied in the course of the motion, may be taken as part of the definition of a uniform film. By the chain rule we then have $0 = \partial\rho/\partial\xi^{\alpha} = \rho_{,\beta}\partial\theta^{\beta}/\partial\xi^{\alpha}$. The presumed invertibility of the relation between the surface-fixed and convected coordinates implies that the matrix $(\partial\theta^{\beta}/\partial\xi^{\alpha})$ is non-singular and hence that $\rho_{,\beta} = 0$. The mass-conservation law (3.59) then yields $\rho_t = 0$, implying that ρ is a fixed constant on the surface ω, independent of θ^{α} and t. If the film is uniform in the sense described, then its response to H and K should be the same at all material points. There is then no explicit coordinate dependence in the areal energy density W. The latter result is modified in the case of films with non-uniform bending properties in which the non-uniformity is induced by a diffusing species, for example (Agrawal and Steigmann 2011). In the next section we show that it is also modified by viscous flow in the surface.

4 A Transport Theory Without Dilation or Distension

A much simpler model emerges in the limit of a dilute concentration of proteins. In this limit it is appropriate to assume that the embedded proteins have a negligible effect on surface dilation, which thus remains fixed at the value $J = 1$ in the absence of lipid distension. This affords the simplest generalization of the Canham/Helfrich framework to accommodate protein diffusion on the surface. The overall energy balance under isothermal conditions, and for negligible inertia, remains as before; namely:

$$D = P - dE/dt, \tag{4.1}$$

where E is the energy content of the lipid–protein system, P is the power supplied to the system by mechanical means and by protein flux, and D is the dissipation, assumed to satisfy $D \geq 0$ in every possible process.

4.1 *Energetics*

Let W again stand for the strain energy per unit area of the evolving membrane. It is conventional, in the standard Canham/Helfrich theory of lipid bilayers, to neglect areal dilation of the surface and thus to impose two-dimensional incompressibility as a constraint. Here we assume the distribution of trans-membrane proteins to be sufficiently dilute that the areal incompressibility constraint remains meaningful. In this case, the rate of change of the energy is

$$dE/dt = \int_{\pi} \dot{W} da, \tag{4.2}$$

where π is an arbitrary subsurface of the membrane surface ω; $\dot{W}$, the *material time derivative*, is given by

$$\dot{W} = W_H \dot{H} + W_K \dot{K} + W_\sigma \dot{\sigma}, \tag{4.3}$$

and σ is the areal protein density (i.e., number of proteins per unit current area of ω). Here we have assumed, in accordance with the standard model, that W is a function of the mean and Gaussian curvatures of the surface, H and K, respectively, in addition to the non-standard dependence on protein concentration. For simplicity's sake we suppress dependence on concentration gradient.

The power supply may be written in the form

$$P = P_{mech} - \int_{\partial\pi} \boldsymbol{q} \cdot \boldsymbol{\nu} ds, \tag{4.4}$$

where P_{mech} is the conventional power supplied by mechanical means and $\boldsymbol{q}$ is the energy flux due to protein transport across the boundary $\partial\pi$, with exterior unit normal $\boldsymbol{\nu}$ lying in the tangent plane of the surface. This is related to the state of the membrane and to the protein flux vector $\boldsymbol{m}$ by

$$\boldsymbol{q} = W_\sigma \boldsymbol{m}. \tag{4.5}$$

The stress power S is the power generated by internal interactions in the membrane. It satisfies (Rangamani et al. 2013)

$$S = dE/dt + \tfrac{1}{2}\int_\pi \pi^{\alpha\beta}\dot{a}_{\alpha\beta}da, \tag{4.6}$$

where $\dot{a}_{\alpha\beta}$, the material derivative of the evolving surface metric $a_{\alpha\beta}$, is the straining of the surface due to flow, and $\pi^{\alpha\beta}$ is the stress arising from viscous resistance to flow. Combining this with (4.1) and (4.4), we obtain

$$D = P - S + \tfrac{1}{2}\int_\pi \pi^{\alpha\beta}\dot{a}_{\alpha\beta}da = P_{mech} - S + \tfrac{1}{2}\int_\pi \pi^{\alpha\beta}\dot{a}_{\alpha\beta}da - \int_{\partial\pi} \boldsymbol{q}\cdot\boldsymbol{\nu}ds. \tag{4.7}$$

Let S_{mech} be the mechanical contribution to the stress power; i.e., the stress power in the absence of diffusion. This is given by (cf. (2.3) and (2.6))

$$S_{mech} = \int_\pi (W_H\dot{H} + W_K\dot{K})da + \tfrac{1}{2}\int_\pi \pi^{\alpha\beta}\dot{a}_{\alpha\beta}da, \tag{4.8}$$

and the total stress power is then given by

$$S = S_{mech} + \int_\pi W_\sigma\dot{\sigma}da. \tag{4.9}$$

As a consequence of the Principle of Virtual work, mechanical equilibrium of the membrane requires that

$$S_{mech} = P_{mech}. \tag{4.10}$$

Accordingly, the expression for the dissipation, due to viscosity and diffusion, reduces to

$$D = \tfrac{1}{2}\int_\pi \pi^{\alpha\beta}\dot{a}_{\alpha\beta}da - \int_{\partial\pi} \boldsymbol{q}\cdot\boldsymbol{\nu}ds - \int_\pi W_\sigma\dot{\sigma}da. \tag{4.11}$$

4.2 Balance Laws

The flux $\boldsymbol{m}$ satisfies the diffusive balance law (cf. (3.22))

$$\dot{\sigma} + m^{\alpha}_{;\alpha} = 0, \tag{4.12}$$

where the second term involving the surface covariant derivative is the divergence of the flux. Moreover, Eq. (4.10) generates three associated balance laws; namely, a normal equation and the two tangential equations of equilibrium (Rangamani et al. 2013)

$$(\partial W/\partial\theta^{\alpha} + \lambda_{,\alpha})a^{\beta\alpha} + \pi^{\beta\alpha}_{;\alpha} = 0, \tag{4.13}$$

where λ is a Lagrange-multiplier field associated with the local constraint of areal incompressibility; $\lambda_{,\alpha}$ is its gradient in the surface; $a^{\beta\alpha}$ is the dual, or reciprocal, surface metric; and the final term is the divergence of the viscous stress. Further, $\partial W/\partial\theta^{\alpha}$, in which θ^{α} are the surface-fixed coordinates, arises from the explicit coordinate dependence in the function W, regarded as a function of H and K. In the present context this is due to the presence of proteins, yielding

$$\partial W/\partial\theta^{\alpha} = W_{\sigma}\sigma_{,\alpha}. \tag{4.14}$$

The normal equation - the so-called *shape equation* - is given by (Rangamani et al. 2013)

$$p = \Delta(\tfrac{1}{2}W_H) + (W_K)_{;\alpha\beta}\tilde{b}^{\alpha\beta} + W_H(2H^2 - K) + 2H(KW_K - W) - 2\lambda H + \pi^{\alpha\beta}b_{\alpha\beta}, \tag{4.15}$$

where p is the lateral pressure on the membrane due, for example, to osmotic pressure, Δ is the surface Laplacian, $b_{\alpha\beta}$ is the 2nd fundamental form (the curvature) of the surface, $\tilde{b}^{\alpha\beta}$ is its cofactor, and $(\cdot)_{;\alpha\beta}$ is the second covariant derivative on the surface. These constitute the partial differential equations to be solved for the surface flow, the surface shape, the Lagrange multiplier and the protein distribution. The system is completed by appending the areal incompressibility constraint. In the present setting this is

$$v^{\alpha}_{;\alpha} = 2Hw, \tag{4.16}$$

where v^{α} is the tangential velocity vector and w is the (scalar) normal velocity.

The diffusive balance law facilitates a useful reduction of the expression for the dissipation. Combining (4.5) with (4.6) and applying Stokes' theorem on the surface, we obtain

$$D = \int_{\pi} [\tfrac{1}{2}\pi^{\alpha\beta}\dot{a}_{\alpha\beta} - m^{\alpha}(W_{\sigma})_{,\alpha}]da. \tag{4.17}$$

4.3 Constitutive Equations

To ensure that membrane processes are dissipative, it is sufficient to adopt the constitutive equations (cf. (3.36) and (3.37))

$$\pi^{\alpha\beta} = \nu a^{\alpha\eta} a^{\beta\mu} \dot{a}_{\eta\mu}, \tag{4.18}$$

where ν, a positive constant, is the intra-membrane surface viscosity, and (cf. (3.39))

$$m^{\alpha} = -c a^{\alpha\beta} (W_{\sigma})_{,\beta}, \tag{4.19}$$

where c, another positive constant, is the diffusivity. These are the relations conventionally adopted in separate studies (Agrawal and Steigmann 2011; Rangamani et al. 2013) concerned either with viscous surface flow or protein diffusion. Other possibilities coupling viscosity and diffusion can be proposed, provided that they conform to the requirement of non-negative dissipation.

Regarding the energy, the appropriate specialization of (3.2) is

$$W(H, K, \sigma) = \alpha(\sigma) + k(\sigma)[H - C(\sigma)]^2 + \bar{k}(\sigma)K. \tag{4.20}$$

The functions α, k and $\bar{k}$ are phenomenological in nature and plausible forms for them must be specified. The foregoing balance laws simplify accordingly, but these simpler forms are not recorded here.

In the limit of a dilute concentration of proteins, σ is small enough to justify use of the approximations

$$\begin{aligned} \alpha(\sigma) &= \alpha(0) + \sigma\alpha'(0) + \tfrac{1}{2}\sigma^2\alpha''(0) + \ldots, \\ k(\sigma) &= k(0) + \sigma k'(0) + \ldots, \\ \bar{k}(\sigma) &= \bar{k}(0) + \sigma\bar{k}'(0) + \ldots \quad \text{and} \\ C(\sigma) &= C(0) + \sigma C'(0) + \ldots . \end{aligned} \tag{4.21}$$

Suppose the membrane forms a natural bilayer in the absence of proteins, so that $C(0) = 0$. Suppose further that the state $\sigma \equiv 0$ is in thermodynamic equilibrium when the surface is flat. Then, W_{σ} vanishes at $\{H, K, \sigma\} = \{0, 0, 0\}$, yielding

$$\alpha(\sigma) \simeq \tfrac{1}{2}\sigma^2\alpha''(0), \quad k(\sigma) \simeq k(0) \equiv k, \quad \bar{k}(\sigma) \simeq \bar{k}(0) \equiv \bar{k} \quad \text{and} \quad C(\sigma) \simeq \sigma C'(0). \tag{4.22}$$

We note that in this limit the constant $\bar{k}$ does not contribute to the equilibrium Eq. (4.15) and may therefore be dropped if the membrane forms a closed surface without boundary. As for the spontaneous curvature, we assume that $C'(0) = \beta\varphi$, where β is an empirical constant and φ is the protein cone angle characterizing the conical geometry of the embedded trans-membrane proteins. Accordingly, $C(\sigma) \simeq 0$

for cylindrical proteins ($\varphi = 0$), regardless of protein density. The expression for the energy reduces to

$$W(H, K, \sigma) = \tfrac{1}{2}\sigma^2\alpha''(0) + k[H - \sigma(\beta\varphi)]^2 + \bar{k}K. \quad (4.23)$$

Numerical solutions obtained using this model are described in (Agrawal and Steigmann 2011).

5 Electromechanics of Polarized Lipid Bilayers

In this final section a model for the electromechanics of lipid bilayers, accounting for flexo-electricity, is obtained as the thin-film limit of the continuum electrodynamics of nematic liquid crystals. A priori restrictions on the polarization field consistent with minimum energy considerations effectively decouple the leading-order membrane problem from the computation of the self field, yielding a substantial simplification vis a vis the three-dimensional theory. While there is a substantial existing literature on this subject, it is marred by the arbitrary suppression of the electric self field on an ad hoc basis, and by the assumption that the polarization field in the membrane may be regarded as a material vector, perpetually aligned along the direction of opposing polar head groups constituting the lipids of the bilayer (Petrov 1999). Neither assumption is compatible with continuum electrodynamics. The latter is used here to derive an asymptotic leading-order-in-thickness model in which the electric self field plays no role. The self field, if desired, can then be computed a posteriori, thereby affording a major simplification of the theory. This, however, requires that the polarization field adjust accordingly.

The idea that lipid bilayers can be regarded as thin liquid crystal films apparently originated in the work of Helfrich (1973). This point of view gave rise to an associated body of work that has been reviewed in (Ou-Yang et al. 1999). As we have seen, the liquid-crystal framework provides a clear conceptual foundation for extensions of the basic purely mechanical theory to coupled-field problems; here this notion is extended to encompass electromechanical interactions. We adapt the three-dimensional electromechanical liquid crystal theory advanced in (DeGennes and Prost 1992; Ericksen 1961, 1962, 1976; Virga 1994) to derive a two-dimensional model for the response of electrically polarized lipid bilayers to applied electric fields generated by a remote source. Our approach differs substantially from those described in (Gao et al. 2008; Mohammidi et al. 2014). For definiteness and for the sake of simplicity, we base our model on the general theory for nematics (Virga 1994), while allowing for the so-called flexo-electric effect (Meyer 1969; DeGennes and Prost 1992).

5.1 Energetics of Three-Dimensional Liquid Crystals in the Presence of a Stationary Applied Field

Numerous variational formulations of electromechanical interactions in deformable media are available in the literature. These have been studied extensively in (Bustamante et al. 2009; Dorfmann and Ogden 2014), to which the interested reader is referred for fuller expositions. There it is shown that Maxwell's equations and the equilibrium equations for a polarized medium in the presence of an applied electric field that is fixed in space, in the absence of applied loads or free electric charges, are such as to render stationary the energy functional

$$\mathcal{E} = \int_R (U - \tfrac{1}{2}\boldsymbol{e}_s \cdot \boldsymbol{p} - \boldsymbol{e}_a \cdot \boldsymbol{p})dv, \tag{5.1}$$

where R is the volume currently occupied by the material in three-space; U is the relevant energy density; $\boldsymbol{p}$ is the polarization per unit volume; $\boldsymbol{e}_s$ is the electric self-field generated by the polarized material; and $\boldsymbol{e}_a$ is the applied electric field, assumed to be assigned as a smooth function in all of three-space, including R. The net electric field is

$$\boldsymbol{e} = \boldsymbol{e}_s + \boldsymbol{e}_a. \tag{5.2}$$

Further, the applied field is a given function of position $\boldsymbol{y}$ in the enveloping three-space. Its variational derivative, associated with a fixed material point (a fixed lipid molecule in the present context), is thus purely convective; i.e.,

$$\dot{\boldsymbol{e}}_a = (grad\boldsymbol{e}_a)\dot{\boldsymbol{y}}, \tag{5.3}$$

where $grad$ is the (spatial) gradient with respect to $\boldsymbol{y}$; and, here and henceforth, superposed dots are used to denote variational derivatives. The field $\boldsymbol{e}_a$ is curl-free; its gradient is therefore symmetric: $grad\boldsymbol{e}_a = (grad\boldsymbol{e}_a)^t$. The self field is also curl free; it is obtained from

$$\boldsymbol{e}_s = -grad V_s, \tag{5.4}$$

where the self-field potential V_s is given by (Kovetz 2000)

$$4\pi\varepsilon_0 V_s(\boldsymbol{y}) = \int_{\partial R} \frac{\boldsymbol{p}' \cdot \boldsymbol{n}'}{|\boldsymbol{y}-\boldsymbol{y}'|} da - \int_R \frac{div'\boldsymbol{p}'}{|\boldsymbol{y}-\boldsymbol{y}'|} dv, \tag{5.5}$$

in which ε_0 is the free-space permittivity; $\boldsymbol{n}'$ is the exterior unit normal to ∂R, expressed as a function of the integration variable $\boldsymbol{y}'$; $\boldsymbol{p}'$ is likewise the polarization in terms of $\boldsymbol{y}'$; and div' is the divergence with respect to $\boldsymbol{y}'$.

The energy density U is a function of the polarization and appropriate deformation variables. In the conventional theory of electro-elasticity the relevant deformation variable is the deformation gradient, the gradient of $\boldsymbol{y} = \chi(\boldsymbol{x})$ with respect to position

$\boldsymbol{x}$ in some fixed reference configuration. Here $\chi(\cdot)$ is a field describing the deformation of material points. In the present application to liquid crystals, the relevant variables are a director field $\boldsymbol{d}(y)$—describing the orientation of the liquid crystal molecules—and its spatial gradient (cf. (1.1))

$$\boldsymbol{D} = grad\boldsymbol{d}. \tag{5.6}$$

We follow the conventional theory and impose $|\boldsymbol{d}(y)| = 1$.

The electric field is given in terms of the polarization by the partial derivative (Toupin 1956; Bustamante et al. 2009)

$$\boldsymbol{e} = U_{\boldsymbol{p}}(\boldsymbol{d}, \boldsymbol{D}, \boldsymbol{p}). \tag{5.7}$$

In applications U is typically assumed to be a quadratic function of $\boldsymbol{D}$, as we have seen. This reflects the notion that the length scale for spatial variations of the director is typically much larger than the local length scale: the molecular length; the dimensionless gradient is then sufficiently small to justify the termination of the Taylor expansion of $U(\boldsymbol{d}, \cdot, \boldsymbol{p})$ at second order. Thus (cf. (1.2)),

$$U = l(\boldsymbol{d}, \boldsymbol{p}) + \boldsymbol{L}(\boldsymbol{d}, \boldsymbol{p}) \cdot \boldsymbol{D} + \tfrac{1}{2}\boldsymbol{D} \cdot \mathcal{L}(\boldsymbol{d}, \boldsymbol{p})[\boldsymbol{D}], \tag{5.8}$$

in which l, $\boldsymbol{L}$ and $\mathcal{L}$ are scalar, second-order tensor and fourth-order tensor-valued functions, respectively, with $\mathcal{L} = \mathcal{L}^t$.

Guided by (Virga 1994), we adopt the specific forms

$$l(\boldsymbol{d}, \boldsymbol{p}) = \tfrac{1}{2}\chi_\perp |\boldsymbol{p}|^2 + \tfrac{1}{2}\chi_a(\boldsymbol{p} \cdot \boldsymbol{d})^2, \tag{5.9}$$

where $\chi_\perp$ and χ_a are the anisotropic dielectric constants, and

$$\boldsymbol{D} \cdot \mathcal{L}[\boldsymbol{D}] = k_1(div\boldsymbol{d})^2 + k_2(\boldsymbol{d} \cdot curl\boldsymbol{d})^2 + k_3 |\boldsymbol{D}\boldsymbol{d}|^2 + (k_2 + k_4)[tr(\boldsymbol{D}^2) - (div\boldsymbol{d})^2], \tag{5.10}$$

in which the latter is independent of $\boldsymbol{p}$, and $k_1 - k_4$ are constants with $2k_1 \geq k_2 + k_4$, $k_2 \geq |k_4|$ and $k_3 \geq 0$, in accordance with the presumed positive-definiteness of $\mathcal{L}$ (Virga 1994). The second expression is the Frank energy for nematic liquid crystals.

To model the flexo-electric effect, we adopt Meyer's proposal (Meyer 1969) in a form similar to that adopted by Ou-Yang et al. (1999). Thus,

$$\boldsymbol{L}(\boldsymbol{d}, \boldsymbol{p}) \cdot \boldsymbol{D} = -\boldsymbol{p} \cdot \boldsymbol{f}(\boldsymbol{d}, \boldsymbol{D}), \quad \text{with} \quad \boldsymbol{f}(\boldsymbol{d}, \boldsymbol{D}) = c_1(div\boldsymbol{d})\boldsymbol{d} + c_2 curl\boldsymbol{d} \times \boldsymbol{d}, \tag{5.11}$$

where c_1 and c_2 are the flexo-electric constants. The relationship (5.7) then furnishes an expression for the electric field:

$$\boldsymbol{e} = \chi_\perp \boldsymbol{p} + \chi_a(\boldsymbol{p} \cdot \boldsymbol{d})\boldsymbol{d} - \boldsymbol{f}(\boldsymbol{d}, \boldsymbol{D}), \tag{5.12}$$

which reduces, in the specialization $\chi_a = \chi_\perp$, to Eq. (2.153) in Ou-Yang et al. (1999) in the case when the electric field vanishes.

Our objective here is to derive the leading-order small-thickness limit of the energy (5.1). This limit is taken to be the energy of a polarized lipid membrane. Stationarity conditions for the limit energy are then identified with the equilibrium equations of a polarized lipid membrane in the presence of an applied field generated by a remote source.

We have in mind a lipid bilayer constituting a membrane structure in a biological cell. Because such a membrane is only one or two molecules across, its thickness is on the order of the local length scale embodied in the constitutive response of the liquid crystal. Accordingly, this is the only length scale arising in the dimension reduction procedure.

5.2 Liquid Crystal Films

In the purely mechanical theory of thin liquid crystal films the leading-order strain-energy density W is associated with the limit (Steigmann 2013)

$$\lim_{t \to 0} t^{-1} \int_R U dv = \int_\omega W da, \tag{5.13}$$

where ω is the interior midsurface of the film, t is the (uniform) thickness of the film, and

$$W = U_{|\omega} \tag{5.14}$$

is the leading-order energy density on ω. This follows by using the volume measure $dv = \mu d\varsigma da$ (Naghdi 1972), where ς is a linear coordinate in the direction of the unit surface normal $\boldsymbol{n}$, regarded as the restriction of $\boldsymbol{d}$ to ω, and $\mu = 1 - 2\varsigma H + \varsigma^2 K$, where H and K, respectively, are the mean and Gaussian curvatures of ω. In effect, then, we suppress misalignment of the lipid molecules with the surface normal - the so-called lipid tilt - as in the classical Canham/Helfrich theory. This is appropriate if the surface density of the lipids is sufficiently high. We have (cf. (1.31))

$$\boldsymbol{n} = \boldsymbol{d}_{|\omega} \quad \text{and} \quad \boldsymbol{D}_{|\omega} = \nabla \boldsymbol{n} + \boldsymbol{\eta} \otimes \boldsymbol{n}, \tag{5.15}$$

where $\nabla(\cdot)$ is the (two-dimensional) surface gradient on ω and $\boldsymbol{\eta}$ is the restriction to ω of the derivative of $\boldsymbol{d}$ in the direction of $\boldsymbol{d}$. Accordingly,

$$\boldsymbol{n} \cdot \boldsymbol{\eta} = 0 \quad \text{on} \quad \omega. \tag{5.16}$$

The extension to the case of polarizable films in the presence of an applied field is immediate. Thus,

$$\lim_{t\to 0} t^{-1}\mathcal{E} = E, \tag{5.17}$$

with

$$E = \int_\omega W da, \tag{5.18}$$

where W is now given by

$$W = (U - \tfrac{1}{2}\boldsymbol{e}_s \cdot \boldsymbol{p} - \boldsymbol{e}_a \cdot \boldsymbol{p})_{|\omega}. \tag{5.19}$$

Remark 1 Quantum mechanical considerations and molecular dynamics simulations (Seelig 1978; Frischleder and Peinel 1982; Warshaviak et al. 1808) indicate that the polarization vector is essentially tangential to the film surface. In this case an estimate based on (5.5)—derived in the mathematically identical context of magneto-statics (Barham et al. 2012)—indicates that the magnitude $|\boldsymbol{e}_s|$ of the self field is of order $O(t\ln t)$; for small t this is negligible compared to unity. It follows that the *leading-order* energy; i.e., the limit of $\mathcal{E}/t$ as $t\to 0$, is given by (5.18) but with W given by

$$W = U_{|\omega} - \boldsymbol{e}_a(\boldsymbol{r})\cdot\boldsymbol{\pi}, \tag{5.20}$$

where $\boldsymbol{\pi} = \boldsymbol{p}_{|\omega}$, $\boldsymbol{r} = \boldsymbol{y}_{|\omega}$ is the position field on ω, and

$$\boldsymbol{n}\cdot\boldsymbol{\pi} = 0 \quad \text{on} \quad \omega. \tag{5.21}$$

Remark 2 The estimate on the self field effectively decouples its computation from the problem of rendering E stationary, implying that it may be evaluated *a posteriori*. This feature affords a major simplification of the theory for thin films *vis a vis* that for bulk continua. Further, in the analogous magnetostatic setting, the condition (5.21), with polarization replaced by magnetization, is known to furnish energetically optimal states of magnetization in thin films (Gioia and James 1997). Thus our approach via dimension reduction provides justification for the suppression of the self field, in the *leading-order* two-dimensional model, under conditions in which the polarization field is tangential to the membrane. In contrast, in (Gao et al. 2008; Mohammidi et al. 2014) no analysis is offered to justify the suppression of the self field.

The self field at points in space remote from the membrane may be evaluated *a posteriori* by applying the divergence theorem to (5.5), for points $\boldsymbol{y}$ not in R. Thus,

$$4\pi\varepsilon_0 V_s(\boldsymbol{y}) = \int_R \boldsymbol{p}'\cdot\frac{\boldsymbol{y}-\boldsymbol{y}'}{|\boldsymbol{y}-\boldsymbol{y}'|^3}dv = t\left[\int_\omega \boldsymbol{\pi}\cdot\frac{\boldsymbol{y}-\boldsymbol{r}}{|\boldsymbol{y}-\boldsymbol{r}|^3}da + o(t)/t\right], \tag{5.22}$$

in which $\boldsymbol{r}$ is the membrane position field. The self field then follows by computing the gradient with respect to $\boldsymbol{y}$ (cf. (5.4)):

$$4\pi\varepsilon_0 \lim_{t\to 0}(t^{-1}\boldsymbol{e}_s) = \int_\omega \boldsymbol{G}\boldsymbol{\pi} da, \tag{5.23}$$

where

$$\boldsymbol{G} = \tfrac{3}{|\boldsymbol{y}-\boldsymbol{r}|^5}(\boldsymbol{y}-\boldsymbol{r}) \otimes (\boldsymbol{y}-\boldsymbol{r}) - \tfrac{1}{|\boldsymbol{y}-\boldsymbol{r}|^3}\boldsymbol{I}. \tag{5.24}$$

The leading-order self field in space is thus delivered by a quadrature over ω after membrane shape has been determined.

In the expression (5.20) we have

$$U_{|\omega} = U(\boldsymbol{n}, -\boldsymbol{b} + \boldsymbol{\eta} \otimes \boldsymbol{n}, \boldsymbol{\pi}), \tag{5.25}$$

where

$$\boldsymbol{b} = -\nabla \boldsymbol{n} \tag{5.26}$$

is the (symmetric) curvature tensor of ω.

The explicit form used here follows from (5.8)–(5.11). For example, the restriction to ω of the function $\boldsymbol{f}(\boldsymbol{d}, \boldsymbol{D})$ in (5.11) is given by

$$\boldsymbol{f}_{|\omega} = c_1(\boldsymbol{n} \cdot \boldsymbol{\eta} - 2H)\boldsymbol{n} + c_2\boldsymbol{\eta}, \tag{5.27}$$

where

$$H = \tfrac{1}{2} tr\boldsymbol{b} \tag{5.28}$$

is the mean curvature of ω. This expression may be simplified by imposing (5.16), but we refrain from doing so for reasons discussed below. Here use has been made of (5.15) and the formula $curl\boldsymbol{d} \times \boldsymbol{d} = \boldsymbol{D}\boldsymbol{d}$, which follows from the fact that $\boldsymbol{d}$ is a field of unit vectors (Virga 1994).

To reduce (5.10) we first introduce a coordinate parametrization $\boldsymbol{r}(\theta^\alpha)$ of ω. This induces the natural tangent basis $\boldsymbol{a}_\alpha = \boldsymbol{r}_{,\alpha}$ and associated dual basis $\boldsymbol{a}^\alpha$, where $(\cdot)_{,\alpha} = \partial(\cdot)/\partial\theta^\alpha$. Then, the restriction of $curl\boldsymbol{d}$ to ω is (Steigmann 2013)

$$(curl\boldsymbol{d})_{|\omega} = \boldsymbol{a}^\alpha \times \boldsymbol{n}_{,\alpha} + \boldsymbol{n} \times \boldsymbol{\eta}, \tag{5.29}$$

where $\boldsymbol{a}^\alpha \times \boldsymbol{n}_{,\alpha} = -b_{\alpha\beta}\boldsymbol{a}^\alpha \times \boldsymbol{a}^\beta$, with $b_{\alpha\beta} = \boldsymbol{a}_\alpha \cdot \boldsymbol{b}\boldsymbol{a}_\beta$, vanishes by virtue of the symmetry of $\boldsymbol{b}$; accordingly, $(\boldsymbol{d} \cdot curl\boldsymbol{d})_{|\omega} = 0$.

Using (5.26) with $\boldsymbol{b}\boldsymbol{n} = \boldsymbol{0}$ we also derive

$$tr(\boldsymbol{D}^2)_{|\omega} = tr(\boldsymbol{b}^2) + (\boldsymbol{n} \cdot \boldsymbol{\eta})^2. \tag{5.30}$$

Applying the Cayley–Hamilton formula

$$\boldsymbol{b}^2 = 2H\boldsymbol{b} + K\mathbf{1}, \tag{5.31}$$

where

$$K = \det \boldsymbol{b} \tag{5.32}$$

is the Gaussian curvature of ω and $\mathbf{1}$ is the identity transformation on T_ω, we then obtain

$$[tr(\boldsymbol{D}^2) - (div\boldsymbol{d})^2]_{|\omega} = 2H\boldsymbol{n} \cdot \boldsymbol{\eta} - 2K, \tag{5.33}$$

which again may be simplified by imposing (5.16).
Remark 3 It is well known (Virga 1994) that the combination $tr(\boldsymbol{D}^2) - (div\boldsymbol{d})^2$ is a null Lagrangian in the three-dimensional theory. It is also well known that the Gaussian curvature is a null Lagrangian in the surface theory; in particular, the total curvature of a closed surface is fixed by its genus and thus contributes only a disposable constant to the energy if the surface topology is fixed. Accordingly, (5.16) implies that the same combination of terms also furnishes a null Lagrangian in the two-dimensional theory.

Altogether, the surface energy (5.8) reduces to

$$\begin{aligned} U_{|\omega} &= \tfrac{1}{2}\chi_\perp |\boldsymbol{\pi}|^2 + \tfrac{1}{2}\chi_a(\boldsymbol{\pi} \cdot \boldsymbol{n})^2 - c_1(\boldsymbol{n} \cdot \boldsymbol{\eta} - 2H)\boldsymbol{n} \cdot \boldsymbol{\pi} - c_2\boldsymbol{\eta} \cdot \boldsymbol{\pi} \\ &\quad + \tfrac{1}{2}k_1(\boldsymbol{n} \cdot \boldsymbol{\eta} - 2H)^2 + \tfrac{1}{2}k_3 |\boldsymbol{\eta}|^2 + (k_2 + k_4)(H\boldsymbol{n} \cdot \boldsymbol{\eta} - K), \end{aligned} \tag{5.34}$$

yielding the net energy density in the form:

$$W = kH^2 + \bar{k}K + \tfrac{1}{2}k_3 |\boldsymbol{\eta}|^2 + \tfrac{1}{2}\chi_\perp |\boldsymbol{\pi}|^2 - c_2\boldsymbol{\eta} \cdot \boldsymbol{\pi} + \tilde{\varphi}\boldsymbol{n} \cdot \boldsymbol{\pi} + \tilde{\psi}\boldsymbol{n} \cdot \boldsymbol{\eta} - \boldsymbol{e}_a(\boldsymbol{r}) \cdot \boldsymbol{\pi}, \tag{5.35}$$

where

$$k = 2k_1, \quad \bar{k} = -(k_2 + k_4) \tag{5.36}$$

and $\tilde{\varphi}$, $\tilde{\psi}$ are certain scalars which will prove to be irrelevant. Then, W may be regarded as a function of the list

$$\{H, K, \boldsymbol{r}, \boldsymbol{n}, \boldsymbol{\eta}, \boldsymbol{\pi}\}, \tag{5.37}$$

subject to the constraints (5.16) and (5.21), in which it is understood that H, K and $\boldsymbol{n}$ are determined by the surface parametrization $\boldsymbol{r}(\theta^\alpha)$. Henceforth we require the doublet $\{\theta^\alpha\}$ to maintain a fixed correspondence with a material point; i.e., a lipid molecule. Thus the coordinates are convected with the lipids.

5.3 Variational Problem and Equilibrium Equations

It is convenient to adopt an extended variational formulation in which the constraints are relaxed. In this formulation we do not impose (5.16) or (5.21), but instead consider the auxiliary energy

$$E^* = \int_{\Omega} [JW + \lambda(J-1) + \bar{\varphi}\boldsymbol{n}\cdot\boldsymbol{\pi} + \bar{\psi}\boldsymbol{n}\cdot\boldsymbol{\eta}]dA, \tag{5.38}$$

where W is given by (5.35), with $\boldsymbol{r}$, $\boldsymbol{\eta}$ and $\boldsymbol{\pi}$ regarded as independent fields; and where $\bar{\varphi}$ and $\bar{\psi}$ are Lagrange-multiplier fields associated with the constraints (5.16) and (5.21). Here Ω is the pre-image of ω in a fixed reference placement, with $da = JdA$. In terms of the convected-coordinate parametrization we have $J = \sqrt{a/A}$, where $a = \det(a_{\alpha\beta})$, $a_{\alpha\beta} = \boldsymbol{a}_\alpha \cdot \boldsymbol{a}_\beta$ is the surface metric, and A is the value of a on Ω. Further, λ is a Lagrange-multiplier field associated with the constraint that the map from any configuration to another preserves local surface area; i.e., that $J = 1$. This restriction is appropriate in the absence of lipid distension, as in the classical Canham/Helfrich theory. As we have seen, the bulk incompressibility of the liquid crystal then implies that area is preserved locally. Generalizations to account for distension in a purely mechanical setting are discussed in Sect. 1 above and in (Kim and Steigmann 2015).

We observe that $\tilde{\varphi}$ and $\tilde{\psi}$ in (5.35) may be absorbed into the Lagrange multipliers and conclude that no generality is lost if the former is replaced by

$$W = U - \boldsymbol{e}_a(\boldsymbol{r})\cdot\boldsymbol{\pi} \tag{5.39}$$

in (5.38), where U is now given by

$$U = kH^2 + \bar{k}K + \tfrac{1}{2}k_3\,|\boldsymbol{\eta}|^2 + \tfrac{1}{2}\chi_\perp\,|\boldsymbol{\pi}|^2 - c_2\boldsymbol{\eta}\cdot\boldsymbol{\pi}. \tag{5.40}$$

We note that the quadratic form involving $\boldsymbol{\eta}$ and $\boldsymbol{\pi}$ is positive definite if and only if $k_3 > 0$, $\chi_\perp > 0$ and $c_2^2 < k_3\chi_\perp$.

The variational derivative of the extended energy, neglecting the variations of the Lagrange multipliers, is

$$\dot{E}^* = \int_{\omega} [\dot{W} + (W+\lambda)\dot{J}/J + \varphi(\dot{\boldsymbol{n}}\cdot\boldsymbol{\pi} + \boldsymbol{n}\cdot\dot{\boldsymbol{\pi}}) + \psi(\dot{\boldsymbol{n}}\cdot\boldsymbol{\eta} + \boldsymbol{n}\cdot\dot{\boldsymbol{\eta}})]da, \tag{5.41}$$

where $\varphi = \bar{\varphi}/J$ and $\psi = \bar{\psi}/J$, and it is understood, having suppressed the variations of the multipliers, that all terms in this expression are to be evaluated *post facto* at states satisfying the constraints (5.16) and (5.21). In the presence of a net lateral pressure p in the direction of the surface normal $\boldsymbol{n}$, the virtual-work statement is given by (Agrawal and Steigmann 2009)

$$\dot{E}^* = \int_{\omega} p\boldsymbol{n}\cdot\dot{\boldsymbol{r}}da + \int_{\partial\omega} \chi ds, \tag{5.42}$$

where χ is the density of edge power. We remark that because of the definition (5.14), the energy in this expression is the actual energy divided by the thickness t; the dimensions of p and χ are affected accordingly. Thus, for example, the actual pressure is tp, and with $p = O(1)$ this is of order t.

We consider the consequences of (5.42) with respect to variations of each variable in turn. The simplest are those associated with the variations $\dot{\boldsymbol{\pi}}$ and $\dot{\boldsymbol{\eta}}$. They are given, respectively, by (cf. (5.7) and (5.15))

$$\boldsymbol{e}_a = \boldsymbol{U}_\pi + \varphi \boldsymbol{n} \tag{5.43}$$

and

$$\boldsymbol{U}_\eta + \psi \boldsymbol{n} = \boldsymbol{0}, \tag{5.44}$$

with

$$\boldsymbol{U}_\pi = \chi_\perp \boldsymbol{\pi} - c_2 \boldsymbol{\eta} \quad \text{and} \quad \boldsymbol{U}_\eta = k_3 \boldsymbol{\eta} - c_2 \boldsymbol{\pi}. \tag{5.45}$$

Accordingly, with the constraints (5.16) and (5.21) in effect we have

$$\psi = 0 \quad \text{and} \quad \boldsymbol{\eta} = (c_2/k_3)\boldsymbol{\pi}; \tag{5.46}$$

with

$$\varphi = \boldsymbol{n} \cdot \boldsymbol{e}_a \quad \text{and} \quad \boldsymbol{U}_\pi = \mathbb{P}\boldsymbol{e}_a, \tag{5.47}$$

where $\mathbb{P} = \boldsymbol{I} - \boldsymbol{n} \otimes \boldsymbol{n}$, in which $\boldsymbol{I}$ is the identity for 3-space, is the projection onto T_ω. We note that $\mathbb{P} = \boldsymbol{1}$, the identity on T_ω. Then, from (5.45),

$$D\boldsymbol{\pi} = \mathbb{P}\boldsymbol{e}_a(\boldsymbol{r}), \quad \text{where} \quad D = \chi_\perp - c_2^2/k_3. \tag{5.48}$$

This furnishes the polarization uniquely in terms of the surface parametrization, provided that $D \neq 0$ and the applied field is assigned as a function in space. When $\chi_\perp > 0$, the sign of D is controlled by the strength of the flexo-electric effect; thus D is positive or negative according as $|c_2|$ is small or large, respectively. These alternatives correspond to the relevant quadratic form in the energy being positive definite or indefinite, respectively.

These results imply that the *equilibrium* value of the energy (5.40) may be regarded as a function of H, K and $\boldsymbol{\pi}$; on combining (5.46) and (5.48), the explicit expression is found to be

$$U = kH^2 + \bar{k}K + \tfrac{1}{2}D\,|\boldsymbol{\pi}|^2. \tag{5.49}$$

With the foregoing in effect, (5.41) and (5.42) furnish the residual virtual-work statement

$$\int_{\omega}[\dot{W} + (W + \lambda)\dot{J}/J + \varphi\boldsymbol{\pi}\cdot\dot{\boldsymbol{n}}]da = \int_{\omega} p\boldsymbol{n}\cdot\dot{\boldsymbol{r}}da + \int_{\partial\omega}\chi ds, \tag{5.50}$$

in which all variations are induced by the virtual velocity

$$\boldsymbol{u} = \dot{\boldsymbol{r}} \tag{5.51}$$

with $\dot{\boldsymbol{\pi}} = \boldsymbol{0}$. In particular,

$$\dot{J}/J = \boldsymbol{a}^{\alpha}\cdot\boldsymbol{u}_{,\alpha} \tag{5.52}$$

and

$$\dot{\boldsymbol{n}} = \varepsilon^{\beta\alpha}\boldsymbol{a}_{\beta}\times\boldsymbol{u}_{,\alpha} - (\dot{J}/J)\boldsymbol{n}, \tag{5.53}$$

where $\varepsilon^{\beta\alpha}$ is the contravariant Levi–Civita permutation tensor. Further,

$$\dot{W} = \dot{U} - \dot{\boldsymbol{e}}_a(\boldsymbol{r})\cdot\boldsymbol{\pi}, \tag{5.54}$$

with

$$\dot{\boldsymbol{e}}_a = (grad\boldsymbol{e}_a)_{|\omega}\boldsymbol{u} \quad \text{and} \quad \dot{U} = U_H\dot{H} + U_K\dot{K}, \tag{5.55}$$

in which we have invoked $U_{\boldsymbol{\eta}} = \boldsymbol{0}$ (cf. (5.44) and $(5.46)_1$). From (5.49),

$$U_H = 2kH \quad \text{and} \quad U_K = \bar{k}. \tag{5.56}$$

Expressions for the variations $\dot{H}$ and $\dot{K}$ given previously will be recalled in the next subsection. To facilitate their representation we use the decomposition (cf. (3.42))

$$\boldsymbol{u} = u^{\alpha}\boldsymbol{a}_{\alpha} + u\boldsymbol{n}, \tag{5.57}$$

where u^{α} and u, respectively, are the *tangential* and *normal* variations of the position field.

Tangential Variations

For tangential variations we have $u = 0$ and (cf. (2.11))

$$\dot{J}/J = u^{\alpha}_{;\alpha}, \quad \dot{H} = u^{\alpha}H_{,\alpha} \quad \text{and} \quad \dot{K} = u^{\alpha}K_{,\alpha}. \tag{5.58}$$

Thus,

$$(W + \lambda)\dot{J}/J = [(W + \lambda)u^{\alpha}]_{;\alpha} - u^{\alpha}(W + \lambda)_{,\alpha}, \tag{5.59}$$

where

$$W_{,\alpha} = U_{,\alpha} - \boldsymbol{a}_\alpha \cdot (grad\boldsymbol{e}_a)_{|\omega}\boldsymbol{\pi} - \boldsymbol{e}_a \cdot \boldsymbol{\pi}_{,\alpha}; \tag{5.60}$$

whereas, with $\dot{\boldsymbol{\pi}} = \mathbf{0}$,

$$\dot{W} = u^\alpha[U_H H_{,\alpha} + U_K K_{,\alpha} - \boldsymbol{a}_\alpha \cdot (grad\boldsymbol{e}_a)_{|\omega}\boldsymbol{\pi}]. \tag{5.61}$$

We thus reach

$$\dot{W} + (W + \lambda)\dot{J}/J = [(W + \lambda)u^\alpha]_{;\alpha} + u^\alpha(U_H H_{,\alpha} + U_K K_{,\alpha} - U_{,\alpha} - \lambda_{,\alpha} + \boldsymbol{e}_a \cdot \boldsymbol{\pi}_{,\alpha}). \tag{5.62}$$

Here we use the fact that U_η vanishes in equilibrium, together with (5.48) and the symmetry of $\mathbb{P}$, to derive

$$U_{,\alpha} = U_H H_{,\alpha} + U_K K_{,\alpha} + \boldsymbol{e}_a \cdot \mathbb{P}\boldsymbol{\pi}_{,\alpha}, \tag{5.63}$$

which furnishes

$$\dot{W} + (W + \lambda)\dot{J}/J = [(W + \lambda)u^\alpha]_{;\alpha} + u^\alpha[(\boldsymbol{e}_a \cdot \boldsymbol{n})(\boldsymbol{n} \cdot \boldsymbol{\pi}_{,\alpha}) - \lambda_{,\alpha}]. \tag{5.64}$$

To reduce the term in (5.50) involving φ, we use (5.53) to obtain

$$\dot{\boldsymbol{n}} = \varepsilon^{\beta\alpha} b_{\lambda\alpha} u^\lambda \boldsymbol{a}_\beta \times \boldsymbol{n}, \tag{5.65}$$

where $\varepsilon_{\beta\lambda}$ is the covariant permutation tensor, together with $\varepsilon^{\beta\alpha}\varepsilon_{\beta\lambda} = \delta^\alpha_\lambda$ (the Kronecker delta). This and

$$\boldsymbol{n} \times \boldsymbol{a}_\beta = \varepsilon_{\beta\gamma}\boldsymbol{a}^\gamma \tag{5.66}$$

yield

$$\dot{\boldsymbol{n}} = -b_{\lambda\alpha} u^\lambda \boldsymbol{a}^\alpha, \tag{5.67}$$

which combines with (5.47) to deliver

$$\varphi\boldsymbol{\pi} \cdot \dot{\boldsymbol{n}} = -(\boldsymbol{n} \cdot \boldsymbol{e}_a) b_{\alpha\beta}\pi^\beta u^\alpha. \tag{5.68}$$

However, the constraint (5.21) implies that

$$\boldsymbol{\pi}_{,\alpha} = \pi^\beta_{;\alpha}\boldsymbol{a}_\beta + b_{\alpha\beta}\pi^\beta\boldsymbol{n}, \tag{5.69}$$

where $(\cdot)_{;\alpha}$ is the covariant derivative on ω. Accordingly,

$$\dot{W} + (W + \lambda)\dot{J}/J + \varphi\boldsymbol{\pi} \cdot \dot{\boldsymbol{n}} = [(W + \lambda)u^{\alpha}]_{;\alpha} - u^{\alpha}\lambda_{,\alpha}. \tag{5.70}$$

Using Stokes' theorem, the surface integral over ω of the first term on the right-hand side may be represented as an integral over the edge $\partial\omega$. Remarkably, the Euler equations emerging from (5.42) under tangential variations are then given simply by $\lambda_{,\alpha} = 0$; i.e.,

$$\lambda = const. \quad \text{on} \quad \omega, \tag{5.71}$$

as in the classical Canham/Helfrich theory for uniform lipid bilayers in the absence of electromagnetic effects (cf. Sect. 2). Appropriate edge conditions, extending those presented in Sect. 2, are developed in (Steigmann and Agrawal 2016).

Normal Variations

In this case $\boldsymbol{u} = u\boldsymbol{n}$, yielding (cf. (2.16))

$$\dot{J}/J = -2Hu, \quad 2\dot{H} = \Delta u + u(4H^2 - 2K) \quad \text{and} \quad \dot{K} = 2KHu + (\tilde{b}^{\alpha\beta}u_{,\alpha})_{;\beta}, \tag{5.72}$$

where, for any scalar field ξ,

$$\Delta\xi = \frac{1}{\sqrt{a}}(\sqrt{a}a^{\alpha\beta}\xi_{,\beta})_{,\alpha} \tag{5.73}$$

is the surface Laplacian in which $a^{\alpha\beta}$ is the dual metric.

Recalling that U_η vanishes at equilibrium and noting that $\boldsymbol{\pi}$ is fixed in the present class of variations, after a lengthy calculation, detailed in (Steigmann et al. 2003), we arrive at

$$\begin{aligned}\dot{W} + (W + \lambda)\dot{J}/J = u[&\Delta(\tfrac{1}{2}U_H) + (U_K)_{;\alpha\beta}\tilde{b}^{\alpha\beta} + U_H(2H^2 - K)\\ &+ 2H(KU_K - W) - 2H\lambda - \boldsymbol{n} \cdot (grad\boldsymbol{e}_a)_{|\omega}\boldsymbol{\pi}]\\ &+ [(\tfrac{1}{2}U_H a^{\alpha\beta} + U_K\tilde{b}^{\alpha\beta})u_{,\alpha}]_{;\beta}\\ &- \{[(U_H)_{,\beta}a^{\alpha\beta} + (U_K)_{,\beta}\tilde{b}^{\alpha\beta}]u\}_{;\alpha}.\end{aligned} \tag{5.74}$$

Here $(\cdot)_{;\alpha\beta}$ is the second covariant derivative on ω and

$$\tilde{\boldsymbol{b}} = 2H\boldsymbol{1} - \boldsymbol{b} \tag{5.75}$$

is the cofactor of the curvature tensor.

For normal variations (5.53) gives

$$\dot{\boldsymbol{n}} = -\boldsymbol{a}^{\alpha}u_{,\alpha}, \tag{5.76}$$

yielding

$$\varphi\boldsymbol{\pi} \cdot \dot{\boldsymbol{n}} = u(\varphi\pi^{\alpha})_{;\alpha} - (\varphi\pi^{\alpha}u)_{;\alpha}. \tag{5.77}$$

Combining this with (5.41) and writing the integrals of the divergences as boundary integrals, from (5.42) the relevant Euler–Lagrange equation is found to be (compare (2.22))

$$\begin{aligned}&\Delta(\tfrac{1}{2}U_H) + (U_K)_{;\alpha\beta}\tilde{b}^{\alpha\beta}U_H(2H^2 - K)\\&+ 2H(\boldsymbol{\pi}\cdot\boldsymbol{e}_a + KU_K - U) - 2H\lambda + (\varphi\pi^\alpha)_{;\alpha}\\&= p + \boldsymbol{n}\cdot(grad\boldsymbol{e}_a)_{|\omega}\boldsymbol{\pi},\end{aligned} \tag{5.78}$$

where

$$(\varphi\pi^\alpha)_{;\alpha} = \frac{1}{\sqrt{a}}(\sqrt{a}\varphi\pi^\alpha)_{,\alpha}. \tag{5.79}$$

This generalizes the well-known *shape equation* of the conventional theory (Ou-Yang et al. 1999; Agrawal and Steigmann 2009). For the particular energy given by (5.49) it reduces to

$$\begin{aligned}&k[\Delta H + 2H(H^2 - K)] - D\,|\boldsymbol{\pi}|^2 H + 2H(\boldsymbol{\pi}\cdot\boldsymbol{e}_a - \lambda) + (\varphi\pi^\alpha)_{;\alpha}\\&= p + \boldsymbol{n}\cdot(grad\boldsymbol{e}_a)_{|\omega}\boldsymbol{\pi}, \quad \text{with} \quad \varphi = \boldsymbol{n}\cdot\boldsymbol{e}_a.\end{aligned} \tag{5.80}$$

This may be simplified by using (5.47) and (5.48) to reach

$$D\boldsymbol{\pi}\cdot\boldsymbol{e}_a = \mathbb{P}\boldsymbol{e}_a\cdot\boldsymbol{e}_a = |\mathbb{P}\boldsymbol{e}_a|^2 = |\boldsymbol{e}_a|^2 - \varphi^2 \tag{5.81}$$

and

$$2H\boldsymbol{\pi}\cdot\boldsymbol{e}_a - D\,|\boldsymbol{\pi}|^2 H = D^{-1}H(|\boldsymbol{e}_a|^2 - \varphi^2). \tag{5.82}$$

Numerical examples are discussed in (Steigmann and Agrawal 2016).

References

A. Agrawal, D.J. Steigmann, Boundary-value problems in the theory of lipid membranes. Contin. Mech. Thermodyn. **21**, 57–82 (2009)

A. Agrawal, D.J. Steigmann, A model for surface diffusion of trans-membrane proteins on lipid bilayers. ZAMP **62**, 549–563 (2011)

R. Aris, *Vectors, Tensors and the Basic Equations of Fluid Mechanics* (Dover, N.Y., 1989)

M. Arroyo, A. DeSimone, Relaxation dynamics of fluid membranes. Phy. Rev. **E79**(031915), 1–17 (2009)

E. Baesu, R.E. Rudd, J. Belak, M. McElfresh, Continuum modeling of cell membranes. Int. J. Non-linear Mech. **39**, 369–377 (2004)

M. Barham, D.J. Steigmann, D. White, Magnetoelasticity of highly deformable thin films: theory and simulation. Int. J. Non-linear Mech. **47**, 185–196 (2012)

R. Bustamante, A. Dorfmann, R.W. Ogden, Nonlinear electroelastostatics: a variational framework. ZAMP **60**, 154–177 (2009)

J.W. Cahn, J.E. Hilliard, Free energy of a nonuniform system. I. Interfacial free energy. J. Chem. Phys. **28**, 258–267 (1958)
P. Canham, The minimum energy of bending as a possible explanation of the biconcave shape of the human red blood cell. J. Theor. Biol. **26**, 61–81 (1970)
R. Capovilla, J. Guven, Geometry of lipid vesicle adhesion. Phys. Rev. E **66**(041604), 1–6 (2002)
P.G. Ciarlet, *Mathematical Elasticity, Vol. 1: Three Dimensional Elasticity* (North-Holland, Amsterdam, 1993)
C.M. Dafermos, Disclinations in liquid crystals. Q. J. Mech. Appl. Math. **23**, S49–S64 (1970). Supplement No. 1: Mechanics of Liquid Crystals
P.G. DeGennes, J. Prost, *The Physics of Liquid Crystals* (Oxford University Press, Oxford, 1992)
L. Deseri, M.D. Piccioni, G. Zurlo, Derivation of a new free energy for biological membranes. Contin. Mech. Thermodyn. **20**, 255–273 (2008)
M. Deserno, M.M. Müller, J. Guven, Contact lines for fluid surface adhesion. Phys. Rev. E **76**(011605), 1–10 (2007)
L. Dorfmann, R.W. Ogden, *Nonlinear Theory of Electroelastic and Magnetoelastic Interactions* (Springer, N.Y., 2014)
A. Embar, J. Dolbow, E. Fried, Microdomain evolution on giant unilamellar vesicles. Biomech. Model. Mechanobiol. **12**, 597–615 (2013)
J.L. Ericksen, Conservation laws for liquid crystals. Trans. Soc. Rheol. **5**, 23–34 (1961)
J.L. Ericksen, Hydrostatic theory of liquid crystals. Arch. Ration. Mech. Anal. **9**, 371–378 (1962)
J.L. Ericksen, Equilibrium theory of liquid crystals, in *Advances in Liquid Crystals*, ed. by G.H. Brown (Academic Press, N.Y., 1976), pp. 233–298
J.L. Ericksen, Theory of Cosserat surfaces and its application to shells, interfaces and cell membranes, in *Proceeding of the International Symposium on Recent Developments in the Theory and Application of Generalized and Oriented Media*, ed. by P.G. Glockner, M. Epstein, D.J. Malcolm (Calgary, Alberta, 1979), pp. 27–39
E.A. Evans, R. Skalak, *Mechanics and Thermodynamics of Biomembranes* (CRC Press, Boca Raton, 1980)
H. Frischleder, G. Peinel, Quantum-chemical and statistical calculations on phospholipids. Chem. Phys. Lipids **30**, 121–158 (1982)
L.T. Gao, X.-Q. Feng, Y.-J. Yin, H. Gao, An electromechanical liquid crystal model of vesicles. J. Mech. Phys. Solids **56**, 2844–2862 (2008)
G. Gioia, R.D. James, Micromagnetics of very thin films. Proc. R. Soc. Lond. **A453**, 213 (1997)
L.M. Graves, The Weierstrass condition for multiple integral variation problems. Duke Math. J. **5**, 656–660 (1939)
W. Helfrich, Elastic properties of lipid bilayers: theory and possible experiments. Z. Naturforsch. **28**, 693–703 (1973)
M.G. Hilgers, A.C. Pipkin, Energy-minimizing deformations of elastic sheets with bending stiffness. J. Elast. **31**, 125–139 (1993)
J.T. Jenkins, The equations of mechanical equilibrium of a model membrane. SIAM J. Appl. Math. **32**(755–764), 00 (1977)
C.-I. Kim, D.J. Steigmann, Distension-induced gradient capillarity in lipid membranes. Contin. Mech. Thermodyn. **27**, 609–621 (2015)
A. Kovetz, *Electromagnetic Theory* (Oxford University Press, Oxford, 2000)
K. Mandadapu, Private communication (2016)
R.B. Meyer, Piezoelectric effects in liquid crystals. Phys. Rev. Lett. **22**, 918–921 (1969)
P. Mohammidi, L.P. Liu, P. Sharma, A theory of flexo-electric membranes and effective properties of heterogeneous membranes. ASME J. Appl. Mech. **81**, 011007-1–011007-11 (2014)
M.M. Müller, M. Deserno, J. Guven, Interface-mediated interactions between particles: A geometrical approach. Phys. Rev. **E72**(061407), 1–17 (2005a)
M.M. Müller, M. Deserno, J. Guven, Geometry of surface-mediated interactions. Europhys. Lett. **69**, 482–488 (2005b)

M.M. Müller, M. Deserno, J. Guven, Balancing torques in membrane-mediated interactions: Exact results and numerical illustrations. Phys. Rev. **E76**(011921), 1–16 (2007)
P.M. Naghdi, Theory of Shells and Plates, in *Handbuch der Physik*, vol. VIa/2, ed. by C. Truesdell (Springer, Berlin, 1972)
Z.-C. Ou-Yang, J.-X. Liu, Y.-Z. Xie, *Geometric Methods in the Elastic Theory of Membranes in Liquid Crystal Phases* (World Scientific, Singapore, 1999)
A.G. Petrov, *The Lyotropic State of Matter* (Gordon and Breach, Amsterdam, 1999)
P. Rangamani, A. Agrawal, K. Mandadapu, G. Oster, D.J. Steigmann, Interaction between surface shape and intra-surface viscous flow on lipid membranes. Biomech. Model. Mechanobiol. **12**, 833–845 (2013)
P. Rangamani, A. Benjamani, A. Agrawal, B. Smit, D.J. Steigmann, G. Oster, Small scale membrane mechanics. Biomech. Model. Mechanobiol. **13**, 697–711 (2014)
P. Rangamani, D.J. Steigmann, Variable tilt on lipid membranes. Proc. R. Soc. Lond. A **470**(20140463) (2014). https://doi.org/10.1098/rspa.2014.0463
L.E. Scriven, Dynamics of a fluid interface. Chem. Eng. Sci. **12**, 98–108 (1960)
J. Seelig, ^{31}P nuclear magnetic resonance and the head group structure of phospholipids in membranes. Biochim. Biophys. Acta **515**, 105–140 (1978)
I.S. Sokolnikoff, *Tensor Analysis: Theory and Applications to Geometry and Mechanics of Continua* (Wiley, N.Y., 1964)
D.J. Steigmann, A note on pressure potentials. J. Elast. **26**, 87–93 (1991)
D.J. Steigmann, On the relationship between the Cosserat and Kirchhoff-Love theories of elastic shells. Math. Mech. Solids **4**, 275–88 (1999a)
D.J. Steigmann, Fluid films with curvature elasticity. Arch. Ration. Mech. Anal. **150**, 127–52 (1999b)
D.J. Steigmann, Applications of polyconvexity and strong ellipticity to nonlinear elasticity and elastic plate theory, in *CISM Course on Applications of Poly-, Quasi-, and Rank-One Convexity in Applied Mechanics*, vol. 516, ed. by J. Schröder, P. Neff (Springer, Wien and New York, 2010), pp. 265–299
D.J. Steigmann, A model for lipid membranes with tilt and distension based on three-dimensional liquid crystal theory. Int. J. Non-linear Mech. **56**, 61–70 (2013)
D.J. Steigmann, A. Agrawal, Electromechanics of polarized lipid bilayers. Math. Mech. Complex Syst. **4–1**, 31–54 (2016)
D.J. Steigmann, E. Baesu, R.E. Rudd, M. McElfresh, J. Belak, On the variational theory of cell-membrane equilibria. Interfaces Free Bound. **5**, 357–366 (2003)
R.A. Toupin, The elastic dielectric. J. Ration. Mech. Anal. **5**, 849–915 (1956)
C. Truesdell, *A First Course in Rational Continuum Mechanics* (Academic Press, N.Y., 1991)
E.G. Virga, *Variational Theories for Liquid Crystals* (Chapman and Hall, London, 1994)
D.T. Warshaviak, M.J. Muellner, M. Chachisvilis, Effect of membrane tension on the electric field and dipole potential of lipid bilayer membrane. Biochim. Biophys. Acta **2608–2617**, 2011 (1808)
Q.-S. Zheng, Two-dimensional tensor function representation for all kinds of material symmetry. Proc. R. Soc. Lond. A **443**, 127–138 (1993)

Elasticity and Hereditariness

Luca Deseri

Abstract This chapter collects the lecture notes of the module "Elasticity and Hereditatiness of Lipid Bilayers" delivered at CISM in July 2016. Such material is based primarily on three papers coauthored by this lecturer, and which have been contributing to shed light on the mechanical behavior of lipid bilayers. In particular, the breakthrough from this research is that the underlying nonlinear elastic response of lipid bilayers is fully determined as long as the membrane energy is obtained. Bending and saddle splay rigidities are shown here to be directly obtainable from the membranal response, as well as the line tension, holding together domains in which lipids are in different phases. The power law hereditariness of lipid membranes strikingly shown through rheometric tests has been analyzed in this work through a suitable energetics obtained by the author and coworkers and penalizing small perturbations of ground configurations of such systems.

L. Deseri (✉)
Department of Mechanical, Aerospace and Civil Engineering-MACE,
Brunel University London, Uxbridge UB8 3PH, UK
e-mail: luca.deseri@unitn.it; deseri.mems@pitt.edu

L. Deseri
Department of Mechanical Engineering and Materials Science-MEMS,
University of Pittsburgh, 636 Benedum Hall 3700 O'Hara Street, Pittsburgh, PA 15261, USA

L. Deseri
Department of Mechanical, Civil and Environmental Engineering-DICAM,
University of Trento, 38123 Trento, Italy

L. Deseri
Department of Mechanical Engineering, Carnegie Mellon University, Pittsburgh,
PA 15213-3890, USA

L. Deseri
Department of Nanomedicine, The Methodist Hospital Research Institute, MS B-490,
Houston, TX 77030, USA

D.J. Steigmann (ed.), *The Role of Mechanics in the Study of Lipid Bilayers*,
CISM International Centre for Mechanical Sciences 577,
DOI 10.1007/978-3-319-56348-0_2

1 Introduction

In Deseri et al. (2008) we obtained an energetics for biomembranes, such as lipid bilayers, which accounts for the through-thickness phase transition exhibited by planar structures and curved closed liposomes, like Giant Unilamellar Vescicles (GUVs) (see e.g., Lipowsky and Sackmann 1995). As pointed out in other chapters of this volume, the average thickness of such structures is of the order of 5 nm while the other two dimensions are several orders of magnitudes higher in size. In the treatment mentioned above no distinction is done between the leaflets of a bilayer, thereby inferring that even the energetics of lipid monolayers is compatible with such derivation.

Ways for controlling the morphologies in planar lipid systems and in GUVs are temperature and osmotic pressure based (see e.g., Baumgart et al. 2003; Veatch et al. 2004). Advanced high-resolution fluorescence imaging techniques employed in Baumgart et al. (2003) in particular have highlighted the coexistence of regions (or *phases*) with completely different features, highlighted in red and blue in Fig. 1 included in the same paper. The main contrast among such zones on the membranes is in terms of "degree of curliness" of the lipids, namely how curly and, hence, how short they get relative to their maximum length. This has an impact on the values of the curvatures in GUVs in regions with different degrees of curliness and also in the redistribution of the species within a lipid membrane with a given chemical composition. Because lipid membranes have the molecules free to move in-plane, namely across the surface, the two phases are called *liquid ordered-*L_o and *liquid disordered-*L_d. In some cases, the presence of "lipid rafts" is detected in lipid bilayers. Basically, glycosphingolipid-enriched domains do form such rafts. For instance, the latter occur in the presence of fully saturated chains of sphingomyelin and glycosphingolipids bond with neighboring active functional glycosyl groups. Obviously, any model owing the L_o–L_d transition can consistently predict lipid rafts. The issue is: can a model at the continuum level be more physically based and predict both the phase transitions and the changes in curvature and shapes? Through the last four decades this has been one of the main tasks in the field and, obviously, there no unique answer to this. Among the most prominent works in this direction

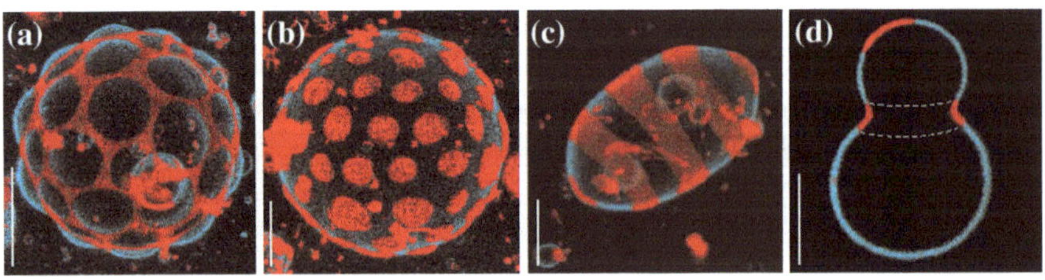

Fig. 1 Images experimentally obtained by Baumgart et al. (2003), showing how phase separation relates to shapes achieved by GUVs. In *red* and *blue* respectively liquid-disordered and liquid-ordered phases. Scale bars 5 μm (Images by courtesy of Baumgart et al. 2003)

one can certainly single out Lipowsky and Sackmann (1995)-Chap. 1. There it was remarked that "here the theory of nonreacting mixtures and the theory of phase transitions are strictly related to the theory of thin, fluid shells". Ultimately, this corroborates the fact that obtaining a physically based model at the continuum level incorporating information regarding the species forming the bilayer, and rendering out the phase transition and the geometrical changes experienced by such structures is an extremely hard task. Contributions focusing on the purely mechanical behavior of such systems can be related to the pioneering work in Canham (1970), Fung (1966), Fung and Tong (1968), although the keystone work in biophysics regarding lipid bilayers can be singled out in Helfrich (1973), where the free energy density per unit area in the case of pure bending was obtained. This led the *Helfrich free energy*, which does coincide with the Kirchhoff–Love bending energy density in the presence of large curvature changes. The latter is well known in Structural Engineering and Solid Mechanics.

A piecewise Helfrich's energy has been employed when different zones of the surface of the bilayer are already known to be occupied by lipids in different phases. Spontaneous curvature in GUVs has also been accounted for in some of the extensions of Helfrich's model.

Along similar lines, a purely mechanical energy for liquid films has been obtained in Keller and Merchant (1991), where the bending stiffness of a liquid surface has been computed. Later, in Steigmann (1999) an expression of the dependence for two-dimensional fluid films exhibiting such stiffness was singled out, thanks to a theory of elastic surfaces. Along similar lines of thinking, in Baesu et al. (2004) it was proposed a stretching–bending energy density.

In all the models above the bilayer was always considered a two-dimensional body, thereby neglecting direct information associated with the thickness of the membrane. This is certainly not what one must do in order to capture the main mechanism of the observed phase transition experienced by the lipids. Indeed, they are seen to be nearly extended in the ordered phase, L_o, whereas they get shorter and curlier in the disordered phase, L_d. Indeed, it is known that a raise in temperature causes the hydrocarbon lipid tails of phospholipids to undergo the phase transition just mentioned above, evidenced by a significant thickness reduction from the liquid-ordered phase L_o to the liquid-disordered phase L_d (see e.g., Falkovitz et al. 1982; Goldstein and Leibler 1988, 1989; Jahnig 1981, 1996; Owicki et al. 1978; Owicki and McConnell 1979; Lipowsky and Sackmann 1995).

The conclusion is that keeping track of thickness changes is essential for lipid membranes and its changes witness the variations of the lipids order. This key issue is addressed in Deseri et al. (2008).

Asymptotic approaches delivering the mechanics of nonlinear elastic shells (see e.g., Koiter 1966) show that the thickness governs the scaling of both the membranal and the bending contributions to the energy density, being the former linear with the thickness while the latter is cubic in this quantity. Henceforth, ignoring the membranal term (as many formulations do) means to neglect an energy contribution to the overall energy which is two orders of magnitude more important than the bending term.

The work done in Deseri et al. (2008) also represents one of the first attempts toward a better understanding of the correlation among lipid order, membrane shape, and chemical composition during either temperature changes or osmotic pressure or both. This has been followed by several contributions in recent years, including Maleki et al. (2013). A related discussion and a derivation of the *line tension*, namely the configurational force arising at between zones of difference phases allowing for zones of finite size, is presented in Deseri and Zurlo (2013). This agrees with results obtained in Trejo and Ben Amar (2011).

The final reduced two-dimensional energetics in Deseri et al. (2008) is consistent with a dimension reduction procedure. This is done in two steps. The first one is to impose a modified Kirchhoff–Love kinematics which accounts for the thickness changes and by enforcing a new symmetry group, introduced in Zurlo (2006), proved Healey et al. (2017) and, eventually in Maleki et al. (2013), thereby delivering a bulk energy density as a function of solely three invariants of the Cauchy–Green strain measure. The second step is to perform an asymptotic expansion of the bulk energy with respect to a reference thickness.

Summing up, the resulting energy density confirms the hierarchy between the membranal and the bending terms described above, although it delivers a uniquely and strikingly revealing expression, explained in Sect. 2. This will eventually lead to deducing the key features of the elastic part of the response of lipid membranes, such as the areal and bending rigidities and the line tension, namely the configurational force holding together zones in different phases.

The main feature of the energy derived in Deseri et al. (2008) is the presence of two turning points in the local stress governing the biological membrane behavior (see Fig. 7a). They are placed in the spinoidal zone for the local part of the energy. Henceforth, whenever the external conditions are such that the aerial stretch, i.e., the reciprocal of the thinning, is enclosed in this region, the response may produce a rapid change of the geometry, i.e., material instabilities can occur. The onset of bifurcated configurations possibly arising from homogeneous configurations characterized by an areal stretch lying in the spinoidal region is studied in Sect. 2.5. The total elastic (Gibbs free) energy expanded upon any ground state in such region will be studied to determine the bifurcated modes and the relationships between the number of nucleated spatial oscillations with the critical values of the areal stretches.

In the sequel we will show that this occurrence is exhibited even when the in-plane viscosity of the lipid membrane is accounted for. In this regard, the experimental observations of lipid viscous behavior showed that the loss and storage moduli are well described by power law functions (Espinosa et al. 2011). This observation suggests that the behavior of the biological membrane is properly described in the framework of fractional hereditariness.

An analysis of the appropriate energetics arising because of viscosity will lead to a new governing functional for studying the influence of the effective viscoelasticity of the lipid membrane on the material instabilities exhibited by the system which is studied in Sect. 3. The resulting viscoelastic free energy has a local and a nonlocal part. There, the power at which stress and hyperstress relax might be different, as diffusion mechanisms may occur at different average speeds depending on whether or not they arise in a boundary layer between different phases or in a given phase.

Exactly like in the purely elastic case, values of the areal stretches for which unknown time evolving bifurcated configurations could occur are sought. This is to investigate the role of the hereditariness on such unstable ground states. To this aim, in full analogy with the elastic case, a variational principle is employed. The Gibbs free energy density prevailing the space-time varying perturbation is taken from Deseri et al. (2014), where a hierarchical rheological model yields the Staverman–Schartzl free energy (extensively studied in Del Piero and Deseri 1996, 1997; Deseri et al. 2006, among many others) as the one for power law materials.

The variational principle yields a non-classical eigenvalue problem. Spatial modes bifurcating from ground states characterized by the areal stretch within the spinoidal zone are of course oscillatory. The period of spatial oscillation is shown to decrease with the ratio of generalized local and nonlocal moduli. Henceforth, the number of oscillations increases with respect to the elastic case. As the ratio just mentioned above increases, for a given number of oscillations the interval of stretches for which bifurcation can occur gets larger if compared with the one determined by the purely elastic behavior. The model then is suggesting that hereditariness increases the chances of nucleating spatially oscillatory bifurcated modes.

Upon exploring the transfer function of the equation governing the eigenvalue problem mentioned above, it is found that, for various values of the local and nonlocal relaxation power, time decay occurs in the response. Hence, spatial oscillations do slowly relax, exhibiting a long tail type response in time.

2 The New Elastic Energy for Lipid Membranes

In this section we briefly recall the main results obtained in Deseri et al. (2008), together with a schematic description of the approach followed in this work. The main result is the derivation of a new surface energy density for the lipid bilayer. This is shown to give the possibility of deducing bending rigidities, line tension, and thickness profile inside the boundary layer during the order–disorder transition from simple experimental data on the stretching behavior of the membrane.

Attention here will be restricted to initially planar membranes, thereby neglecting the effects of spontaneous curvature. An orthonormal basis $(\boldsymbol{e}_1, \boldsymbol{e}_2, \boldsymbol{e}_3)$ is introduced to describe material points in the reference configuration and geometrical changes with respect to that. A simply connected region $\mathscr{B}_0$ of constant thickness h_0 in the direction of $\boldsymbol{e}_3$ and with a flat mid-surface Ω in the plane spanned by $(\boldsymbol{e}_1, \boldsymbol{e}_2)$ depicts the reference configuration for the membrane, thereby not distinguishing between the upper and the lower leaflet of the membrane. Points of $\mathscr{B}_0$ are denoted by

$$\boldsymbol{x} = \mathbf{x} + z\boldsymbol{e}_3,$$

where

$$\mathbf{x} = x\mathbf{e}_1 + y\mathbf{e}_2$$

and $z \in (-h_0/2, h_0/2)$.

In the sequel $\boldsymbol{f}$ represents the deformation map of $\mathscr{B}_0$ and $\mathbf{F} = \nabla \boldsymbol{f}$ its 3×3 gradient. The energy $\mathscr{E}$ stored in the membrane is symbolically expressed as follows:

$$\mathscr{E} = \int_{\mathscr{B}_0} W(\mathbf{F})\, dV = \int_{\Omega} \int_{-h_0/2}^{h_0/2} W(\mathbf{F})\, dz\, d\Omega, \tag{1}$$

where W is the purely elastic Helmholtz energy density per unit volume. Evidently, the energy density per unit surface in the reference configuration reads as follows:

$$\psi = \int_{-h_0/2}^{h_0/2} W(\mathbf{F})\, dz. \tag{2}$$

In-plane fluidity is the main features of lipid membranes at room-to-body temperature. This entails the impossibility of sustaining shear stresses in planes perpendicular to $\boldsymbol{e}_3$, unless viscosity is accounted for. This has been used to restrict the dependence of W on three suitable invariants of $\mathbf{F}$ (see Zurlo 2006; Healey et al. 2017 and Maleki et al. 2013), namely

$$\mathscr{I}(\boldsymbol{x}) = \{\bar{J}(\boldsymbol{x}), \det \mathbf{F}(\boldsymbol{x}), \bar{\phi}(\boldsymbol{x})\}, \tag{3}$$

representing the areal stretch of planes perpendicular to the direction $\boldsymbol{e}_3$, the volume change and the stretch across the thickness, which ultimately will deliver the order parameter for the degree of curliness of the lipids, respectively.

With the aim of catching the out-of-plane kinematics as well as thickness changes, the following ansatz is assumed for the 3D deformation (see Fig. 2):

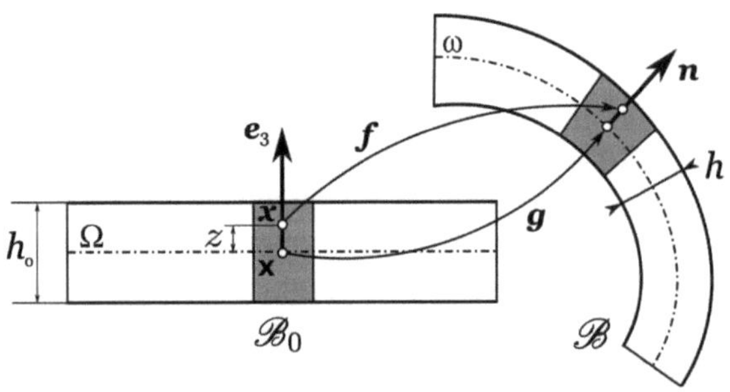

Fig. 2 Schematic representation of the deformation (4) of a plate-like reference configuration $\mathscr{B}_0$ into the current configuration $\mathscr{B}$. The gray box depicts the space occupied by two lipid molecules, their volume being conserved during the deformation (courtesy of Deseri and Zurlo 2013)

$$f(x) = g(\mathbf{x}) + z\,\phi(\mathbf{x})n(\mathbf{x}), \tag{4}$$

where $g(\mathbf{x}) = g(x, y, 0)$ defines the current mid-surface of the membrane, that is $\omega = g(\Omega)$, where n is the outward normal to ω and where

$$\phi(\mathbf{x}) = h(\mathbf{x})/h_0$$

is the *thickness stretch*, with h the current thickness. Such ansatz permits to make explicit the dependence of the invariants $\mathscr{I}$ on z and, ultimately, to perform the expansion of (2) in powers of the reference thickness h_0.

The molecular volume of biological membranes can be shown to stay almost constant in a broad range of temperature (see e.g., Goldstein and Leibler 1989; Owicki et al. 1978). This condition can be made explicit through a *quasi-incompressibility* constraint, namely

$$\det \mathbf{F}(\mathbf{x}, 0) = \bar{J}(\mathbf{x}, 0)\phi(\mathbf{x}) = 1. \tag{5}$$

The gray area in Fig. 2 relates with neighboring lipid molecules across the upper and lower leaflets with respect to the film mid-surface. The constraint (5) is actually a first-order approximation of the exact incompressibility constraint, as $\det \mathbf{F}(x) = \det \mathbf{F}(\mathbf{x}, 0) + O(z)$. In all planar deformations, namely whenever ω deforms in the plane $z = 0$, (5) yields that $\det \mathbf{F}(x) = 1$ is exact. This is the special case considered in the sequel, thereby focusing on planar lipid membranes. It is not difficult to show that the 3D energy density W reduces as follows:

$$w(J) = W(\bar{J}, \det \mathbf{F}, \bar{\phi})\Big|_{z=0} = W(J, 1, J^{-1}), \tag{6}$$

where

$$J(\mathbf{x}) = \bar{J}(\mathbf{x}, 0).$$

The ansatz (4) and the assumption of in-plane fluidity yield the following expansion of (2) up to terms of order h_0^3:

$$\psi = \varphi(J) + \kappa(J)H^2 + \kappa_g(J)K + \alpha(J)||\mathrm{grad}_\omega \widehat{J}||^2, \tag{7}$$

where H and K are, respectively, the mean and Gaussian curvatures of the mid-surface ω, where

$$\varphi(J) = h_0\ w(J) \tag{8}$$

is the stretching energy density of the membrane, scaling with h_0, and where bending rigidities are found to be the following:

$$k(J) = \frac{h_0^2}{6}\varphi''(J) = \frac{h_0^3}{6}w''(J), \qquad k_G(J) = \frac{h_0^2}{12J}\varphi'(J) = \frac{h_0^3}{12J}w'(J), \tag{9}$$

where $' = d/dJ$. It is worth emphasizing that the latter scale with h_0^3, as expected. The last term in (7) reads as follows:

$$\alpha(J) = \frac{h_0^2}{24J^3}\varphi'(J), \tag{10}$$

and it penalizes the gradients of $J = h/h_0$, namely the presence of boundary layers between zones where the lipids are in either of the two possible phases. It is worth emphasizing that $\widehat{J}$ represents the spatial description $\widehat{J} \circ \boldsymbol{g} = J$ of J, and grad_ω is the spatial gradient, with respect to points of the current mid-surface ω.

Often times the bending energy is calculated relative to the current surface ω and, hence, bending rigidities must be expressed relative to the same configuration (see e.g., Baumgart et al. 2003), i.e.,

$$\kappa(J) = \frac{h_0^2}{6J}\varphi''(J), \qquad \kappa_G(J) = \frac{h_0^2}{12J^2}\varphi'(J). \tag{11}$$

The expression (7) is consistent with several models previously introduced in the literature of biological membranes. Indeed, Helfrich's model

$$\psi = kH^2 + k_G K$$

is recovered whenever one considers fixed value of J fixed.

The new energy (7) enables one to predict thickness phase transitions even for planar lipid membranes, including Langmuir films, that remain flat under external inputs, like temperature changes. Such situations are retrieved by (7) by setting $H = K = 0$. This energetics reminds of the resulting energy for cold drawing of polymeric films obtained in (see e.g., Coleman and Newman 1988). If the term factoring α is neglected, (7) agrees with the one determined in Baesu et al. 2004.

It is worth noting that the strategy followed in Deseri et al. (2008) and Zurlo (2006) to deliver (7) accounts for fairly general constitutive assumptions on the 3D energy W, and also accounts for chemical composition, temperature dependence and, potentially, for the presence of spontaneous curvature.

2.1 Stretching Energy

As it is clear from the structure of (7) and from (11) and (10), the pivot information governing the whole energetics is the surface Helmholtz energy $\varphi(J)$. This regulates the in-plane stretching behavior of the membrane and allows for predicting the phase transition phenomena observed in lipid membranes (Fig. 3).

The experimental evidence clearly shows that for a given chemical composition there may exist a temperature range where the L_o and L_d phases coexist, organizing themselves in domains called rafts. In closed membranes, these domains are

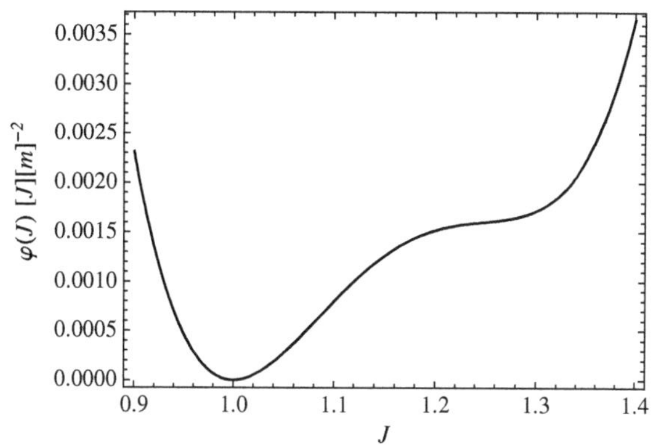

Fig. 3 The stretching energy $\varphi(J)$ adapted from Goldstein and Leibler (1989), for a temperature $T \sim 30°$. The areal stretch $J_o = 1$ corresponds to the unstressed, reference configuration $\mathscr{B}_0$ (courtesy of Deseri and Zurlo 2013)

typically detectable by curvature inhomogeneities, reflecting the occurrence of different bending rigidities (Baumgart et al. 2003). The expressions (11) for the bending moduli enlighten how the order–disorder transition, described by the stretching energy $\varphi(J)$, is connected with bending behavior of the membrane. Furthermore, we will prove that the knowledge of $\varphi(J)$ also determines the line tension occurring at the phase boundary.

Several works, such as Falkovitz et al. (1982), Goldstein and Leibler (1989), Komura et al. (2004), Owicki et al. (1978), Owicki and McConnell (1979), show that in order to provide a suitable expression for $\varphi(J)$ a Landau expansion in terms of the powers of either the thinning field $\phi = h/h_0$ or the areal stretch J is provided. This has the advantage that its (temperature dependent) coefficients are connected to the latent heat and the order parameter jump (e.g., Goldstein and Leibler 1989 and Lipowsky and Sackmann 1995). For the sake of convenience, in the sequel we assume that the natural planar configuration $\mathscr{B}_0$ of the lipid membrane is precisely the ordered phase L_o, where $J = J_o = 1$, so that the stretching energy takes the form

$$\varphi(J) = a_0 + a_1 J + a_2 J^2 + a_3 J^3 + a_4 J^4, \tag{12}$$

where the coefficients a_i $(i = 0, \ldots, 4)$ depend on temperature and chemical composition. Lacking of more experimental information leads one to tune such parameters, thanks to experimentals provided in Goldstein and Leibler (1989), actually also utilized in Komura et al. (2004), to connect with the thinning transition experienced by the lipids. At room temperature $T \sim 30°$, one record the following coefficients for $\varphi(J)$:

$$\begin{gathered} a_0 = 2.03, \quad a_1 = -7.1, \quad a_2 = 9.23, \\ a_3 = -5.3, \quad a_4 = 1.13. \end{gathered} \tag{13}$$

It is worth noting that their dimension is $[J][m]^{-2}$. The choice (13) has been pursued with the aim to show the feasibility of the proposed treatment. Specific data on the bilayer chemical composition and the temperature are required in order to get realistic pictures of the geometrical changes during the expected phase transition.

Summing up we conclude that:

- the membrane energy density $\varphi(J)$ can be completely determined experimentally: this is a local term within the energy and depends on temperature, chemical composition (of the specific lipids), and it scales with the linear power of the reference thickness of the bilayer;
- bending and spatial changes of either the thickness change gradient or of the related areal stretch are detected by the energy, thanks to the arising nonlocal terms;
- the latter coincides with the Helfrich's energy when the gradient term is negligible with respect to the bending one and the elastic moduli do not significantly change with areal stretch and concentration;
- like in the case of Coleman and Newman (1988), the penalization of the gradient of the areal stretch spontaneously arises from the dimension reduction procedure;
- besides prescribing the mean and the Gaussian curvatures, the resulting bending energy density is completely determined by the sole membrane energy density $\varphi(J)$: this relates to the chemical composition of the membrane is the only needed constitutive information of the model.

2.2 *Thinning Transition in Flat Lipid Layers*

A planar membrane in the reference configuration $\mathscr{B}_0$ is displayed in Fig. 4. Its homogeneous thickness in the direction of $\boldsymbol{e}_3$ is denoted by h_0, while its width in the direction of $\boldsymbol{e}_2$ is labeled by B and its length is denoted by L. At $z = 0$ the reference mid-surface Ω is set, while the sides of the planar bilayer are situated $x = \pm L/2$ and $y = \pm B/2$.

Plane strain deformations are considered to explore the main features of the thinning phase transition. Hence, the kinematics reads as follows:

$$\varphi(\boldsymbol{x}) = g(x)\boldsymbol{e}_1 + y\boldsymbol{e}_2 + z\phi(x)\boldsymbol{e}_3, \tag{14}$$

where x is the variable in the direction $\boldsymbol{e}_1$. The deformation gradient of such φ reads as follows:

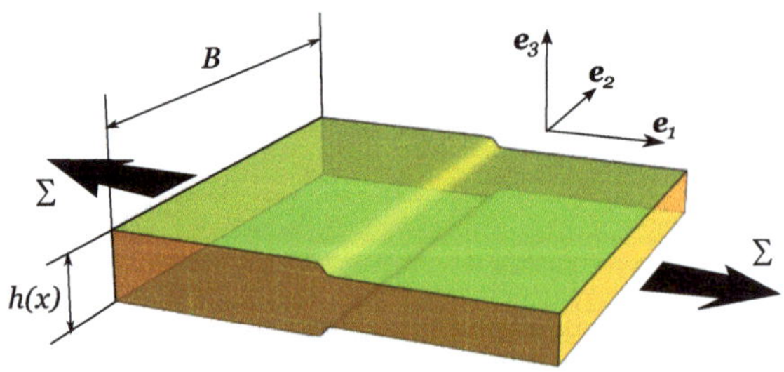

Fig. 4 Plane strain lipid bilayer undergoing phase transition from the thicker L_o domain to the thinner L_d domain under a traction Σ in the $\boldsymbol{e}_1$ direction (courtesy of Deseri and Zurlo 2013)

$$\mathbf{F} = \nabla\varphi = \begin{bmatrix} g_x & 0 & 0 \\ 0 & 1 & 0 \\ z\phi_x & 0 & \phi \end{bmatrix}, \tag{15}$$

where $_x$ denotes differentiation with respect to that (only) variable. The displacement component along $\boldsymbol{e}_1$ is $u(x) = g(x) - x$. The stretch in direction of the length of the bilayer is introduced in the sequel

$$\lambda(x) = g_x(x). \tag{16}$$

The 3D quasi-incompressibility reduces to $\phi = \lambda^{-1}$ on Ω, so that the membrane deformation is completely determined by λ.

We note that

$$||\text{grad}_\omega \hat{J}||^2 = ||\text{grad}_\omega \hat{\lambda}||^2 = \lambda_x^2 \lambda^{-2}, \tag{17}$$

after setting $\hat{\lambda} = \lambda \circ g^{-1}$, representing the spatial description of λ,

The resulting energy density per unit area (7) reads as follows:

$$\psi(\lambda, \lambda_x) = \varphi(\lambda) + \frac{h_0^2}{24}\lambda^{-5}\varphi'(\lambda)\lambda_x^2 \tag{18}$$

where $' = d/d\lambda$ (here $J = \lambda$). Upon introducing

$$\gamma(\lambda) = -\frac{h_0^2}{12}\lambda^{-5}\varphi'(\lambda), \quad \beta(\lambda) = \frac{1}{2}\gamma'(\lambda), \tag{19}$$

(see Coleman and Newman 1988), the energy density above can be rewritten as follows:

$$\psi(\lambda, \lambda_x) = \varphi(\lambda) - \frac{1}{2}\gamma(\lambda)\lambda_x^2. \tag{20}$$

It is worth noting that if γ would be replaced by a negative constant, the energy (20) coincides to the Cahn–Hilliard functional (Cahn and Hilliard 1958). The fact that the constant $\gamma < 0$ in such a model is required for stability purposes. In (19) the fact that γ depends on λ makes (20) to resemble the energy density deduced in Coleman and Newman (1988). Even in this case the condition $\gamma(\lambda) < 0$ is required for nucleating phase boundaries. This is in fact the case for the energy density (12).

For the sake of argument, opposite tractions Σ (force per reference length) are taken to arise on the edges $x = \pm L/2$. Due to the presence of λ_x hypertractions Γ performing work against u_x must be accounted for. Henceforth, the work performed on the bar reads as follows:

$$\mathscr{W}(u, u_x) = B\,[\Sigma u]_{-L/2}^{+L/2} + B\,[\Gamma u_x]_{-L/2}^{+L/2}. \tag{21}$$

Evidently, the total potential energy for any g is the sum of the total strain energy, obtained by integrating (20) across the membrane, minus the work (21), i.e.,

$$\mathscr{E}(\gamma) = B\int_{-L/2}^{L/2} \psi(\lambda, \lambda_x)\,dx - \mathscr{W}(u, u_x). \tag{22}$$

Stationarity of (22) yields the Euler–Lagrange equation for u and the associated boundary conditions. A perturbation $\eta(x)$ is imposed on the underlying g, namely

$$g_\varepsilon(x) := g(x) + \varepsilon\eta(x), \tag{23}$$

to deliver those information from (22). The arbitrariness of η leads to the first variation $\delta E = dE(g_\varepsilon)/d\varepsilon|_{\varepsilon=0}$ of (22), thereby delivering the Euler–Lagrange equation. Upon integrating such relation, the following condition is obtained:

$$\Sigma = \varphi'(\lambda) + \beta(\lambda)\lambda_x^2 + \gamma(\lambda)\lambda_{xx} = const., \tag{24}$$

holding in the open interval $(-L/2, L/2)$, with the boundary conditions

$$[\Gamma + \gamma(\lambda)\lambda_x]_{-L/2} = [\Gamma + \gamma(\lambda)\lambda_x]_{+L/2} = 0. \tag{25}$$

Strain localizations are investigated to explore the possible coexistence of ordered regions, in the L_o phase, and disordered zones, the thinner L_d phase: this transition may be connected through a boundary layer. With the aim of getting rid of edge effects induced by the boundary (Coleman and Newman 1988), the length L is considered unbounded relative to the reference thickness h_0. Henceforth, $-\infty < x < \infty$. The particular case in which $\Gamma = 0$ at the boundaries is explored in the sequel, so that (25) implies $\lambda_x \to 0$ as $x \to \pm\infty$. Nontrivial and bounded solutions of (24) are sought. In Coleman and Newman (1988) it is shown that they verify the equation

$$x - \bar{x} = \int_{\lambda(\bar{x})}^{\lambda(x)} \left(\frac{-2}{\gamma(\lambda)}\int_{\lambda_a}^{\lambda}\left[\varphi'(\zeta) - \Sigma\right]d\zeta\right)^{-\frac{1}{2}} d\lambda, \tag{26}$$

where $\bar{x}$ is arbitrary and where λ_a is either the value of λ at a specific location or a limiting value at which $\lambda_x = 0$. The derivation of (26) is detailed in Deseri and Zurlo (2013).

Whenever $\gamma(\lambda) < 0$, nontrivial bounded solutions of (24) have been completely characterized in Coleman and Newman (1988) for given tractions Σ. Depending on the number of locations at which $\lambda_x = 0$, the solutions of the problem are shown to fall in one of the following classes:

1. λ is strictly monotone, if $\lambda_x \neq 0$ for any finite location;
2. λ exhibits either a bulge or a neck, if there exists precisely one location x at which $\lambda_x = 0$;
3. λ is periodic, if there is more than one finite value of x at which $\lambda_x = 0$.

Strictly monotone solutions are analyzed in the sequel. In such cases the following relations hold:

$$\begin{array}{ll} \lim_{x\to-\infty}\lambda=\lambda_*, & \lim_{x\to+\infty}\lambda=\lambda^* \\ \lim_{x\to\pm\infty}\lambda_x=0, & \lim_{x\to\pm\infty}\lambda_{xx}=0. \end{array} \tag{27}$$

Coleman and Newman (1988) show that such conditions can be attained provided that the applied traction equals the Maxwell stress Σ_M, which is determined by the equal area rule

$$\int_{\lambda_*}^{\lambda^*}[\varphi'(\lambda)-\Sigma_M]\,d\lambda=0, \tag{28}$$

with

$$\varphi'(\lambda_*)=\varphi'(\lambda^*)=\Sigma_M,$$

bearing in mind that these solutions are uniquely determined to within a reflection or translation. The fact that $\lambda(x)$ is monotonic allows for determining the location map x in terms of λ from (26), with $\lambda_a\equiv\lambda_*$ and $\bar{x}$ arbitrary, such that $\lambda_*<\lambda(\bar{x})<\lambda^*$.

For the specific energy (12), it turns out that

$$\Sigma_M=5.92\,\mathrm{mN\,m^{-1}},\quad \lambda_*=1.025,\quad \lambda^*=1.308, \tag{29}$$

which is consistent with what it is displayed in (see Fig. 5). For the sake of illustration, one may take $h_0=45.5$ Å for the reference thickness of the ordered phase (see Deseri and Zurlo 2013) and its reference to Goldstein and Leibler (1989) and by making use of (12, 19), the numerical integration of (26) yields $\lambda(x)$ within the range (λ_*,λ^*).

The boundary layer is displayed in Fig. 6 as a result of the solutions of the Euler–Lagrange equation mentioned above. Evidently, with $\lambda(x)$ strictly monotonic, the limit values (λ_*,λ^*) are asymptotic values at infinity. Nonetheless, the solution depicted in Fig. 6 is characterized by a strong strain localization inside a boundary layer of length $\simeq$15 Å. It is between λ_* and λ^* where such a boundary layer is almost completely localized. As it was expected, the length of the boundary layer and the membrane thickness are of the same order. This is in agreement with

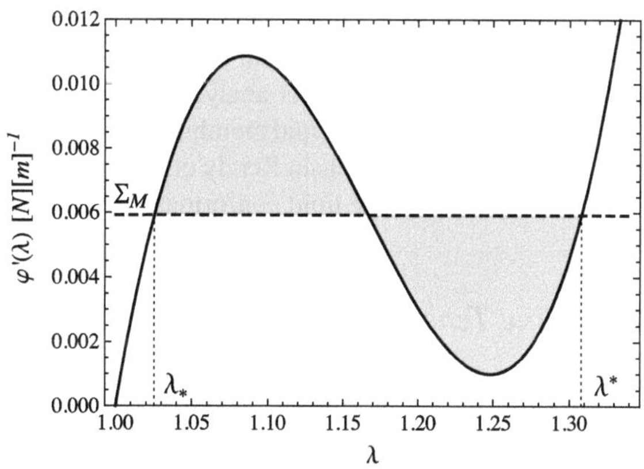

Fig. 5 The function $\varphi'(J)$ and the value of the Maxwell stress $\Sigma_M=5.92\,\mathrm{mN\,m^{-1}}$, resulting from the equal area construction (*gray regions*) (courtesy of Deseri and Zurlo 2013)

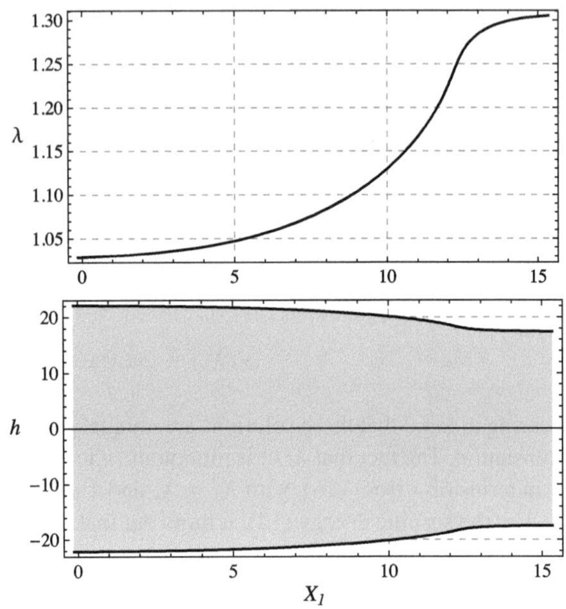

Fig. 6 The function $\lambda(x)$ (*up*) and the thickness profile $h(x)$ (*down*) in correspondence of $\Sigma = \Sigma_M$. Lengths expressed in Å (courtesy of Deseri and Zurlo 2013)

previously obtained estimates, such as the one obtained in Akimov et al. (2004). It goes without saying that the stretch is almost constant outside the boundary layer. The two domains where the stretch is practically equal to λ_* and λ^* are the L_o and L_d phases, respectively. From (24), the (Piola) stress (per reference length) in both phases equals Σ_M, whereas the Cauchy stress (per current length) in the two domains amounts to

$$\begin{aligned} t^{L_o} &= t_* = \Sigma_M \lambda_* = 6.07\,\mathrm{mN\,m^{-1}} \\ t^{L_d} &= t^* = \Sigma_M \lambda^* = 7.74\,\mathrm{mN\,m^{-1}}. \end{aligned} \tag{30}$$

Of course such values strongly depend on the form of $\varphi(J)$ taken in (12). Although this is certainly the case, such values are consistent with estimates of surface stress in ordered and disordered domains inferred through experimental investigations (see e.g., Semrau et al. 2008). In the latter paper it is shown that the stress in the disordered phase is significantly higher than in the ordered one. Furthermore, the values of surface stress obtained in this analysis are within the range of values of tension physiologically intrinsic of lipid membranes, namely ($0-15\,\mathrm{mN\,m^{-1}}$). The estimates above agree with the results in Reddy et al. (2012), where the role played by surface tension in changes of the lipid conformational order has been investigated.

2.3 Line Tension Holding Zones in a Given Phase

Before introducing the line tension, as the configurational force capable to hold zones in one phase surrounded by others in a different phase, we prove that (26) is a global minimizer of the total potential energy $\mathscr{E}$ in the class of smooth solutions fulfilling the boundary conditions (27). Furthermore, one can also show that this profile delivers an optimal value of the line tension.

In order to do so, we recall that phase coexistence follows two different approaches: the gradient theory and the sharp interface approximation.

The gradient theory does not allow discontinuities in the field. In either case, the analysis leading to phase transition between two different zones relies upon minimizing the total potential energy introduced earlier in the text, namely:

$$\mathscr{E} = \int_{\Omega} \left[\varphi(J) + \alpha(J) \| \mathrm{grad}_{\omega} \hat{J} \|^2 \right] d\Omega - \mathscr{W}. \tag{31}$$

The sharp interface approximation allows for the order parameter J to be subject to discontinuities; in this case the total potential energy reads

$$\mathscr{F} = \int_{\Omega} \varphi(J)\, d\Omega + \sigma\, \ell([\![J]\!]) - \mathscr{W}, \tag{32}$$

where σ is the *line tension* between the two phases which, from the dimensional standpoint, is a force. Here ℓ is the length of the interface, which in this approximation is a jump set, i.e., the union of regions across which J can tolerate jumps.

In Deseri and Zurlo (2013) a rigorous analysis demonstrates the strict connection between the sharp interface approach and the gradient theory. Indeed, it is proved that minimizers of $\mathscr{E}$ converge (in a suitable sense) to minimizers of $\mathscr{F}$ (see e.g., Alberti 2000 for explanations). An optimal value for the line tension can be deduced by evaluating the global minima of $\mathscr{E}$ in the class of solutions fulfilling the boundary conditions (27).

Because of compatibility we recall that $u_x(x) = \lambda(x) - 1$. Henceforth, the work can be rewritten as follows:

$$\mathscr{W} = B \int_{-L/2}^{L/2} \Sigma_M \lambda\, dx - B \Sigma_M L. \tag{33}$$

It is worth noting that the following quantity, essentially representing a Gibbs free energy density for the lipid membrane, remains constant at the minimizer, i.e.,

$$\varphi(\lambda_*) - \Sigma_M \lambda_* = \varphi(\lambda^*) - \Sigma_M \lambda^*.$$

This suggests to consider the energy

$$\tilde{\varphi}(\lambda) = \varphi(\lambda) + c,$$

where

$$c = \Sigma_M \lambda_* - \varphi(\lambda_*) = \Sigma_M \lambda^* - \varphi(\lambda^*), \tag{34}$$

so that

$$\tilde{\varphi}(\lambda_*) - \Sigma_M \lambda_* = \tilde{\varphi}(\lambda^*) - \Sigma_M \lambda^* = 0. \tag{35}$$

We also note that away from the characteristic stretches λ_* and λ^*, i.e., for $\lambda \neq \lambda_*$ and $\lambda \neq \lambda^*$, the following inequality holds:

$$\tilde{\varphi}(\lambda) - \Sigma_M \lambda \geq 0. \tag{36}$$

After discussing the sharp interface approximation, the gradient functional is now analyzed. Outcomes from the latter will be compared with the former. Indeed the film is subject to a traction Σ_M, and we consider a monotonic stretch profile $\lambda(x)$ within the interval $(-L/2, L/2)$. Assume that $\lambda \to \lambda_*$ as $x \to -L/2$ and that $\lambda \to \lambda^*$ as $x \to L/2$. Obviously, Σ_M is the Maxwell value introduced in Sect. 2.2.

Consider the total potential energy per unit length $\mathscr{E}/L$. By utilizing ((33), (35)), for any thickness profile satisfying the boundary conditions (27), the relation below follows:

$$\frac{\mathscr{E}}{L} = \frac{B}{L} \int_{-L/2}^{L/2} \left[(\tilde{\varphi}(\lambda) - \Sigma_M \lambda) - \frac{\gamma(\lambda)}{2} \lambda_x^2 \right] dx + d, \tag{37}$$

where $d = B(\Sigma_M - c)$ is a constant.

On closing, the profile characterized by (26) and verifying stationarity is now shown to be a minimizer for $\mathscr{E}/L$. This is based on a result in Alberti (2000). By $\tilde{\varphi}(\lambda) - \Sigma_M \lambda \geq 0$, by $-\gamma(\lambda)\lambda_x^2 \geq 0$, by the monotonicity of λ and by the inequality $a^2 + b^2 \geq 2ab$, it follows the following inequality:

$$\frac{\mathscr{E}}{L} \geq \frac{B}{L} \int_{\lambda_*}^{\lambda^*} \sqrt{-2\gamma(\lambda)\,(\tilde{\varphi}(\lambda) - \Sigma_M \lambda)}\; d\lambda + d, \tag{38}$$

and equality holds if and only if $a = b$, namely if and only if

$$\tilde{\varphi}(\lambda) - \Sigma_M \lambda = -\frac{\gamma(\lambda)}{2} \lambda_x^2. \tag{39}$$

If one now simply recognizes that

$$\tilde{\varphi}(\lambda) - \Sigma_M \lambda = \int_{\lambda_*}^{\lambda} \left(\varphi'(\zeta) - \Sigma_M \right) d\zeta \tag{40}$$

from (35), (26) is obtained in exact form by integrating (39). Finally, we just showed that the following minimum is actually attained:

$$\min \left(\frac{\mathscr{E}}{L} \right) = $$

$$= \frac{B}{L} \int_{\lambda_*}^{\lambda^*} \sqrt{-2\gamma(\lambda)\,(\tilde{\varphi}(\lambda) - \Sigma_M \lambda)}\; d\lambda + d, \tag{41}$$

within the functions verifying the boundary conditions (27), provided that $\lambda(x)$ is given by (26).

Consider now any configuration characterized by $\lambda = \lambda_*$ for $x < x_0$ and $\lambda = \lambda^*$ for $x > x_0$, so that in $x = x_0$ there is a sharp interface. The location x_0 is an arbitrary finite point. In this configuration, one can show that the total potential energy per unit length (32) becomes

$$\frac{\mathscr{F}}{L} = \frac{B}{L}\sigma + d. \tag{42}$$

By comparing (37), (41) and (42) the line tension of the sharp interface model remains determined as follows:

$$\sigma = \int_{\lambda_*}^{\lambda^*} \sqrt{-2\gamma(\lambda)\,(\tilde{\varphi}(\lambda) - \Sigma_M \lambda)}\; d\lambda. \tag{43}$$

Numerical data (13) and integration of (43) owe the following number for the line tension:

$$\sigma = 3.88 \cdot 10^{-13}\mathrm{N} \tag{44}$$

which is consistent with the experimentally found value $9 \pm 0.3 \cdot 10^{-13}$N (see e.g., Baumgart et al. 2003; Semrau et al. 2008). The predicted thickness profile and line tension are then consistent with pre-existing analyses for lipid membranes, that account for the competition of stretching and tilt elasticity. This latter phenomenon is due to the fact that lipid molecules can deviate from the mid-surface normal (see e.g., Akimov et al. 2004; Hamm and Kozlov 2000).

2.4 *Elastic Properties of the Lipid Membrane*

In the sequel we explore values for the elastic moduli in a lipid bilayer undergoing a traction Σ_M. Here each pure phase is characterized by a specific value of the stretch λ, namely $\lambda = \lambda_*$ for the liquid-ordered phase L_o and $\lambda = \lambda^*$ for the liquid-disordered one L_d.

Area compressibility A tangent area compressibility modulus

$$K_A(\lambda) := \varphi''(\lambda) \tag{45}$$

is defined as the change of surface stress, $\varphi'(\lambda)$, induced by a change in stretch. As the membrane energy $\varphi(\lambda)$ is a fourth-order polynomial, the compressibility stiffness K_A is nonconstant and takes the form

$$K_A(\lambda_*) = K_A(\lambda^*) = 181\,\mathrm{mN\,m^{-1}}, \tag{46}$$

in λ_* and λ^*, and for the unstretched membrane ($\lambda = 1$)

$$K_A(1) = 288\,\text{mN}\,\text{m}^{-1}. \tag{47}$$

Henceforth, K_A manifests softening. The obtained values are consistent with measurements available in the literature. In particular, the highest areal stretch is $\delta A/A_0 = \lambda_* - 1 = 0.025$ and it agrees with the value of rupture stretches found in Lipowsky and Sackmann (1995).

Bending stiffness Relation $(11)_1$ yields values in agreement with previous results (see e.g., Bermúdez et al. 2004; Evans 1974; Norouzi et al. 2006; Pan et al. 2009; Rawicz et al. 2000). Specifically, in the ordered and disordered phases the following values are obtained:

$$\kappa^{L_o} = \kappa(\lambda_*) = 6.10 \cdot 10^{-19}\,\text{J}, \tag{48}$$

$$\kappa^{L_d} = \kappa(\lambda^*) = 4.78 \cdot 10^{-19}\,\text{J}. \tag{49}$$

It is worth noting that the ratio of these rigidities is

$$\frac{\kappa^{L_o}}{\kappa^{L_d}} = 1.27 \tag{50}$$

in agreement with the experimental findings (see e.g., Baumgart et al. 2003; Semrau et al. 2008).

Gaussian stiffness Normally the evaluation of this quantity refers to the spontaneous curvature of each leaflet (Hu et al. 2012; Siegel and Kozlov 2004), while in Deseri and Zurlo (2013), Zurlo (2006) these values are not accounted for. There each leaflet has no spontaneous curvature and the resulting κ_G is of the order of 10^{-21}J. This is then turns out to be two orders of magnitude lower than existing estimates available in Norouzi et al. 2006, Semrau et al. (2008). This discrepancy could be solved either incorporating those spontaneous curvatures of each monolayer of by incorporating the lateral (and highly nonconstant through thickness) pressure profile within the bilayer. This is actually under investigation.

Keeping the approach of Deseri and Zurlo (2013), Zurlo (2006), relations $(9)_2$ and (10) yield

$$\alpha(J) = \frac{k_G(J)}{2J^2}, \tag{51}$$

highlighting the connection between changes of the Gaussian rigidity with changes in the gradient of thinning and, ultimately, of the areal stretch. This connection is actually not surprising. Indeed, thanks to the Gauss–Bonnet Theorem, k_G emerges at the boundaries of each region characterized by constant values of J. Namely, k_G appears at the phase boundaries between the L_o and the L_d phases. The role of $\alpha(J)$ emerges instead while trying to evaluate the line tension inside the boundary layer, as highlighted in Sect. 2.3. Such instances are consistent with the relation established in Eq. (51).

2.5 The Onset of Change of Elastic Phase

In this section we obtain the linearized equation of lipid membrane under the plane strain geometry (14) with $g_x = \bar{J}$ and $\phi = \bar{\phi}$ (hence $\phi_x = 0$). In this regard let us denote with ε the strain field perturbing uniformly the stretched configuration just described. The elastic free energy density (20) for the membrane is then evaluated at the perturbed configuration $J = \bar{J} + \varepsilon$, and takes the following form:

$$\begin{aligned}\psi(\varepsilon, \varepsilon_x) &= \varphi\left(\bar{J} + \varepsilon\right) + \alpha(\bar{J} + \varepsilon)|| \left(\bar{J} + \varepsilon\right)_x ||^2 \\ &\approx \varphi(\bar{J}) + \varphi'(\bar{J})\,\varepsilon + \frac{\varphi''(\bar{J})}{2}\varepsilon^2 + \alpha(\bar{J})\,||\varepsilon_x||^2\,,\end{aligned} \tag{52}$$

where we neglected higher order contributions in ε^2 to define ψ_{DZ}. Then the free energy takes the following form:

$$\Psi_{DZ} = \int_\Omega \psi_{DZ}(\varepsilon, \varepsilon_x) dx, \tag{53}$$

where $\Omega \in [-L/2, L/2]$, and

$$\psi_{DZ}(\varepsilon, \varepsilon_x) = \varphi(\bar{J}) + \varphi'(\bar{J})\,\varepsilon + \frac{\varphi''(\bar{J})}{2}\varepsilon^2 + \alpha(\bar{J})\,\varepsilon_x^2\,. \tag{54}$$

The (in-plane) displacement field is expressed through a perturbation v such that $u = \bar{u} + v$, and $\varepsilon(x) = v_x(x)$.

Due to the presence of nonlocal terms ε_x, we recall that the hypertractions Γ perform work against displacement gradient v_x at the boundary. Henceforth, the total energy $\mathscr{E}$ change in a neighborhood of the "ground" (homogeneous) configuration reads as follows:

$$\mathscr{E} = B\,\Psi_{DZ} - \mathscr{W}(v, v_x), \tag{55}$$

where the external work reads now as follows:

$$\mathscr{W}(v, v_x) = B\,[\Sigma\ \times\ (\bar{u} + v) + \Gamma\ \times\ (\bar{u}_x + v_x)]_{\partial\Omega}\,, \tag{56}$$

where $\bar{u} = \bar{J}_x$ is zero if the ground configuration is homogeneously stretched. For the sake of conciseness, nonhomogeneous ground configurations will not be analyzed here, although the issue is addressed in Deseri et al. (2016). By substituting (52) and (55) in (56) the total energy change takes the following form:

$$\begin{aligned}\mathscr{E} = B &\int_\Omega \left(\varphi + \varphi'(\bar{J})\,v_x + \frac{\varphi''(\bar{J})}{2}v_x^2 + \alpha(\bar{J})\,v_{xx}^2\right) dx \\ &- B\,[\Sigma\,v + \Gamma\,v_x]_{\partial\Omega} + \bar{\mathscr{E}},\end{aligned} \tag{57}$$

where

$$\bar{\mathscr{E}} = B \int_{\Omega} \varphi(\bar{J})dx - [\Sigma\,\bar{u} + \Gamma\,\bar{u}_x]_{\partial\Omega}\,. \tag{58}$$

From now on, every item with the over-bar is calculated on the ground configuration (e.g., $\bar{\varphi} = \varphi(\bar{J})$ etc.), and we will denote with $'$ the derivative with respect to the spatial variable x.

The linear Euler–Lagrange equation for the perturbations of planar membranes is derived through stationarity of $\mathscr{E}$ (see Appendix A1 in Deseri et al. 2016 for details). Such equation together with its boundary conditions reads as follows:

$$\begin{cases} 2\bar{\alpha}\,v'''' - \bar{\varphi}''\,v'' = 0 & \text{in } \Omega \\ \text{either } \bar{\varphi}''\,v' - 2\bar{\alpha}\,v''' = \Sigma - \bar{\varphi} \text{ or } \delta v = 0 & \text{in } \partial\Omega \\ \text{either } 2\bar{\alpha}\,v'' = \Gamma \text{ or } \delta v' = 0 & \text{in } \partial\Omega \end{cases} \tag{59}$$

It is worth noting that homogeneous configurations, and hence their corresponding values $\bar{J}$, from which oscillatory perturbations could arise are still not known at this point. In order to find them, a parameter ω is introduced as follows:

$$\omega^2 := \begin{cases} +\dfrac{\bar{\varphi}''}{2\bar{\alpha}} & \text{if } \bar{\varphi}'' > 0 \\ -\dfrac{\bar{\varphi}''}{2\bar{\alpha}} & \text{if } \bar{\varphi}'' < 0, \end{cases} \tag{60}$$

where, because of (10) and (9), we have

$$\frac{\bar{\varphi}''}{2\bar{\alpha}} = \frac{12}{h_0^2}\frac{\bar{\varphi}''}{\bar{\varphi}'}\bar{J}^5. \tag{61}$$

Relation (59) can then be rewritten as follows:

$$\begin{cases} v'''' \mp w^2\,v'' = 0 & \text{in } \Omega \\ \text{either } \pm\,\omega^2 v' - v''' = \dfrac{\Sigma - \bar{\varphi}}{2\bar{\alpha}} \text{ or } \delta v = 0 & \text{in } \partial\Omega \\ \text{either } 2\bar{\alpha}\,v'' = \Gamma \text{ or } \delta v' = 0 & \text{in } \partial\Omega. \end{cases} \tag{62}$$

Boundary conditions yield obviously several cases. For the sake of illustration, we choose the case in which the displacement is constrained and the hypertractions are imposed at the boundary, i.e., $v = 0$ and $2\bar{\alpha}\,v'' = \Gamma$.

The value of ω^2 does determine the type of solution arising from this analysis. In particular, the phase changes start to be seen from the onset arising, thanks to the specific value of ω^2. In order to investigate such onset, subcases are identified depending on $\bar{J}$ relative to the landscape of the membrane energy φ in Fig. 3. Indeed, because such a function has at most one stationary point J_0 unless the lipid bilayer is at its transition temperature, inspection of Fig. 7 below shows that there are four values of J besides $\bar{J}$ to be accounted for, i.e., $J_* \le J_{max} \le J_{min} \le J^*$. Here J_{max}

and J_{min} are points of turning curvature for $\varphi(J)$, whereas J_* and J^* are the values of the two points sharing the value of the tangent to the graph of $\varphi(J)$.

Two alternative situations may arise depending on the sign of $\bar{\varphi}''$, depending on whether or not the ground state $\bar{J}$ belongs to the spinoidal, hence unstable, zone of $\varphi(J)$.

2.6 Unstable Region: $\bar{\varphi}'' < 0$

The case just mentioned is investigated in this section. Here, $\bar{J}$ is then such that $J_{max} < \bar{J} < J_{min}$, corresponding to a region of negative tangent for the membrane stress $\tau(J) = \varphi'(J)$ (see Fig. 7). The Euler–Lagrange equation (62) takes then the following form:

$$v^{''''} + \omega^2 v^{''} = 0, \tag{63}$$

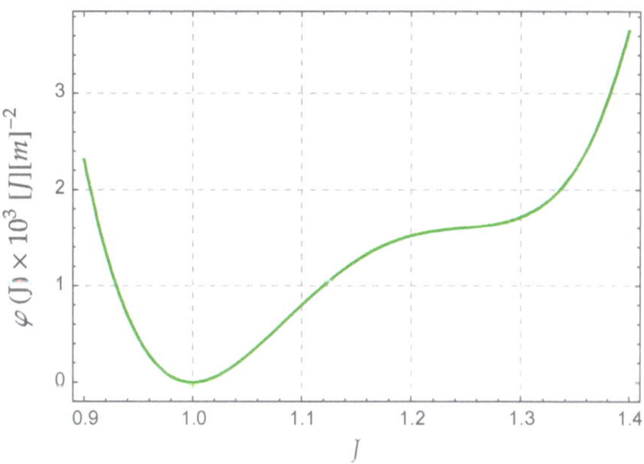

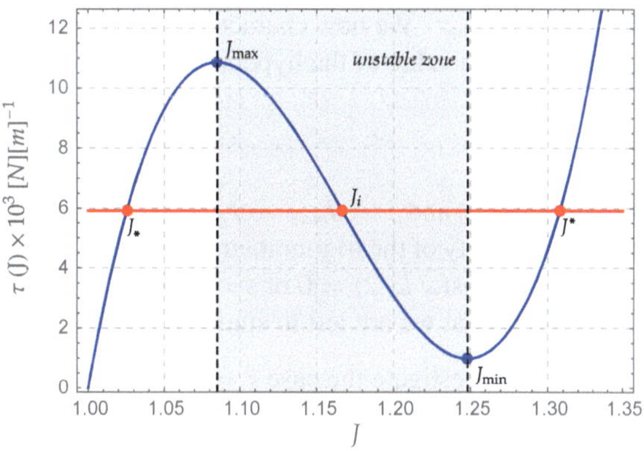

Fig. 7 The membrane energy $\varphi(J)$ for a temperature $T \sim 30°$ and related local stress $\tau(J)$. The value $J_o = 1$ corresponds to the unstressed, reference configuration $\mathscr{B}_0$ (courtesy of Deseri and Zurlo 2013; Deseri et al. 2016)

which general solution reads as follows:

$$v(x) = A_1 \cos(\omega x) + A_2 \sin(\omega x) + A_3 x + A_4. \tag{64}$$

The primary interest here is to investigate the influence of the boundary conditions below:

$$v\Big|_{\partial\Omega^-} = 0 \quad v\Big|_{\partial\Omega^+} = 0 \quad 2\bar{\alpha}v''\Big|_{\partial\Omega^-} = \hat{\Gamma}_L \quad 2\bar{\alpha}v''\Big|_{\partial\Omega^+} = \hat{\Gamma}_R \tag{65}$$

where $\hat{\Gamma}_R = \Gamma\Big|_{\partial\Omega^+}$ and $\hat{\Gamma}_L = \Gamma\Big|_{\partial\Omega^-}$. For the sake of brevity we set

$$c = \cos(\omega L/2) \;\; \text{and} \;\; s = \sin(\omega L/2).$$

The boundary conditions assume can be then recast in the following form:

$$\begin{cases} A_1 c - A_2 s - A_3 \dfrac{L}{2} + A_4 = 0 \\ 2\bar{\alpha}\omega^2 (-A_1 c + A_2 s) = \hat{\Gamma}_L \end{cases} \text{at } x = -\frac{L}{2}$$

$$\begin{cases} A_1 c + A_2 s + A_3 \dfrac{L}{2} + A_4 = 0 \\ 2\bar{\alpha}\omega^2 (-A_1 c - A_2 s) = \hat{\Gamma}_R \end{cases} \text{at } x = +\frac{L}{2}$$

We further choose a constant hyperstress at the boundary, namely $\hat{\Gamma}_L = \hat{\Gamma}_R = \hat{\Gamma}$, leading to the simplified set of algebraic conditions below:

$$\begin{bmatrix} 0 & s & \frac{L}{2} & 0 \\ c & 0 & 0 & 1 \\ 0 & s & 0 & 0 \\ -2\bar{\alpha}\,\omega^2 c & 0 & 0 & 0 \end{bmatrix} \begin{pmatrix} A_1 \\ A_2 \\ A_3 \\ A_4 \end{pmatrix} = \begin{pmatrix} 0 \\ 0 \\ 0 \\ \hat{\Gamma} \end{pmatrix}. \tag{66}$$

We record that the determinant of the coefficient matrix of such system reads as follows $\bar{\alpha}\, c\, s\, L\, \omega^2$. We now characterize the nontrivial modes (64) of the system no matter what the value of the hyperstress, namely we investigate the solutions of

$$\bar{\alpha}\, c\, s\, L\, \omega^2 = 0. \tag{67}$$

Because of (10) and $1 < J_{max} < \bar{J} < J_{min}$, we note that $\bar{\alpha} > 0$ for all $\bar{J} > 1$. Then, the orthogonality of the trigonometric functions imposes that the equation is satisfied if either $c = \cos(\omega L/2) = 0$ or $s = \sin(\omega L/2) = 0$.

It follows that we are left to study only two subcases.

Case 1. We investigate the case $s = 0$ and $c = \pm 1$. Such instance implies that

$$\omega = \frac{2\, n\, \pi}{L} \tag{68}$$

and relation (68) allows for showing that this circumstance occurs whenever the ground state solves the nonlinear algebraic equation below:

$$\frac{\bar{\varphi}''}{\bar{\varphi}'}\bar{J}^5 = -\frac{n^2\pi^2}{3}\left(\frac{h_0}{L}\right)^2. \tag{69}$$

It is worth noting that the ratio $(h_0/L)^2$ measures the thinness of the bilayer and it is of the order 10^{-8} or smaller. Henceforth, from (69) it follows that a large finite number n of oscillations arise in the onset of bifurcation starting from ground states solving (69). Indeed this is possible just by noting that for J such that $\bar{\varphi}'' \to 0^-$, namely right after the change on convexity of φ. The solution of the resulting system permits to get the amplitudes of the nth mode, i.e.,

$$\begin{bmatrix} 0 & 0 \ \frac{L}{2} \ 0 \\ \pm 1 & 0 \ 0 \ 1 \\ 0 & 0 \ 0 \ 0 \\ \mp 2\bar{\alpha}\,\omega^2 & 0 \ 0 \ 0 \end{bmatrix} \begin{pmatrix} A_1 \\ A_2 \\ A_3 \\ A_4 \end{pmatrix} = \begin{pmatrix} 0 \\ 0 \\ 0 \\ \hat{\Gamma} \end{pmatrix}$$

then

$$\begin{cases} A_1 = \mp\dfrac{\hat{\Gamma}}{2\bar{\alpha}\,\omega^2} \\ A_3 = 0 \\ A_4 = \mp A_1 \end{cases}.$$

Then, the corresponding buckled solution of order n reads

$$\begin{aligned} v_n(x) = {} & \pm\frac{\hat{\Gamma}}{8\,\bar{\alpha}\,n^2\,\pi^2}\left[\cos\left(2n\pi\frac{x}{L}\right) - 1\right] \\ & + A_2\sin\left(2n\pi\frac{x}{L}\right). \end{aligned} \tag{70}$$

It goes without saying that even if no hyperstress $\hat{\Gamma}$ is present at the boundary, (70) guarantees that a bifurcated mode $v_n = A_2\sin\left(2n\pi\frac{x}{L}\right)$ does occur.

Case 2. We now instead explore the following situation:

$$s = \pm 1 \ \text{ and } \ c = 0.$$

In this case we have

$$\omega = \frac{(1+2\,n)\,\pi}{L} \tag{71}$$

and

$$\frac{\bar{\varphi}''}{\bar{\varphi}'}\bar{J}^5 = -\frac{(1+2n)^2\pi^2}{12}\left(\frac{h_0}{L}\right)^2, \tag{72}$$

which definitely has solutions for $\bar{J}$ so that $\bar{\varphi}'' \to 0^-$ for the same very reason discussed for case 1. Boundary conditions lead to $A_2 = A_3 = A_4 = 0$. It follows that solutions exist if and only if $\hat{\Gamma} = 0$ and they take the form

$$v_n(x) = A_1 \cos(\omega x) = A_1 \cos\left((1+2n)\pi \frac{x}{L}\right). \tag{73}$$

2.7 Stability Region: $\bar{\varphi}'' > 0$

If the ground state $\bar{J}$ is not in the spinoidal zone, namely there $\bar{\varphi}'' > 0$, and either $1 < \bar{J} < J_{max}$ or $\bar{J} > J_{min}$, the balance equation reduces to

$$v'''' - \omega^2 v'' = 0, \tag{74}$$

and its general solution becomes

$$v(x) = A_1 \cosh(\omega x) + A_2 \sinh(\omega x) + A_3 x + A_4, \tag{75}$$

hence no oscillations arise.

2.8 Singular Ground States: $\bar{\varphi}'' = 0$

Singular values for the ground states are $\bar{J} = J_{max}$ and $\bar{J} = J_{min}$. There, the first derivative of the local stress with respect to J is zero and, hence, $\bar{\varphi}'' = 0$. This immediately tells that $\omega = 0$, and the resulting governing equation, $v'''' = 0$, admits

$$v(x) = A_0 + A_1 x + A_2 x^2 + A_3 x^3 \tag{76}$$

as solution. If (65) are imposed at the boundary with $\hat{\Gamma}_R = \hat{\Gamma}_L = \hat{\Gamma}$, the constants in the previous relation become as follows:

$$A_0 = -\frac{\hat{\Gamma} L^2}{16\bar{\alpha}} \quad A_1 = 0 \quad A_2 = \frac{\hat{\Gamma}}{4\bar{\alpha}} \quad A3 = 0, \tag{77}$$

thereby leading to a unique solution. In other words, no bifurcations arise from singular ground states and perturbations do not arise in the absence of hyperstress at the boundary.

3 Hereditariness of Lipid Membranes

Available experimental data Harland et al. (2010), Espinosa et al. (2011), Craiem and Magin (2010) show that lipid bilayers present an *anomalous* rate-dependent behavior within broad ranges of temperature. Anomaly means that if the loss and storage moduli[1] in any rheometric test are plotted against frequency, such quantity scale with a noninteger power of the frequency itself. Indeed, Harland et al. (2010) showed that the storage and loss modulus are proportional to the frequency through a power law of fractional order, i.e., $G^{'}(\omega) \propto \omega^{\beta}$ and $G^{''}(\omega) \propto \omega^{\beta+1}$, where the exponent β depends on temperature and specific chemical composition of the biological structure. This justifies the term "fractional" for such a kind of response. Fractional hereditariness is then an intrinsic feature of lipid membranes. Perturbations of the ground states from which bifurcations of phases occur are nucleated and then evolve in time according to such behavior.

Results in Harland et al. (2010) show that lipid membranes are not purely elastic and this is in fact only an asymptotic condition. Nevertheless, such structures have been predominantly modeled as hyperelastic surfaces. Physiological conditions of cells are in fact characterized by intracellular and extracellular viscous fluid compartments cooperating to vary the areal stretch several times during cell lifetimes. The corresponding membrane stress therefore changes in time and can achieve significantly higher values than the ones evaluated by utilizing nonlinear elasticity. The time change of such stress can even evolve to the extent of either causing rupture of the cell membrane or to modify toward ceramide phase, and then to cell apoptosis, the lipids across the membrane (Craiem and Magin 2010).

3.1 *The Physics of Hereditariness in Lipid Structures*

As pointed out before, lipid systems forming cytoplasmatic membranes present time-hereditary properties (Espinosa et al. 2011). Storage and loss moduli $G^{'}(\omega)$, $G^{''}(\omega)$ of lipid membrane depend on the type of lipids. The presence of very common lipids like phosphatidylcholine (PODC) and sphingomyelin (SM) do heavily influence the rate behavior of lipid layers, thereby showing various morphologies ultimately affecting the resulting effective viscosity of the membrane. The phases can be either liquid-ordered or gel-phase, for temperatures over or below the melting temperatures of the PODC. For SM the liquid-disordered or the solid phase (ceramide) can be involved depending on the temperature of the system.

From the point of view of modeling, it is obvious that the use of Maxwell rheological elements to describe storage and loss moduli of the material does not provide a suitable representation for the behavior of lipid membranes for the simple reason that

[1] For the reader who is not familiar with this standard terminology, we recall that the right-handed Fourier transform of a given relaxation function represents the "complex modulus" of a viscoelastic material; its real part is the "storage modulus", while its imaginary part is its "loss modulus".

Maxwell models yield $G^{'}(\omega) \propto \omega$ and $G^{''}(\omega) \propto \omega^2$, never observed in experiments (see e.g., Espinosa et al. 2011).

It is then obvious that the only way to account for hereditary behavior of lipid membranes must contain fractional-order features, where creep and relaxation are described as power laws so that $J(t) \propto t^{\beta}$ and $G(t) \propto t^{-\beta}$, respectively. Small perturbations arising from homogeneous ground states must then be studied by making use of the Boltzmann–Volterra convolution integral. This allows for keeping track the stress evolution at any x depending on the strain history $\epsilon(x, t)$, namely

$$\sigma(x, t) = \frac{C_\beta}{\Gamma[1-\beta]} \int_{-\infty}^{t} (t-\tau)^{-\beta} \, \dot{\epsilon}(x, \tau) \, d\tau. \tag{78}$$

The right-hand side of the latter relation relates with the Caputo fractional-order derivative $\mathcal{D}_t^{\beta}$ defined as follows:

$$\mathcal{D}_t^{\beta} f(t) = \frac{1}{\Gamma(\beta)} \int_{-\infty}^{t} (t-\tau)^{-\beta} \dot{f}(x, \tau) d\tau, \tag{79}$$

introduced in Caputo (1969) and explored in several papers ever since (see e.g., Podlubny 1998; Magin 2010; Samko et al. 1987; Kilbas et al. 2006). The *springpot element* introduced in Scott-Blair (1974) is a rheological element associated to (79). This detects an intermediate behavior between a linear elastic spring and a viscous dashpot, which are then limiting cases obtained for $\beta = 0$ and $\beta = 1$, respectively.

When it comes to considering more complex studies of nucleations of phase perturbations in the presence of elasticity and viscosity, one needs to provide an expression of the free energy, delivering the key element of a variational principle suitable for the desired investigations. The free energy provided by Deseri et al. (2014) for power law hereditary systems is then used in the sequel. This can be further specialized to characterize the non-dissipated part of the power performed in a given springpot by an underlying stress, thereby allowing for a powerful tool suitable for handling lipid membrane hereditariness.

3.2 *The Free Energy for Small Perturbations of Planar Lipid Structures*

In this section we aim to obtain and solve the balance equations governing the nucleation and evolution of small perturbations of homogeneous ground states in hereditary and planar lipid membranes.

The limiting elastic case is well described through (54), containing both the local term, $\varepsilon(x\, t)$, and a nonlocal one, $\varepsilon_x(x,\ t)$. Henceforth, when it comes to accounting for fractional hereditariness of our systems, the expression of the free energy function is then the sum of contributions related to the local and the nonlocal state variables (for the notion of state in hereditary systems see e.g., Del Piero and Deseri 1997; Deseri et al. 1999, 2006).

It is then reasonable to infer that nucleation and evolution of small perturbations from homogeneous ground states are determined by the local and nonlocal stresses $\sigma_L(x,t)$ and $\sigma_N(x,t)$ respectively, i.e.,

$$\sigma_L(x,t) = \int_0^t G_L(t-\tau)\dot{\varepsilon}(x,\tau)\,\mathrm{d}\tau, \tag{80a}$$

$$\sigma_N(x,t) = \int_0^t G_N(t-\tau)\dot{\varepsilon}_x(x,\tau)\,\mathrm{d}\tau, \tag{80b}$$

where G_L and G_N represent the local and nonlocal relaxation functions (relative to the given ground state $\bar{J}$), respectively, defined as follows:

$$G_L(t) = \bar{\varphi}'' + f_L(t),$$
$$G_N(t) = 2\bar{\alpha} + f_N(t).$$

Asymptotically, we require the following relations to hold:

$$\lim_{t\to\infty} f_L(t) = \lim_{t\to\infty} f_N(t) = 0, \tag{81}$$

as the elastic case must be retrieved as limit. The analytic dependence of both $f_L(t)$ and $f_N(t)$ on time can be determined by experimental observations of the evolution of the phases as well as of their transition zone. The striking experimental evidence discussed in the section above induces us to utilize a power law relaxation function to model both local and nonlocal evolution of the constitutive response. In general two different laws for describing the local and the nonlocal contributions have to be considered; here we assume

$$G_L(t) = \bar{\varphi}'' + C_L t^{-\lambda}, \tag{82a}$$

$$G_N(t) = 2\bar{\alpha} + C_N t^{-\nu}, \tag{82b}$$

where C_L and C_N are generalized moduli of the local and nonlocal relaxations, λ and ν are the decay exponents of the relaxations, chosen in the (open) interval (0, 1). Relations (82) yield a fractional-order rheological element introduced in (79).

The free energy function $\Psi(x,\, t)$ is chosen to be additive in two distinguished terms:

$$\Psi(x,\, t) = \Psi_{DZ}(x,\, t) + \Psi_V(x,\, t), \tag{83}$$

where $\Psi_{DZ}(x,\, t)$ is defined by (53) and represents the elastic contribution to the free energy at equilibrium (see Del Piero and Deseri 1996), while $\Psi_V(x,\, t)$ is the free energy characterizing the hereditary response of the system. The latter has been obtained in Deseri et al. (2014). There it has been shown that a multiscale procedure across the spectrum of observation scales of a fractal material does deliver (i) a power law relaxation function and (ii) a Staverman–Schartzl free energy, which is indeed utilized here for Ψ_V. Studies on Staverman–Schartzl free energies can be found in Breuer and Onat (1964), Del Piero and Deseri (1996), Del Piero and Deseri (1997), among other works. The results in Deseri et al. (2014) and formulas (80), (82) yield $\Psi(x,\, t)$ as follows:

$$\Psi(x,\, t) = \Psi_L(\varepsilon(x,\, t)) + \Psi_N(\varepsilon_x(x,\, t)), \tag{84}$$

where the subscripts L and NL stand for local and nonlocal, respectively. The former term depends upon the strain, while the latter one is a functional of its gradient. Results in Breuer and Onat (1964) and Deseri et al. (2014) suggest to introduce a kernel $K(\circ, \circ)$, symmetric in its arguments, namely such that $K(\circ, \circ) \geq 0$ and $K(\tau_1, \tau_2) = K(\tau_2, \tau_1)$ hold. Specifically, each contribution is taken as follows:

$$\begin{aligned}\Psi_L(x,\, t) &= \frac{1}{2} K_L(0, 0)\varepsilon(x, t)^2 \\ &+ \varepsilon(x, t)\int_{-\infty}^{t} \dot{K}_L(0, t-\tau)\varepsilon(x, \tau)\mathrm{d}\tau \\ &+ \frac{1}{2}\int_{-\infty}^{t}\int_{-\infty}^{t} \ddot{K}_L(t-\tau_1, t-\tau_2)\varepsilon(x, \tau_1)\varepsilon(x, \tau_2)\mathrm{d}\tau_1\mathrm{d}\tau_2,\end{aligned} \tag{85a}$$

$$\begin{aligned}\Psi_N(x,\, t) &= \frac{1}{2} K_N(0, 0)\varepsilon_x(x, t)^2 \\ &+ \varepsilon_x(x, t)\int_{-\infty}^{t} \dot{K}_N(0, t-\tau)\varepsilon_x(x, \tau)\mathrm{d}\tau + \\ &+ \frac{1}{2}\int_{-\infty}^{t}\int_{-\infty}^{t} \ddot{K}_N(t-\tau_1, t-\tau_2)\varepsilon_x(x, \tau_1)\varepsilon_x(x, \tau_2)\mathrm{d}\tau_1\mathrm{d}\tau_2,\end{aligned} \tag{85b}$$

where

$$K_L(t, 0) = \bar{\varphi}'' + \frac{C_L}{\Gamma(1-\lambda)}(t+\delta)^{-\lambda} = G_L^{\delta}(t), \tag{86a}$$

$$K_N(t, 0) = 2\bar{\alpha} + \frac{C_N}{\Gamma(1-\nu)}(t+\delta)^{-\nu} = G_N^{\delta}(t), \tag{86b}$$

where δ is a preloading time. This comes from the fact that no strain process starts with abrupt jump and, instead, it does require some time, δ, to reach a desired value.

The relations $K_L(0, t) = K_L(t, 0)$ and $K_N(0, t) = K_N(t, 0)$ also do hold. This result, together with (82) and the considerations addressed in Eqs. (17–22) in Deseri et al. (2014), permits to write the free energy as follows:

$$\Psi_L(x,\ t) = \frac{1}{2} G_L^\delta(0)\varepsilon^2(x,t) + \varepsilon(x,t)\int_{-\infty}^{t} \dot{G}_L^\delta(t-\tau)\varepsilon(x,\tau)\mathrm{d}\tau + \frac{1}{2}\int_{-\infty}^{t}\int_{-\infty}^{t} \ddot{G}_L^\delta(2t-\tau_1-\tau_2)\varepsilon(x,\tau_1)\varepsilon(x,\tau_2)\mathrm{d}\tau_1\mathrm{d}\tau_2, \tag{87a}$$

$$\Psi_N(x,\ t) = \frac{1}{2} G_N^\delta(0)\varepsilon_x^2(x,t) + \varepsilon_x(x,t)\int_{-\infty}^{t} \dot{G}_N^\delta(t-\tau)\varepsilon_x(x,\tau)\mathrm{d}\tau + \frac{1}{2}\int_{-\infty}^{t}\int_{-\infty}^{t} \ddot{G}_N^\delta(2t-\tau_1-\tau_2)\varepsilon_x(x,\tau_1)\varepsilon_x(x,\tau_2)\mathrm{d}\tau_1\mathrm{d}\tau_2, \tag{87b}$$

where $\varepsilon(x,\ t) = v_x(x,\ t)$, and $v(x,\ t)$ is the space-time perturbation process of the underlying ground state of the membrane. Ultimately, the free energy associated with the perturbation process $v(x,t)$ becomes the following:

$$\mathscr{E} = B\int_{t_1}^{t_2}\left(\int_{\Omega}[\Psi_L(x,\ t) + \Psi_N(x,\ t)]\,\mathrm{d}x\right)\mathrm{d}t - B\ [\Sigma\ v(x,t) + \Gamma\ v_x(x,t)]_{\partial\Omega}\,, \tag{88}$$

where t_1 and $t_2 > t_1$ are two subsequent times during which the time evolution of the membrane is investigated.

3.3 *Time Evolution of Phase Perturbations*

The governing equation for the evolution of small perturbations v is sought by imposing the stationarity of $\mathscr{E}$ within the class of synchronous variations, i.e., such that $\delta v(\circ, t_1) = \delta v(\circ, t_2)$. This leads to the Euler–Lagrange equation in the following form (see Deseri et al. 2016 for details):

$$2\bar{\alpha}\,\frac{\partial^4}{\partial x^4}\left(v + C_N^* \mathcal{D}_t^\nu v\right) - \bar{\varphi}''\,\frac{\partial^2}{\partial x^2}\left(v + C_L^* \mathcal{D}_t^\lambda v\right) = y(x), \tag{89}$$

where $C_L^* = C_L/\bar{\varphi}''$ and $C_N^* = C_N/2\bar{\alpha}$ represent the normalized local and nonlocal moduli of the membrane, respectively, and the forcing term $y(x)$ is defined as follows:

$$y(x) = 2\bar{\alpha}\frac{\partial^4 v_0}{\partial x^4} - \bar{\varphi}''\frac{\partial^2 v_0}{\partial x^2}, \tag{90}$$

where $v_0(x)$ is an initial perturbation induced on the system. This represents the initial perturbation of the ground state before the relaxation takes place. The balance equation (89) is endowed with the boundary conditions to be retrieved from the following conditions:

$$\begin{cases} \text{either} \\ \bar{\varphi}''\left(v'+\bar{C}_L\mathcal{D}_t^{\lambda}v'\right)-2\bar{\alpha}\left(v'''+\bar{C}_N\mathcal{D}_t^{\nu}v'''\right)=\Sigma+\Sigma_0 \\ \text{or} \\ \delta v=0 \end{cases} \tag{91a}$$

$$\begin{cases} \text{either} \\ 2\bar{\alpha}\left(v''+\bar{C}_N\mathcal{D}_t^{\nu}v''\right)=\Gamma+2\bar{\alpha}\,\varepsilon_0' \\ \text{or} \\ \delta v'=0 \end{cases} \tag{91b}$$

Here $\Sigma_0=\bar{\varphi}''\varepsilon_0+2\,\bar{\alpha}\,\varepsilon_0''$ is the initial stress arising on the bilayer associated with the initially perturbed configuration. Obviously, whenever the membrane is initially perturbation-free then (89) and its boundary conditions give us an eigenvalue problem: this will be solved in Sect. 3.6.

Separation of variables is employed here to solve (89), namely we seek for solutions in the form

$$v(x,t)=f(x)\,q(t), \tag{92}$$

where $q(t)$ describes the time change of the perturbation and $f(x)$ describes the shape of the mode. Substitution of (92) in (89) leads to the following pair of equations

$$\frac{2\bar{\alpha}}{\bar{\varphi}''}\frac{f''''(x)}{f''(x)}=\frac{q(t)+C_L^*\,\mathcal{D}_t^{\lambda}q(t)}{q(t)+C_N^*\,\mathcal{D}_t^{\nu}q(t)}=k^2, \tag{93}$$

where k^2 is a constant to be determined. We remind that the expression (60) relating $\frac{2\bar{\alpha}}{\bar{\varphi}''}$ to the spatial frequency (squared) ω^2 does hold. Because here we focus on the circumstances for which spatial oscillations can occur, the only case of interest is when $\bar{\varphi}''<0$. Henceforth, we will solve the following equations:

$$-\frac{1}{\omega^2}\frac{f''''(x)}{f''(x)}=\frac{q(t)+C_L^*\,\mathcal{D}_t^{\lambda}q(t)}{q(t)+C_N^*\,\mathcal{D}_t^{\nu}q(t)}=k^2. \tag{94}$$

The very same boundary conditions assumed for the elastic case (65) will be considered for the viscoelastic problem, namely:

$$\begin{cases} v\Big|_{\partial\Omega^-}=v\Big|_{\partial\Omega^+}=0 \\ 2\bar{\alpha}\left[v''+C_N^*\,\mathcal{D}_t^{\nu}v''\right]\Big|_{\partial\Omega^-}=2\bar{\alpha}\left[v''+C_N^*\,\mathcal{D}_t^{\nu}v''\right]\Big|_{\partial\Omega^+}=\hat{\Gamma} \end{cases} \tag{95}$$

which by (92) yield the following relations:

$$\begin{cases} f(x)\Big|_{\partial\Omega} = 0 \\ 2\bar{\alpha} f''\left[q(t) + C_N^* \mathcal{D}_t^\nu q(t)\right]\Big|_{\partial\Omega} = \hat{\Gamma} \end{cases} \tag{96}$$

3.4 Spatial Modes for the Perturbations

The spatial mode $f(x)$ verifies (93), namely

$$f''''(x) + k^2\omega^2 f''(x) = 0. \tag{97}$$

the solution of (97) reads as

$$f(x) = A_1\cos(\zeta x) + A_2\sin(\zeta x) + A_3 x + A_4, \tag{98}$$

after setting

$$\zeta^2 = k^2\omega^2. \tag{99}$$

Boundary conditions (96) allow for determining the coefficients A_i, $i = 1 \div 4$. In particular, the second boundary condition yields

$$2\bar{\alpha} f''\Big|_{\partial\Omega}\left[q(t) + C_N^* \mathcal{D}_t^\nu q(t)\right] = \hat{\Gamma} \quad \forall t,$$

to be satisfied if either $\hat{\Gamma}$ is a prescribed function of time or if it is constant. Whenever this is the case, then

$$q(t) + C_N^* \mathcal{D}_t^\nu q(t) = \kappa_n, \tag{100}$$

where κ_n is a constant. Consequently, the boundary condition under exam reads as follows:

$$2\bar{\alpha} f''\Big|_{\partial\Omega}\kappa_n = \hat{\Gamma}. \tag{101}$$

Moreover, this condition at the edge highlights that the second derivative evaluated in such location $v''(x, t)\Big|_{\partial\Omega}$ can be zero for whatever value of κ_n if and only if no hyperstress arises at the edges, i.e.,

$$f''\Big|_{\partial\Omega} = 0 \iff \hat{\Gamma} = 0. \tag{102}$$

For such a case, Eq. (100) is irrelevant. After setting $s = \sin(\zeta L/2)$ and $c = \cos(\zeta L/2)$, the boundary conditions can be written explicitly in the following form:

$$\begin{cases} A_1\, c - A_2\, s - A_3 \dfrac{L}{2} + A_4 = 0 \\ 2\bar{\alpha}\zeta^2\, (-A_1\, c + A_2\, s)\, \kappa_n = \hat{\Gamma} \end{cases} \text{at } x = -\frac{L}{2}$$

$$\begin{cases} A_1\, c + A_2\, s + A_3 \dfrac{L}{2} + A_4 = 0 \\ 2\bar{\alpha}\zeta^2\, (-A_1\, c - A_2\, s)\, \kappa_n = \hat{\Gamma} \end{cases} \text{at } x = +\frac{L}{2}$$

Such a system is the analog of (66):

$$\begin{bmatrix} 0 & s & \frac{L}{2} & 0 \\ c & 0 & 0 & 1 \\ 0 & s & 0 & 0 \\ -2\bar{\alpha}\, \kappa_n \zeta^2 c & 0 & 0 & 0 \end{bmatrix} \begin{pmatrix} A_1 \\ A_2 \\ A_3 \\ A_4 \end{pmatrix} = \begin{pmatrix} 0 \\ 0 \\ 0 \\ \hat{\Gamma} \end{pmatrix} \tag{103}$$

whose nontrivial solutions can be found by studying the roots of the determinant, namely after solving:

$$\bar{\alpha}\, c\, s\, L\, \kappa_n\, \zeta^2 = 0. \tag{104}$$

The ground states $\bar{J}$ from which bifurcations may occur are given by the latter equation no matter what κ_n, and because the constants $\bar{\alpha}$, L are always nonzero, only two possibilities are left.

Case 1. Because $\zeta^2 = k^2\, \omega^2$ with $k > 0$ (although still unknown at this stage), if $s = 0$ we have

$$k^2\, \omega^2 = \frac{4n^2\pi^2}{L^2}, \tag{105}$$

and

$$-\frac{\bar{\varphi}''}{\bar{\varphi}'}\bar{J}^5 = \frac{n^2\pi^2}{3\, k^2}\left(\frac{h_0}{L}\right)^2. \tag{106}$$

Case 2. If $c = 0$ then $\hat{\Gamma} = 0$. As highlighted in (102), this happens if and only if $f''\,(\partial\Omega) = 0$.

3.5 *Time Evolutions of the Perturbations*

The expression of $q(t)$ can be traced back to the solution of the equation coming from the boundary condition (101).

Whenever in (96) the boundary condition on the second derivative of the displacement is nonzero, the presence of a hyperstress $\hat{\Gamma}$ at the edges implies that the time-dependent term is constant, assuring that relation (100) holds. This equation is solved in Deseri et al. (2016) and delivers the following expression:

$$q(t) = \frac{\kappa_n}{C_N^*} t^\nu E_{\nu,\nu+1}\left(-\frac{1}{C_N^*} t^\nu\right) + q_0 E_\nu \left(-\frac{1}{C_N^*} t^\nu\right), \tag{107}$$

where $E_{\alpha,\beta}(z)$ is the Mittag-Leffler function of two parameters.

Nonetheless, separation of variables imposes (93) to be fulfilled. This, together with relation (100), delivers the following differential equation:

$$q(t) + C_L^* \mathcal{D}_t^\lambda q(t) = k^2 \kappa_n, \tag{108}$$

again solved in Deseri et al. (2016) by means of the same method, delivering the following expression for q:

$$q(t) = \frac{k^2 \kappa_n}{C_L^*} t^\lambda E_{\lambda,\lambda+1}\left(-\frac{1}{C_L^*} t^\lambda\right) + h_0 E_\lambda \left(-\frac{1}{C_L^*} t^\lambda\right). \tag{109}$$

Obviously the two obtained expressions for q must agree at all times. This is certainly true in the trivial case for which the local and nonlocal terms have both the same relaxation exponent $\lambda = \nu$ and the same normalized parameters $C_L^* = -C_N^*$, namely $k^2 = 1$, recalling that $C_L^* < 0$ has been rendered nondimensional by taking C_L and dividing it by $\bar{\varphi}'' < 0$.

3.6 Eigenvalue Problem Governing the Time Dependence of the Perturbations

Because Eqs. (93) and (100) both govern the evolution function q a complete study of such a requirement is needed. Indeed, those two equations deliver the following fractional-order eigenvalue problem:

$$C_L^* \mathcal{D}_t^\lambda q(t) - C_N^* k^2 \mathcal{D}_t^\nu q(t) + (1 - k^2) q(t) = 0. \tag{110}$$

The solution method of such a problem is here based on the right-sided Fourier transform $Q(p)$

$$Q(p) := \int_0^{+\infty} e^{-i\,p\,t} q(t)\, dt \qquad p \in \mathbb{R}. \tag{111}$$

By Fourier transforming both sides of (110) we obtain

$$\left[C_L^* (-i\,p)^\lambda - C_N^* k^2 (-i\,p)^\nu + (1 - k^2)\right] Q(p) = 0. \tag{112}$$

The zeros of the function inside square brackets provide the eigenvalues of the fractional differential equation (110) no matter what $Q(p)$ is. It is worth noting that the constant k^2 appearing in (93) for the first time must be a real number. The algebraic equation (112) can actually be manipulated by separating the real and the imaginary parts as follows:

$$
\begin{aligned}
k^2 &= \frac{1 + C^*_{L}\, p^{\lambda}\, (c_{\lambda} - i\, s_{\lambda})}{1 + C^*_{N}\, p^{\nu}\, (c_{\nu} - i\, s_{\nu})} \\
&= \frac{\left(1 + C^*_{L}\, p^{\lambda}\, c_{\lambda}\right) - i\, \left(C^*_{L}\, p^{\lambda}\, s_{\lambda}\right)}{\left(1 + C^*_{N}\, p^{\nu}\, c_{\nu}\right) - i\, \left(C^*_{N}\, p^{\nu}\, s_{\nu}\right)} = \frac{a - i\, b}{c - i\, d} \\
&= \frac{a - i\, b}{c - i\, d}\frac{c + i\, d}{c + i\, d} = \frac{a\, c + b\, d}{c^2 + d^2} + i\, \frac{a\, d - b\, c}{c^2 + d^2},
\end{aligned}
$$

after setting

$$
\begin{cases} a = 1 + C^*_{L}\, p^{\lambda}\, c_{\lambda} \\ b = C^*_{L}\, p^{\lambda}\, s_{\lambda} \end{cases} \qquad \begin{cases} c = 1 + C^*_{N}\, p^{\nu}\, c_{\nu} \\ d = C^*_{N}\, p^{\nu}\, s_{\nu} \end{cases},
$$

$$
c_{\alpha} = \cos(\alpha\, \pi/2)
$$

$$
s_{\alpha} = \sin(\alpha\, \pi/2),
$$

$\alpha = \lambda,\ \nu$. Because k is real, the former complex algebraic equation delivers the following real-valued conditions to be verified, namely,

$$
k^2 = \frac{a\, c + b\, d}{c^2 + d^2} \tag{113a}
$$

$$
a\, d - b\, c = 0. \tag{113b}
$$

Equation (113b) can be rewritten as follows:

$$
C^*_{N}\, p^{\nu}\, s_{\nu} - C^*_{L}\, p^{\lambda}\, s_{\lambda} + C^*_{L}\, C^*_{N}\, p^{\lambda+\nu}\, (s_{\nu} c_{\lambda} - c_{\nu} s_{\lambda}) = 0
$$

and, through the transformation formulas for the difference of two angles, it becomes

$$
\begin{aligned}
C^*_{N}\, p^{\nu}\, \sin\left(\nu\, \frac{\pi}{2}\right) - C^*_{L}\, p^{\lambda}\, \sin\left(\lambda\, \frac{\pi}{2}\right) + \\
+ C^*_{L}\, C^*_{N}\, p^{\lambda+\nu} \sin\left((\nu - \lambda)\, \frac{\pi}{2}\right) = 0.
\end{aligned} \tag{114}
$$

Finally, a relationship for k^2 is found in the following form:

$$
k^2 = \frac{\left(1 + C^*_{L}\, p^{\lambda}\, c_{\lambda}\right)\left(1 + C^*_{N}\, p^{\nu}\, c_{\nu}\right) + \left(C^*_{L}\, p^{\lambda}\, s_{\lambda}\right)\left(C^*_{N}\, p^{\nu}\, s_{\nu}\right)}{\left(1 + C^*_{N}\, p^{\nu}\, c_{\nu}\right)^2 + \left(C^*_{N}\, p^{\nu}\, s_{\nu}\right)^2}. \tag{115}
$$

Whenever the trivial case $\lambda = \nu$ and $C^*_{L} = C^*_{N}$ is considered, Eq. (114) has solution $p = 0$, that implies $k^2 = 1$, as noticed qualitatively above. The solution of (115) cannot be found in closed form. In Figs. 8 and 9 some numerical results are represented whenever the moduli C^*_{L}, C^*_{N} and both the exponents are known.

The ratio

$$
R = -C^*_{L}/C^*_{N}
$$

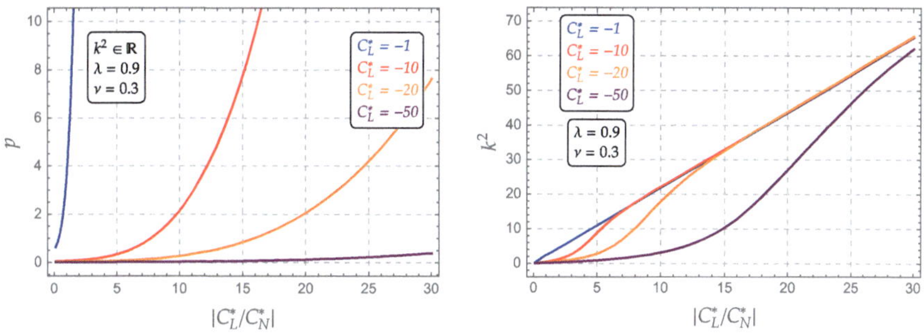

Fig. 8 Locus of the real values of p and correspondent eigenvalues k^2 as function of the ratio $R = -C_N^*/C_L^*$ whenever $\lambda = 0.9$ and $\nu = 0.3$ (see (114) and (115)) (courtesy of Deseri et al. 2016)

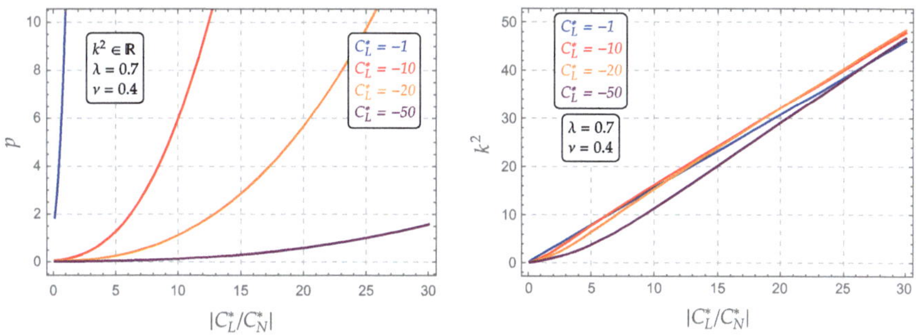

Fig. 9 Locus of the real values of p and correspondent eigenvalues k^2 as function of the ratio $R = -C_N^*/C_L^*$ whenever $\lambda = 0.7$ and $\nu = 0.4$ (see (114) and (115)) (courtesy of Deseri et al. 2016)

shows that the eigenvalues are bijections of p. Hence, there is also a one-to-one correspondence between R and k^2. Of course, each bifurcation is characterized by a value of k^2 which modifies the left and right branch of the ratio $\bar{\varphi}''/\bar{\varphi}'$:

$$-k^2 \frac{\bar{\varphi}''}{\bar{\varphi}'} \bar{J}^5 = \frac{n^2 \pi^2}{3} \left(\frac{h_0}{L} \right)^2, \tag{116}$$

which is the viscoelastic analog of (69).

A numerical example based on the very same energetics utilized in the elastic case is displayed in Fig. 10. This diagram shows that k^2 acts as a rescaling parameter, thereby amplifying the ratio $\bar{\varphi}''/\bar{\varphi}'$ as k increases. While the values of J_n are not modified by such rescaling, the upper bound of the curve is highly influenced by such parameter. This has an impact on the maximum number of oscillations, n_{max}, as displayed in Fig. 10. Henceforth, by plotting in Fig. 11 the values of the critical J in terms of the number of oscillations, one can notice that the left (blue color) and right (red color) branches do have different shapes, thereby modifying their intersections with any given $\bar{J}$.

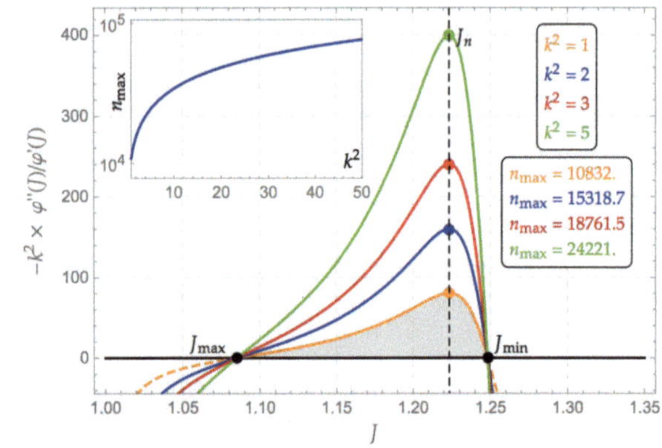

Fig. 10 Left-hand side of Eq. (116) in terms of J. It is highlighted the influence of of k^2 and the corresponding maximum number of spatial oscillations n_{max} is displayed (courtesy of Deseri et al. 2016)

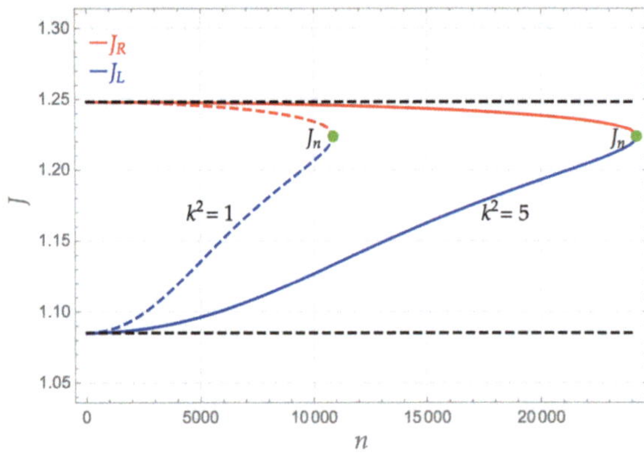

Fig. 11 Modification of the left and right intersections depending on k^2 (courtesy of Deseri et al. 2016)

3.7 *Influence of the Initial Conditions*

The "full" fractional differential equation (110) with inhomogeneous initial conditions is analyzed in this section, namely,

$$\begin{cases} C_L^* \mathcal{D}_t^\lambda q(t) - C_N^* k \mathcal{D}_t^\nu q(t) + (1-k^2) q(t) = 0, \\ q(0) = q_0. \end{cases}$$

The right-handed Fourier transform is again employed here to account the initial condition, i.e.,

$$C_L^* \left[(i\,p)^\lambda \hat{Q} - (i\,p)^{\lambda-1} q_0 \right] - C_N^* k^2 \left[(i\,p)^\nu \hat{Q} + \right.$$
$$\left. - (i\,p)^{\nu-1} q_0 \right] + \hat{Q}\,(1-k^2) = 0,$$

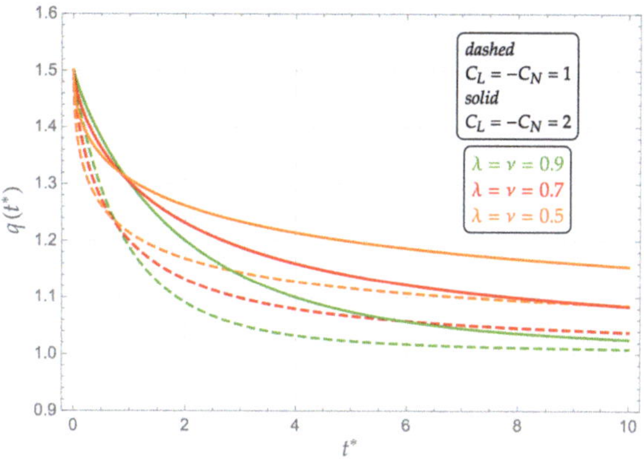

Fig. 12 Time-dependent transfer function for two chosen values of $C_L^* = -C_N^*$ and $h_0 = 1.5$. Here $t^* = \sqrt[\nu]{\frac{t^\nu}{C_N^*}}$ is a dimensionless time (courtesy of Deseri et al. 2016)

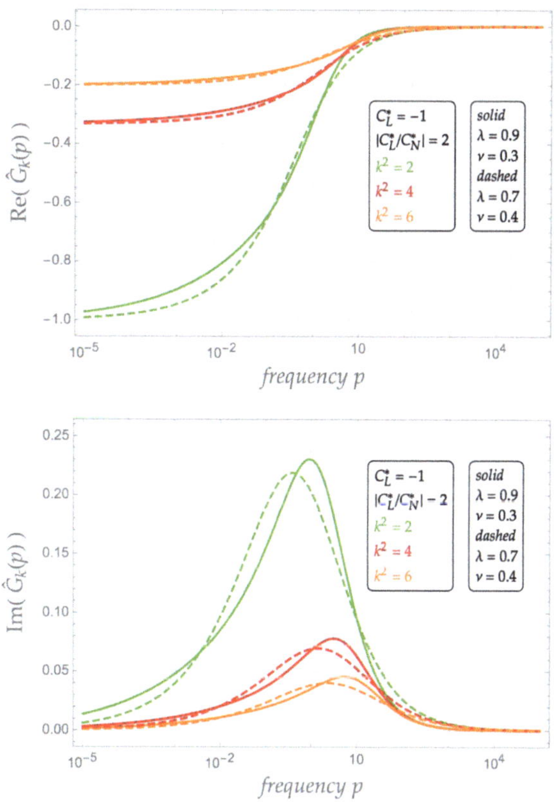

Fig. 13 Transfer function $\hat{G}_k(p)$: real and imaginary parts (courtesy of Deseri et al. 2016)

whose solution $\hat{Q}(p)$ reads as follows:

$$\hat{Q}_k(p) = \hat{G}_k(p)\, q_0 \left(C_L^*\,(i\,p)^{\lambda-1} - C_N^*\,k^2\,(i\,p)^{\nu-1}\right), \tag{117}$$

where

$$\hat{G}_k(p) = \frac{1}{C_L^*\,(i\,p)^{\lambda} - C_N^*\,k^2\,(i\,p)^{\nu} + (1-k^2)} \tag{118}$$

is the transfer function for this problem. It is worth noting that this function strictly depends on the order of the eigenvalue, k^2.

From Podlubny (1998), Eqs. 5.22–5.25, p. 155 (where $a = C_L^*$, $\beta = \lambda$, $b = -C_N^*\,k^2$, $\alpha = \nu$ and $c = 1 - k^2$), we find the anti-right-handed Fourier transform of such a function, which reads as follows:

$$\begin{aligned} G_k(t) &= \mathcal{F}^{-1}\left\{\hat{G}_k(p); t\right\} = \\ &= \frac{1}{C_L^*}\sum_{z=0}^{\infty}(-1)^z\left(\frac{1-k^2}{C_L^*}\right)^{z+1} t^{\lambda(z+1)-1} E^{(z)}_{\lambda-\nu,\lambda+z\nu}\left(\frac{C_N^*}{C_L^*}k^2\,t^{\lambda-\nu}\right). \end{aligned} \tag{119}$$

The obtained result is then represented by a series of Mittag-Leffler functions with two parameters. This plays the role of modulating the membrane response no matter what the initial data is. For the sake of illustration, the transfer function is numerically explored in Fig. 12 whenever two subcases of $C_L^* = -C_N^*$ are considered, by assuming several values of the exponential decay $\lambda = \nu$. Similarly, in Fig. 13 the real and imaginary parts of the transfer function are analyzed whenever different exponents of the decay $\lambda \neq \nu$ are chosen for some values of k^2. The Mittag-Leffler function drives the evolution of the membrane stretch, determining changes in the amplitude of the membrane response, as expected from the analysis with a separation of variables.

4 Conclusions

The mechanical behavior of biological membranes is regulated by the interaction of an extremely rich list of features, such as their thinness, their special constitutive nature which enables them to sustain bending moments but not in-plane shear stresses unless their viscosity is accounted for, their chemical composition and, furthermore, their capability of undergoing ordering–disordering phenomena. The resulting effects of this interaction are evidenced by a strong variety of configurations that can be achieved and kept by biological membranes at equilibrium for given values of overall chemical composition, controlled temperature, or applied osmotic pressure.

Within this framework, a remarkable issue is the analysis of line tension at the boundary of ordered–disordered domains: it is now recognized that, together with bending rigidities, line tension plays a major role in maintaining nonspherical configurations observed in experiments (see e.g., Akimov et al. 2004). In the effort of deducing a physically based model of lipid membranes where the bending behavior, the order–disorder transition, and the chemical composition are consistently considered, in Deseri et al. (2008), Deseri and Zurlo (2013), Zurlo (2006) the expression of the energetics regulating the thermo-chemo-mechanical behavior of biological membranes was derived, within the framework of a formal asymptotic 3D-to-2D reduction, based on thinness assumptions. This model reveals the possibility of describing the geometrical (shape) and conformational (state of order) behavior of the lipid bilayer on the basis of one single ingredient: the in-plane membrane stretching elasticity, regulating the material response with respect to local area changes on the membrane mid-surface. A confirmation of these possibilities is given in Choksi et al. (2012), where a model energy obtainable from the one deduced in Deseri et al. (2008) is proved to exhibit two-phase global minimizers resembling observed configurations in Baumgart et al. (2003). In essence, the major point in Deseri et al. (2008), Deseri and Zurlo (2013), Zurlo (2006) is that the bilayer stretching elasticity is enough to describe its order–disorder transition (together with the influence of chemical composition), to determine the profile and the length of the boundary layer where the membrane thickness passes from a thicker domain (ordered phase) to a thinner one (disordered phase), to evaluate the corresponding line tension and finally to determine the bending rigidities in both phases.

A prototypical planar problem has been studied in Deseri and Zurlo (2013) with the aim of elucidating the potentials of the model described above and summarized in the present work. On the basis of a Landau expansion of the stretching energy density, calibrated, thanks to the experimental results in Goldstein and Leibler (1989), the line tension, the thickness profile inside the boundary layer and the area compressibility and bending moduli are obtained. Those calculated quantities show a satisfactory comparison with the data known in the literature.

Lipid phase transition arising in planar membrane and triggered by material instabilities and their linearized evolution are studied in Deseri et al. (2016) and summarized in this work. There, the effective viscoelastic behavior inherited by their exhibited power law in-plane viscosity (Espinosa et al. 2011) is accounted for. At first it is shown that the critical set of areal stretches is determined in the limiting case of elasticity and for two sets of boundary conditions. Spatial oscillations corresponding to the nucleated configurations arising from any of such critical stretches are investigated. Perturbations of the phase ordering of lipids are predicted to form bifurcated shapes, sometimes of large periods relative to the reference thickness of the bilayer. The corresponding membrane stress changes are also oscillatory. Then, the influence of the effective viscoelasticity of the membrane on its material instabilities is investigated. A variational principle based on the search of stationary points of a Gibbs free energy in the class of synchronous perturbation is employed for such analysis. The resulting Euler–Lagrange equation is a fractional-order partial differential equation yielding a non-classical eigenvalue initial boundary value problem. The eigenvalues are found to be amplified with respect to their elastic counterpart. Spatial modes and transfer functions characterizing the resulting admissible perturbations of the underlying ground configurations are determined. It is found that while the range of critical areal stretches not get affected, the number of oscillations per given critical stretch significantly increases, thereby drastically reducing the period of oscillations of the bifurcated configurations. Nevertheless, a “long tail” type relaxation of the bifurcated configurations is shown to occur. Furthermore, whenever the same power law applies both for the local and the nonlocal response, the explicit time decay is displayed, while in all of the other cases the frequency dependence of the real and imaginary parts of the transfer function reveal that fading memory in time occurs as well (see Fig. 13).

Acknowledgements The author wishes to thank the organizer of this course, David Steigmann, for his invitation to contribute to this course. The other lecturers are also acknowledged for the nice and extended discussions that allowed for exchange of ideas on the topic of this course.

The author is extremely grateful to Giuseppe Zurlo (National University of Galway, Ireland), formerly his Ph.D. student, for the very extensive discussions and long-standing collaboration from his early days in 2002. His key contribution to this research has had huge impact in its assessment and development. Timothy J. Healey (Cornell University) and Roberto Paroni (University of Sassari, Italy) also gratefully acknowledged for the very extended discussions on the early stages of the 2008 work.

Grateful acknowledgements go to Massimiliano Zingales (University of Palermo), Kaushik Dayal (Carnegie Mellon University) as collaborators on key aspects related to the hereditary response of lipid bilayers. Massimiliano Fraldi (University of Napoli-Federico II) is also gratefully ackowledged for his illuminating remarks and insights on biological tissues and biomechanics, as well as Valentina Piccolo (University of Trento), a graduate student working with myself and other people on various topics, who also provided new perspectives on the applications of Fractional Analysis to lipid membranes and helped a lot to edit this work.

The author is grateful to the financial support provided by (i) the NSF Grant no.DMS-0635983 of the Center for Nonlinear Analysis, Carnegie Mellon University, (ii) for the direct financial support of the Dept. of Mechanical Engineering and Materials Science-MEMS of the University of Pittsburgh for, and also to (iii) the support of the EU Grant “INSTABILITIES” ERC-2013-ADG Instabilities and nonlocal multiscale modelling of materials held by Prof. Davide Bigoni from the University of Trento.

References

S.A. Akimov, P.I. Kuzmin, J. Zimmerberg, An elastic theory for line tension at a boundary separating two lipid monolayer regions of different thickness. J. Electroanal. Chem. **564**, 13–18 (2004)

G. Alberti, An approach via Γ−convergence, in *Calculus of Variations and Partial Differential Equations*, Topics on Geometrical Evolution Problems and Degree Theory, ed. by L. Ambrosio, N. Dancer (Springer, Berlin, 2000)

E. Baesu, R.E. Rudd, J. Belak, M. McElfresh, Continuum modeling of cell membranes. Int. J. Non-Linear Mech. **39**(3), 369–377 (2004)

T. Baumgart, W.W. Webb, S.T. Hess, Imaging coexisting domains in biomembrane models coupling curvature and line tension. Nature **423**, 821–824 (2003)

H. Bermúdez, D.A. Hammer, D.E. Discher, Effect of bilayer thickness on membrane bending rigidity. Langmuir **20**, 540–543 (2004)

S. Breuer, E. Onat, On the determination of free energies in linear viscoelastic solids. ZAMP **15**, 184–191 (1964)

J.W. Cahn, J.E. Hilliard, Free energy of a non-uniform system i - interfacial free energy. J. Chem. Phys. **28**, 258–267 (1958)

P.B. Canham, The minimum energy as possible explanation of the biconcave shape of the human red blood cell. J. Theor. Biol. **26**(1), 61–81 (1970)
M. Caputo, *Elasticità e Dissipazione* (Zanichelli, Bologna, 1969)
R. Choksi, M. Morandotti, M. Veneroni, Global Minimizers for Axisymmetric Multiphase Membranes, arXiv preprint (2012), arXiv:1204.6673
B.D. Coleman, D.C. Newman, On the rheology of cold drawing. i. elastic materials. J. Polym. Sci.: Part B: Polym. Phys. **26**, 1801–1822 (1988)
D. Craiem, R.L. Magin, Fractional order models of viscoelasticity as an alternative in the analysis of red blood cell (rbc) membrane mechanics. Phys. Biol. **7**(1), 13001 (2010)
G. Del Piero, L. Deseri, On the analytic expression of the free energy in linear viscoelasticity. J. Elast. **43**, 247–278 (1996)
G. Del Piero, L. Deseri, On the concepts of state and free energy in linear viscoelasticity. Arch. Ration. Mech. Anal. **138**, 1–35 (1997)
L. Deseri, G. Zurlo, The stretching elasticity of biomembranes determines their line tension and bending rigidity. Biomech. Model. Mechanobiol. **12**, 1233–1242 (2013)
L. Deseri, G. Gentili, M.J. Golden, An expression for the minimal free energy in linear viscoelasticity. J. Elast. **54**, 141–185 (1999)
L. Deseri, M.J. Golden, M. Fabrizio, The concept of a minimal state in viscoelasticity: new free energies and applications to pdes. Arch. Ration. Mech. Anal. **181**, 43–96 (2006)
L. Deseri, M. Piccioni, G. Zurlo, Derivation of a new free energy for biological membranes. Contin. Mech. Term **20**(5), 255–273 (2008)
L. Deseri, M. Di Paola, M. Zingales, Free energy and states of fractional-order hereditariness. Int. J. Solids Struct. **51**, 3156–3167 (2014)
L. Deseri, P. Pollaci, M. Zingales, K. Dayal, Fractional hereditariness of lipid membranes: Instabilities and linearized evolution. J. Mech. Behav. Biomed. Mater. **58**, 11–27 (2016)
G. Espinosa, I. López-Montero, F. Monroy, D. Langevin, Shear rheology of lipid monolayers and insights on membrane fluidity. PNAS **108**(15), 6008–6013 (2011)
E.A. Evans, Bending resistance and chemically induced moments in membrane bilayers. Biophys. J. **14**, 923–931 (1974)
M.S. Falkovitz, M. Seul, H.L. Frisch, H.M. McConnell, Theory of periodic structures in lipid bilayer membranes. Proc. Natl. Acad. Sci. USA **79**, 3918–3921 (1982)
Y.C. Fung, Theoretical considerations of the elasticity of red blood cells and small blood vessels. Proc. Fed. Am. Soc. Exp. Biol. **25**(6), 1761–1772 (1966)
Y.C. Fung, P. Tong, Theory of sphering of red blood cells. Biophys. J. **8**, 175–198 (1968)
R.E. Goldstein, S. Leibler, Model for lamellar phases of interacting lipid membranes. Phys. Rev. Let. **61**(19), 2213–2216 (1988)
R.E. Goldstein, S. Leibler, Structural phase transitions of interacting membranes. Phys. Rev. A. **40**(2) (1989)
M. Hamm, M.M. Kozlov, Elastic energy of tilt and bending of fluid membranes. Eur. Phys. J. E **3**, 323–335 (2000)
C.W. Harland, M.J. Bradley, R. Parthasarathy, Phospholipid bilayers are viscoelastic. PNAS **107**(45), 19146–19150 (2010)
T.J. Healey, R. Paroni, L. Deseri, Material gamma-limits for biological in-plane fluid plates. (2017)
W. Helfrich, Elastic properties of lipid bilayers: theory and possible experiments. Z. Naturforsch [C], **28**(11), 693–703 (1973)
M. Hu, J.J. Briguglio, M. Deserno, Determining the gaussian curvature modulus of lipid membranes in simulations. Biophys. J. **102**, 1403–1410 (2012)
F. Jahnig, Critical effects from lipid-protein interaction in membranes. Biophys. J. **36**, 329–345 (1981)
F. Jahnig, What is the surface tension of a lipid bilayer membrane? Biophys. J. **71**, 1348–1349 (1996)
J.B. Keller, G.J. Merchant, Flexural rigidity of a liquid surface. J. Stat. Phys. **63**(5–6), 1039–1051 (1991)

A.A. Kilbas, H.M. Srivastava, J.J. Trujillo, *Theory and Applications of Fractional Differential Equations* (Elsevier, Amsterdam, 2006)
W.T. Koiter, On the nonlinear theory of thin elastic shells. Proc. K. Ned. Akad. Wet. B **69**, 1–54 (1966)
S. Komura, H. Shirotori, P.D. Olmsted, D. Andelman. Lateral phase separation in mixtures of lipids and cholesterol. Europhys. Lett. **67**(2) (2004)
R. Lipowsky, E. Sackmann (eds.), *Handbook of Biological Physics-Structure and Dynamics of Membranes*, vol. 1 (Elsevier Science B.V, Amsterdam, 1995)
R.L. Magin, Fractional calculus models of complex dynamics in biological tissues. Comput. Math. Appl. **59**(5), 1586–1593 (2010)
M. Maleki, B. Seguin, E. Fried, Kinematics, material symmetry, and energy densities for lipid bilayers with spontaneous curvature. Biomech. Model. Mechanobiol. **12**(5), 997–1017 (2013)
D. Norouzi, M.M. Müller, M. Deserno, How to determine local elastic properties of lipid bilayer membranes from atomic-force-microscope measurements: a theoretical analysis. Phys. Rev. E, **74** (2006)
J.C. Owicki, H.M. McConnell, Theory of protein-lipid and protein-protein interactions in bilayer membranes. Proc. Natl. Acad. Sci. USA **76**, 4750–4754 (1979)
J.C. Owicki, M.W. Springgate, H.M. McConnell, Theoretical study of protein-lipid interactions in bilayer membranes. Proc. Natl. Acad. Sci. USA **75**, 1616–1619 (1978)
J. Pan, S. Tristram-Nagle, J.F. Nagle, Effect of cholesterol on structural and mechanical properties of membranes depends on lipid chain saturation. Phys. Rev. E: Stat. Nonlinear **80**(021931) (2009)
I. Podlubny, *Fractional Differential Equation* (Academic, New York, 1998)
W. Rawicz, K.C. Olbrich, T. McIntosh, D. Needham, E. Evans, Effect of chain length and unsaturation on elasticity of lipid bilayers. Biophys. J. **79**, 328–339 (2000)
A.S. Reddy, D. Toledo Warshaviak, M. Chachisvilis, Effect of membrane tension on the physical properties of dopc lipid bilayer membrane. Bioch. Biophys. Acta **1818**, 2271–2281 (2012)
S.G. Samko, A.A. Kilbas, O.I. Marichev, *Fractional Integrals and Derivatives. Theory and Applications* (Gordon & Breach Science Publishers, London, 1987)
G.W. Scott-Blair, Psychoreology: links between the past and the present. J. Texture Stud. **5**, 3–12 (1974)
S. Semrau, T. Idema, L. Holtzer, T. Schmict, C. Storm, Accurate determination of elastic parameters for multicomponent membranes. PRL **100**(088101) (2008)
D.P. Siegel, M.M. Kozlov, The gaussian curvature elastic modulus of n-monomethylated dioleoylphosphatidylethanolamine: Relevance to membrane fusion and lipid phase behavior. Biophys. J. **87**, 366–374 (2004)
D.J. Steigmann, Fluid films with curvature elasticity. Arch. Ration. Mech. Anal. **150**, 127–152 (1999)
M. Trejo, M. Ben, Amar. Effective line tension and contact angles between membrane domains in biphasic vesicles. Eur. Phys. J. E **34**(8), 2–14 (2011)
S.L. Veatch, V.I. Polozov, K. Gawrisch, S.L. Keller, Liquid domains in vescicles investigated by nmr and fluorescence microscopy. Biophys. J. **86**, 2910–2922 (2004)
G. Zurlo. Material and geometric phase transitions in biological membranes. Dissertation for the Fulfillment of the Doctorate of Philosophy in Structural Engineering, University of Pisa, etd-11142006-173408 (2006)

Lipid Membranes: From Self-assembly to Elasticity

M. Mert Terzi and Markus Deserno

Abstract In aqueous solution, lipid molecules spontaneously assemble into macroscopic bilayer membranes, which have highly interesting mechanical properties. In this chapter, we first discuss some basic aspects of this self-assembly process. In the second part, we then revisit and slightly expand a well-known continuum-level theory that describes the elastic properties pertaining to membrane geometry and lipid tilt. We then illustrate in part three several conceptually different strategies for how one of the emerging elastic parameters—the bending modulus—can be obtained in computer simulations.

1 Surfactant Self Assembly: Morphology and Statistical thermodynamics

Surfactant molecules are amphiphiles: they comprise different chemical moieties which are soluble in different solvents. Since they are linked together chemically, this requires nature to grapple with an interesting problem: how best to lower the free energy, given that no matter what the solvent conditions are, some chemical moieties will likely be "unhappy." Nature's solution to this is *self assembly*—a process by which larger scale structures form cooperatively, such that unfavorable solvent contact is largely avoided. Self assembly is an amazing and hugely important example of an emergent phenomenon, in that it creates new physical entities (namely, the aggregates) which can be much bigger than the individual molecules they are made of. This transition in relevant scale is the primary reason why we can deal with these aggregates using physical tools that are quite removed from atomistic modeling—such as continuum elasticity. How self-assembly works, is the topic of our first section.

M.M. Terzi · M. Deserno (✉)
Department of Physics, Carnegie Mellon University,
5000 Forbes Avenue, Pittsburgh, PA 15213, USA
e-mail: deserno@andrew.cmu.edu

D.J. Steigmann (ed.), *The Role of Mechanics in the Study of Lipid Bilayers*,
CISM International Centre for Mechanical Sciences 577,
DOI 10.1007/978-3-319-56348-0_3

The classical work on this topic is the groundbreaking paper by Israelachvili et al. (1976), from which the present section picks some of the most beautiful results and elaborates them in a bit more detail. Good discussions can also be found in textbooks on soft condensed matter physics, such as Jones (2002) or Witten and Pincus (2004).

1.1 Morphology

Lipids, or more generally surfactants, are molecules which are typically divided into a "head" and a "tail." The head is hydrophilic (water soluble), for instance because it has polar groups (e.g., hydroxyl or carbonyl groups), or because it is charged (e.g., amino, carboxyl, or phosphate groups). The tail, on the other hand, is hydrophobic (water insoluble), and for lipids generally consists of two aliphatic chains. They typically contain between 12 and 22 carbon atoms, usually connected by single bonds, but sometimes with one or more double bonds (in the latter case one speaks of "unsaturated lipids"). Figure 1 gives a simple illustration of this by showing pictures of lipids using some commonly employed computational models for studying them. Notice that only one of these models strives for a full chemical resolution. The others simplify the chemical architecture more or less drastically, but they all keep one key aspect: lipids are amphiphiles.

The key effect on which self assembly relies is a cooperative aggregation of surfactants that tries to bury the water-insoluble tails in the interior of the aggregate, shielding them from the aqueous solvent by a layer of hydrophilic head groups.

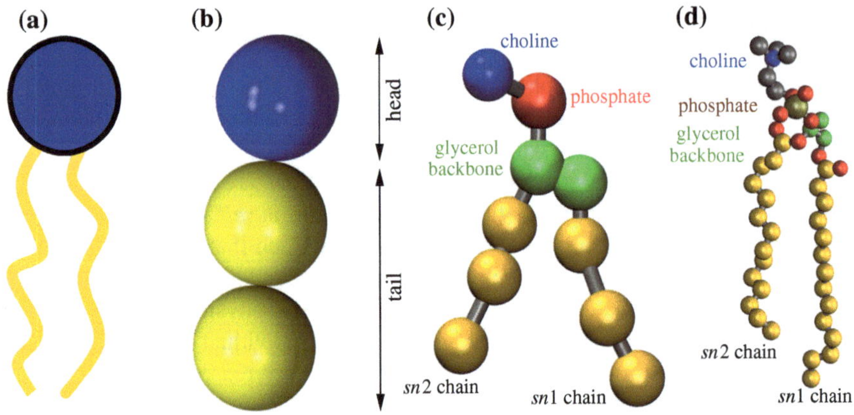

Fig. 1 Illustration of the morphology of a lipid molecule. Panel **a** shows a typical physicist's cartoon—a hydrophilic head group with two schematic tails; panel **b** takes this sketch serious and translates it into a highly coarse grained model (Cooke et al. 2005); panel **c** illustrates a lipid on the MARTINI level (Marrink et al. 2007), where the number of beads is increased, but still each bead accounts for approximately 3–4 heavy atoms; and panel **d** displays a united-atom lipid model of DMPC (dimyristoylphosphatidylcholine) (Berger et al. 1997), in which every atom (except non-polar hydrogens) are explicitly accounted for. Adapted from (Wang and Deserno 2016)

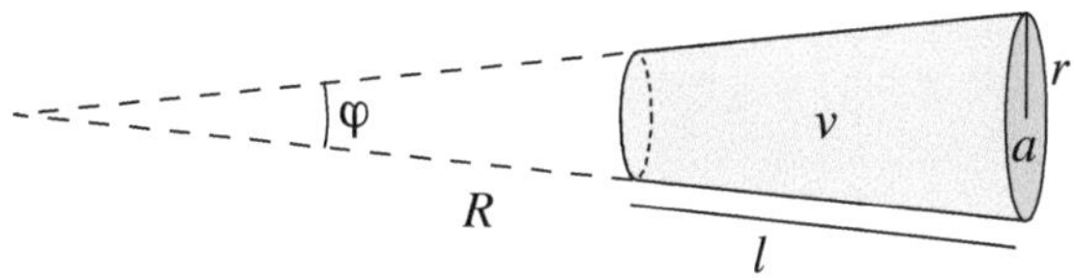

Fig. 2 Simplified shape-description of a surfactant as a blunted cone

Interestingly, there are numerous different morphologies in which that could happen, and this depends on the *shape* of the surfactant. For instance, if the lipid has a relatively large head group and a thin tail—if it looks like an ice cream cone—then we can imagine these surfactants packing together to form little spheres. But if the shape of a lipid is less obviously pointed, then lower curved structures seem more likely—such as cylindrical aggregates or even planar sheets. As we will now see, Israelachvili et al. (1976) have developed a beautifully simple way to make this intuition quantitative.

Let us represent a lipid schematically as a building block that is approximately cylindrical, but with a somewhat tapered tail region, as illustrated in Fig. 2, so that it looks like a blunted cone. The area of its head-group surface is $a = \pi r^2$, its volume is v, and its length is l. Imagine we need N of these object to piece them together into a sphere of radius $R_{\rm sph}$. It is then obvious that we must have

$$Nv = V_{\rm sph} = \frac{4}{3}\pi R_{\rm sph}^3 \,, \tag{1a}$$

$$Na = A_{\rm sph} = 4\pi R_{\rm sph}^2 \,. \tag{1b}$$

Dividing these two equations, N cancels, and we get an equation for the radius of that sphere:

$$\frac{v}{a} = \frac{1}{3} R_{\rm sph} \,. \tag{2}$$

At the center of the sphere we cannot have any empty space. Hence the radius $R_{\rm sph}$ which we found cannot be larger than the length l of the amphiphile—imagine for instance that there is a largest length to which the tails can stretch, and that limits the sphere's radius: $R_{\rm sph} \le l$. This results in the condition

$$\text{spheres:} \qquad \frac{v}{al} =: P \le \frac{1}{3} \,, \tag{3}$$

where we defined the so-called *packing parameter* P. We hence find that if this condition on P is satisfied, these lipid building blocks will indeed like to aggregate into spherical objects, which go under the name *spherical micelles*.

We can repeat this argument, but now instead consider packing the building blocks into a cylinder of radius $R_{\rm cyl}$ and length $L_{\rm cyl}$; Assuming that $L_{\rm cyl}$ is large enough to ignore end effects, we then get

$$Nv = V_{\rm cyl} = \pi R_{\rm cyl}^2 L_{\rm cyl} \; , \tag{4a}$$

$$Na = A_{\rm cyl} = 2\pi R_{\rm cyl} L_{\rm cyl} \; . \tag{4b}$$

Again dividing these two equations yields

$$\frac{v}{a} = \frac{1}{2} R_{\rm cyl} \; . \tag{5}$$

Once more, requiring that the resulting value for the cylinder's radius is not larger than the lipid length l leads to the condition

$$\text{cylinders:} \qquad \frac{1}{3} \le P \le \frac{1}{2} \; , \tag{6}$$

where the lower cutoff comes from the previous case: if P is even smaller than $\frac{1}{3}$, we already know that we get spheres.

We can again repeat this argument, but now we pack the amphiphiles into a planar bilayer structure of area $A_{\rm bil}$ and thickness $b_{\rm bil}$, leading to

$$Nv = V_{\rm bil} = b_{\rm bil} A_{\rm bil} \; , \tag{7a}$$

$$Na = A = 2A_{\rm bil} \; , \tag{7b}$$

and dividing these two equations gives

$$\frac{v}{a} = \frac{1}{2} b_{\rm bil} \; . \tag{8}$$

Again, the thickness of each individual leaflet (i.e., half the bilayer's thickness) cannot exceed the length l to which the lipid can stretch, $\frac{1}{2} b_{\rm bil} \le l$, and so we find

$$\text{bilayers:} \qquad \frac{1}{2} \le P \le 1 \; . \tag{9}$$

The argument, as presented, is remarkably simple; Israelachvili et al. (1976) look at the situation in a fair bit more detail, but the key findings nevertheless hold up. In fact, this line of reasoning works well even for building blocks which are very simple and not very pliable–such as the lipid model from Fig. 1b. Cooke and Deserno (2006) showed that by simply changing the head-group size of the three-bead lipid, one can drive the entire morphological transition from spheres over cylinders to bilayers; if one pushes the packing parameter even larger, the lamellar phase becomes unstable. This is illustrated in Fig. 3.

Of course, the transitions *themselves* do not yet tell whether the simple packing-parameter theory works; but this theory makes a prediction that can be tested. Taking the area per lipid from a flat bilayer as the value for a, and using one of the transitions (say, spheres to cylinders) to pinpoint v/l, one can write the packing parameter as

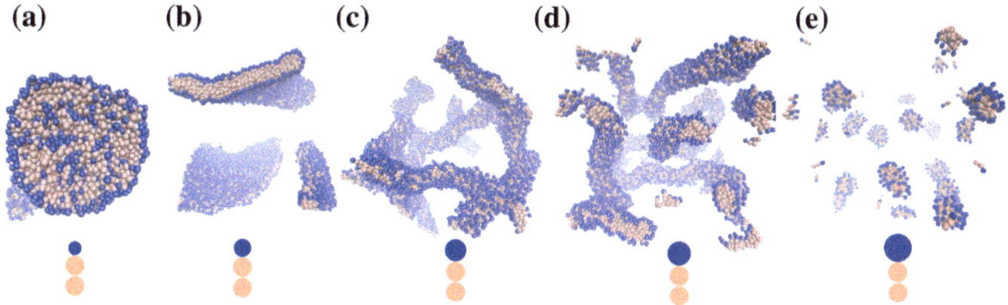

Fig. 3 The different morphologies of amphiphilic aggregates are controlled by amphiphile shape, even for models as simple as that from Fig. 1b. Reprinted from Cooke and Deserno (2006) with permission from Elsevier

a function of the head-group size of the lipid. This then gives a prediction for the head-group size where the other transition (cylinders to bilayer) happens. Cooke and Deserno (2006) show that this prediction indeed works.

The geometrical picture we have in mind by now is that a smaller packing parameter P corresponds to a more cone-like shape, while for a larger P the lipid becomes more cylindrical. This intuition can be verified (and made more precise) by a simple calculation: if Ω is the solid angle of the blunted cone, then its volume can be written as

$$v = \frac{1}{3}\Omega\left[R^3 - (R-l)^3\right] = \Omega\left[R^2 l - R l^2 + \frac{1}{3}l^3\right] . \tag{10}$$

Since its head surface is $a = \Omega R^2$, we find $P = 1 - \frac{l}{R} + \frac{1}{3}\left(\frac{l}{R}\right)^2$, a quadratic equation that can be solved for R, from which we then get the solid angle. Since, furthermore, $\Omega = 2\pi\left(1 - \cos\frac{\varphi}{2}\right) \approx \frac{1}{4}\pi\varphi^2$, where the last approximation is good for small φ, we arrive at the opening angle

$$\frac{\varphi}{r/l} \approx 3\left[1 - \sqrt{1 - \frac{4}{3}(1-P)}\right] . \tag{11}$$

This relation is illustrated in Fig. 4. The characteristic ratio r/l defines an angle, and the actual opening angle φ is some multiple of that—twice as big for cones at the boundary between spheres and cylinders, and about 1.3 times as big at the boundary between cylinders and planes. Of course, the angle vanishes at $P = 1$. Notice that we can alternatively also calculate the lipid spontaneous curvature, defined as $K_{0,\mathrm{m}} = 2/R$. For P close to 1 we find for this parameter

$$K_{0,\mathrm{m}} l \approx 2(1-P) + \frac{2}{3}(1-P)^2 + \cdots \tag{12}$$

This provides a link between a parameter from continuum Helfrich theory, $K_{0,\mathrm{m}}$, and a parameter from the self assembly problem, P.

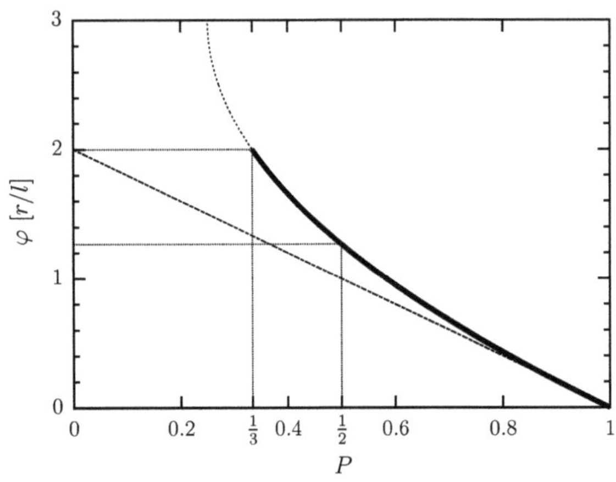

Fig. 4 Relation between the opening angle of the blunted cone from Fig. 2 (measured in units of r/l) and the packing parameter P. Around $P = 1$ we have $\varphi \approx \frac{2r}{l}(1 - P)$

1.2 Statistical Thermodynamics

Knowing the shape of the aggregate is only the beginning. We surely also want to know, under what conditions such aggregates form, and if they come in different sizes (say, what's the length of a cylindrical micelle?), we want to know what that is, too.

The problem is interesting, because entropy plays a key role. Were it only a matter of energy, any kind of amphiphile would aggregate to any other amphiphile, no matter how weak any attractive interaction is. But when we consider entropy, we realize that aggregation strongly reduces the translational entropy of amphiphiles. To understand this energy–entropy balance better we again follow Israelachvili et al. (1976). Let us therefore define

$$\varepsilon_n : \text{energy per molecule in } n \text{ -aggregate} \tag{13a}$$

$$\phi_n : \text{concentration of } n\text{-aggregates} \tag{13b}$$

$$X_n : \text{concentration of monomers in } n\text{-aggregates}, \ = n\phi_n \ , \tag{13c}$$

where an "n-aggregate" is a self-assembled aggregate of molecules consisting exactly of n molecules (or monomers or 1-aggregates). You may think of X_n in the following way: consider only the n-aggregates in solution (mentally remove all the others) and now ask, what is the overall concentration of all amphiphiles left in the system?

The total energy of one n-aggregate is of course $E_n = n\varepsilon_n$. Observe that this does *not* imply that $E_n \propto n$, since ε_n also depends on n. The energy density due to n-aggregates is therefore

$$e_n = \phi_n E_n = \phi_n n \varepsilon_n = X_n \varepsilon_n \ . \tag{14}$$

For the (translational) entropy density of n-aggregates we will simply assume an ideal gas law, so that we get

$$s_n = -k_B\,\phi_n\big(\log\phi_n - 1\big)\,. \tag{15}$$

The total free energy density is then the sum of the energetic and entropic terms over all aggregate sizes:

$$f = \sum_{n=1}^{N}\{e_n - Ts_n\} = \sum_{n=1}^{N}\left\{X_n\varepsilon_n + k_BT\,\frac{X_n}{n}\left(\log\frac{X_n}{n} - 1\right)\right\}, \tag{16}$$

where N is the total number of molecules, and hence also the biggest aggregate we can get.

We are interested in the distribution function of aggregate sizes, X_n, *subject to the constraint* that the total amount of material in the system is fixed, meaning

$$\sum_{n=1}^{N} X_n =: X = \text{fixed}\,, \tag{17}$$

where X is the total monomer concentration in the system. We can calculate this distribution function by minimizing Eq. (16) subject to the constraint, which we enforce by means of a Lagrange multiplier μ:

$$0 \stackrel{!}{=} \frac{\partial}{\partial X_n}\left\{f[X_n] - \mu\left[X - \sum_{m=1}^{N} X_m\right]\right\}. \tag{18}$$

This readily gives

$$\phi_n = \mathrm{e}^{-\beta n(\varepsilon_n - \mu)}\,, \tag{19}$$

where as usual $\beta = 1/k_BT$. From this we in particular also get the monomer concentration ϕ_1, and so we can eliminate the Lagrange multiplier μ from the expression:

$$\phi_n = \left[\phi_1\,\mathrm{e}^{\beta(\varepsilon_1 - \varepsilon_n)}\right]^n\,. \tag{20}$$

This is a very important general result. How it plays out in reality depends entirely on ε_n, which in turn depends crucially on the geometry of the aggregate—spherical cylindrical, or planar. Regardless: we see that if $\varepsilon_n < \varepsilon_1$, meaning that it is favorable for a monomer to be in an n-aggregate compared to being isolated in solution, the exponential factor becomes large and the concentration of n-aggregates goes up. But let us now specifically look at the individual geometries.

Spherical micelles. What is the energy of a monomer in a micelle consisting of n monomers? This is potentially a difficult question, but we will circumvent it by looking at the physics: packing monomers of some particular curvature into a spherical aggregate will likely result in some particular size—say, m—at which they fit best, and deviations away from that size will be suboptimal. Let us hence assume that, to lowest order, the energy is simply quadratic in the deviation from that particular optimal state:

$$\varepsilon_n = \varepsilon_m + \frac{1}{2}\varepsilon^*(n-m)^2 \,. \tag{21}$$

Inserting this into Eq. (19) leads to

$$\phi_n = \exp\left\{-\beta n\left(\varepsilon_m + \frac{1}{2}\varepsilon^*(n-m)^2 - \mu\right)\right\} \,, \tag{22}$$

where n needs to be determined from the normalization condition (17). Notice that this distribution is *cubic* in the exponent. However, we can simplify it by expanding the exponent around its maximum, up to quadratic order, and hence find an approximate Gaussian distribution that describes ϕ_n reasonably well. To do so, we need to calculate

$$0 \stackrel{!}{=} \frac{\partial}{\partial n}\left[-\beta n\left(\varepsilon_m + \frac{1}{2}\varepsilon^*(n-m)^2 - \mu\right)\right] \,, \tag{23}$$

which leads to the solution n^* at which the function peaks:

$$n^* = \frac{m}{3}\left[2 + \sqrt{1 - \frac{6(\varepsilon_m - \mu)}{\epsilon^* m^2}}\right] \approx m - \frac{\epsilon_m - \mu}{\epsilon^* m} \,, \tag{24}$$

where the approximation results from expanding the square root to first-order, since the term $6(\varepsilon_m - \mu)/\epsilon^* m^2$ is small. We then find the quadratic expansion

$$n\left(\varepsilon_m + \frac{1}{2}\varepsilon^*(n-m)^2 - \mu\right) \approx \text{const.} + \frac{1}{2}\epsilon^* m\sqrt{1 - \frac{6(\varepsilon_m - \mu)}{\epsilon^* m^2}}(n-n^*)^2 \,. \tag{25}$$

This shows that the micelle distribution can be approximated as a Gaussian,

$$\phi_n \approx \text{const.} \times \exp\left\{-\frac{(n-n^*)^2}{2\sigma^2}\right\} \,, \tag{26}$$

with the mean value n^* given in Eq. (24) and the variance given by

$$\sigma^2 = \frac{k_B T}{\epsilon^* m} \,. \tag{27}$$

This shows that the distribution widens at larger temperature, and is narrower for bigger micelles.

The effects on the structure on a single micelle are curious but minor in the spherical case; what is truly remarkable and very important is the overall aggregation thermodynamics which this model implies. In order to not get bogged down in tedious math (chiefly from dealing with the normalization condition (17)), let us instead look at a *two-state system*, in which we only have monomers coexisting with m-aggregates, and the normalization condition becomes $X = \phi_1 + m\phi_m$. Furthermore, we have

$$\phi_m \stackrel{(20)}{=} (\phi_1 e^{\beta(\varepsilon_1 - \varepsilon_m)})^m = (\phi_1 e^{\alpha})^m , \tag{28}$$

where we defined $\alpha = \beta(\varepsilon_1 - \varepsilon_m) > 0$ (we know the sign because we know that it is energetically favorable to form an m-aggregate). The normalization condition then becomes

$$X = \phi_1 + m e^{\alpha m} \phi_1^m . \tag{29}$$

This must be solved for ϕ_1, but notice that this is an mth order polynomial equation. This looks exceedingly troublesome, but it in fact becomes simple to get an approximate solution if we remember that m is likely large: recall from Sect. 1.1 and Fig. 2 that the number of surfactants in a spherical micelle can be written as $N = 4\pi l^2/a = 4\pi l^2/\pi r^2 = (2l/r)^2$, and with a reasonable estimate of $a \approx 0.5\,\text{nm}^2$ (and hence $r \approx 0.4$ nm) and $\ell \approx 2$ nm, we find $N \approx 100$. We then see that the second term in Eq. (29) stays extremely small for large ϕ_1 and then very rapidly picks up and completely dominates the value of X—see the left hand graph in Fig. 5. The crossover happens where the two terms on the right hand side are approximately equal, leading to

$$\phi_1 = m e^{\alpha m} \phi_1^m \quad \Longrightarrow \quad \phi_1 = \left(\frac{1}{m}\right)^{\frac{1}{m-1}} e^{-\frac{\alpha m}{m-1}} \approx e^{-\alpha} , \tag{30}$$

where the approximation is very good because $m \gg 1$ (recall in particular that $(1/m)^{1/m} \approx 1 - (\ln m)\, m^{-1} + \mathcal{O}(m^{-2})$). This shows that a critical concentration exists, $\phi_{\text{cmc}} = e^{-\alpha}$, at which something startling happens: up to that concentration, the normalization condition is dominated by ϕ_1, and this means that the solution exists almost exclusively of monomers. But at ϕ_{cmc} the second term takes over, and from now on adding extra material will almost exclusively go into aggregates. This is very visible if we plot the inverse of the normalization condition—see the right hand side of Fig. 5: the concentration of monomers initially grows linearly with the amount of added material, but it levels off quite abruptly at ϕ_{cmc}, meaning that from now on any additional material will form micelles, which so far did not exist. The concentration ϕ_{cmc} is called the *critical micelle concentration*, usually abbreviated as "cmc," and it is a fundamentally important quantity for any aggregation problem. We will soon see that the concept remains relevant beyond the case of spherical micelles we have discussed just now. Notice that $\alpha = \beta(\varepsilon_1 - \varepsilon_m)$ is not just positive, but can be a fair amount bigger than 1, since the energy which an amphiphile gains in an aggregate compared to being in isolation can be many $k_B T$. This implies, in turn, that the cmc can be very low: not much material needs to be added before micelles form. For instance, the cmc for the standard surfactant

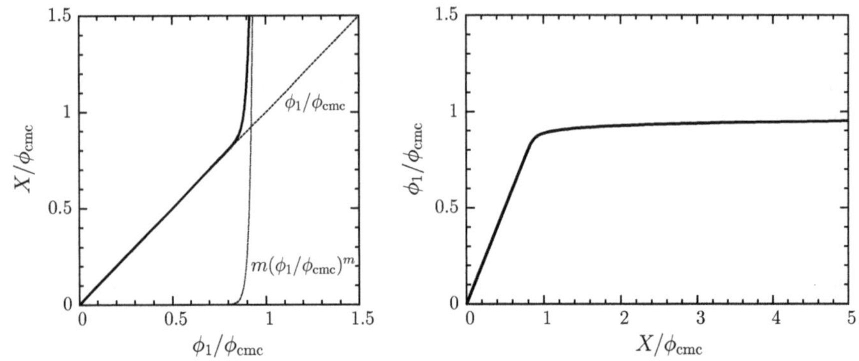

Fig. 5 The *left plot* shows the total monomer concentration in all aggregates combined, X, as a function of the concentration of *single* monomers, ϕ_1. Since X emerges as a sum of ϕ_1 and a second term $m(\phi_1/\phi_{\rm cmc})^m$ with a large m (in the graph we chose $m = 50$), there is a sharp crossover near $\phi_1 = \phi_{\rm cmc} = {\rm e}^{-\alpha}$. The right picture simply flips the axes and shows the monomer concentration ϕ_1 as a function of the total concentration of added amphiphiles. Initially, the monomer concentration grows linearly with the amount of added amphiphiles—up to the concentration $\phi_{\rm cmc}$, at which point it essentially stays constant

sodiumdodecylsulfate (SDS) is about 8 mM in water at 25 °C, at which point the aggregation number of the micelles is $m \approx 60$ (Turro and Yekta 1978).

It should be noted that the micellization transition is not a phase transition in the classical sense: there is no discontinuity or non-analyticity in any of the thermodynamic functions; the transition is always rounded, since m is large but finite. Regardless, it is a very pronounced change in the system's behavior, and as such it dominates aggregation physics.

Cylindrical micelles. The difference between the spherical and the cylindrical case enters via the energy per monomer in an aggregate, ε_n. For spheres we made the reasonable assumption in Eq. (21) that there is a typical size for a micelle, and that the energy will deviate quadratically as we move away from that value. This cannot be true for cylindrical micelles, though, since they have an unspecified length: we can easily make cylindrical micelles longer by simply adding more amphiphiles to the linear part. The aggregation energy of these amphiphiles will be always the same, for they cannot know how long the cylindrical aggregate is of which they are a part. However, amphiphiles at the two end caps of the micelle must have a different energy, and it must be *larger* than the energy of amphiphiles in the wormlike middle, for if that were not so, spherical micelles would form in the first place. It is hence reasonable to write the total energy of a cylindrical micelle of n monomers as $E_n = n\varepsilon_\infty + 2E_{\rm cap}$, and hence the energy per monomer is

$$\varepsilon_n = \varepsilon_\infty + \frac{2E_{\rm cap}}{n} =: \varepsilon_\infty + \frac{\alpha\, k_{\rm B}T}{n} \,. \tag{31}$$

Notice that the dimensionless number α must be large: it is the excess energy (in units of $k_B T$) of all end-cap monomers. Since these caps consist of two semi-spheres, they together make up essentially one full *spherical* micelle, whose aggregation number is $\mathcal{O}(100)$, and it seems fair to estimate that the excess energy for each monomer stuck in the wrong local geometry is at least a sizable fraction of $k_B T$.

Inserting this ansatz for ε_n into Eq. (20), we get

$$\phi_n = \left[\phi_1 \, e^{\beta(\varepsilon_1 - \varepsilon_\infty - \alpha k_B T/n)}\right]^n = \left[\phi_1 \, e^{\beta(\varepsilon_1 - \varepsilon_\infty)}\right]^n e^{-\alpha} = \left[\phi_1 \, e^{\alpha}\right]^n e^{-\alpha} , \tag{32}$$

where the last step follows since this equation must be true also for $n = 1$.

It is now highly useful to define the scaled concentrations $\tilde{\phi}_n = \phi_n e^{\alpha}$, because in these variables Eq. (32) becomes

$$\tilde{\phi}_n = \tilde{\phi}_1^n . \tag{33}$$

The distribution of the $\tilde{\phi}_n$ is *exponential*, which is remarkably wide (we will make this more precise below) and very different from the spherical case, where the distribution was sharply peaked around an optimal size. Notice that in order for it to be normalizable, we must have $\tilde{\phi}_1 < 1$, implying that the monomer concentration can never exceed $e^{-\alpha}$—a concentration we will soon recognize as the cmc for the cylindrical case.

If we define the scaled total concentration of monomers as $\tilde{X} = X e^{\alpha}$, the normalization condition (17) becomes

$$\tilde{X} = \sum_{n=1}^{N} n \, \tilde{\phi}_n = \sum_{n=1}^{N} n \, \tilde{\phi}_1^n . \tag{34}$$

Sums of this type can be done by the following elegant trick:

$$\sum_{n=1}^{N} n^b x^n = \sum_{n=1}^{N} \left(x \frac{\partial}{\partial x}\right)^b x^n = \left(x \frac{\partial}{\partial x}\right)^b \sum_{n=1}^{N} x^n = \left(x \frac{\partial}{\partial x}\right)^b \frac{x - x^{N+1}}{1 - x} , \tag{35}$$

where in the last step we summed the well-known geometric series. Moreover, since we know that in our case $x < 1$ and N is *very* large, we can drop the x^{N+1} term (or, equivalently, set $N \to \infty$), and so we for instance find

$$\sum_{n=1}^{\infty} n \, x^n = \left(x \frac{\partial}{\partial x}\right) \frac{x}{1 - x} = \frac{x}{(1 - x)^2} , \tag{36a}$$

$$\sum_{n=1}^{\infty} n^2 \, x^n = \left(x \frac{\partial}{\partial x}\right)^2 \frac{x}{1 - x} = \frac{x(1 + x)}{(1 - x)^3} , \tag{36b}$$

$$\sum_{n=1}^{\infty} n^3 x^n = \left(x\frac{\partial}{\partial x}\right)^3 \frac{x}{1-x} = \frac{x(1+x(4+x))}{(1-x)^4} \,. \tag{36c}$$

Hence, using Eq. (36a), the normalization condition (34) becomes a quadratic equation for $\tilde{\phi}_1$ that is easy to solve

$$\tilde{X} = \frac{\tilde{\phi}_1}{(1-\tilde{\phi}_1)^2} \quad \Longrightarrow \quad \tilde{\phi}_{1\pm} = \frac{1+2\tilde{X} \pm \sqrt{1+4\tilde{X}}}{2\tilde{X}} \,. \tag{37}$$

Since we know $\tilde{\phi}_1 < 1$, the minus sign is the correct choice. Expanding the solution for small and large $\tilde{X}$, we find

$$\tilde{\phi}_1 = \begin{cases} \tilde{X} + \mathcal{O}(1) & : \tilde{X} \ll 1 \\ 1 - 1/\sqrt{\tilde{X}} + \mathcal{O}(\tilde{X}^{-1}) & : \tilde{X} \gg 1 \end{cases} \,. \tag{38}$$

As promised, we can again define a cmc, $\phi_{\rm cmc} = {\rm e}^{-\alpha}$, such that below the cmc the monomer concentration in our solution is proportional to the amount of added material, while for concentrations larger than the cmc any added material goes into micelles, leaving the monomer concentration below $\phi_{\rm cmc}$, and approaching it with a very slow $1/\sqrt{X}$ asymptotics. This is illustrated in Fig. 6.

We already know that the distribution of micelle sizes is exponential, but we might also want to know what the mean and the variance are. These are easily calculated by working out (weight-averaged) moments of n. For the first one, we find

$$\langle n \rangle = \frac{\sum_{n=1}^{\infty} n\tilde{X}_n}{\sum_{n=1}^{\infty} \tilde{X}_n} = \frac{\sum_{n=1}^{\infty} n^2\tilde{\phi}_n}{\sum_{n=1}^{\infty} n\tilde{\phi}_n} \overset{*}{=} \frac{1+\tilde{\phi}_1}{1-\tilde{\phi}_1} \overset{\#}{=} \sqrt{1+4\tilde{X}} \,, \tag{39}$$

where at $*$ we used Eqs. (36a) and (36b) and at # we inserted the solution (37). Hence, the average micelle length grows like the square root of the concentration: $\langle n \rangle \approx 2\sqrt{X/\phi_{\rm cmc}}$.

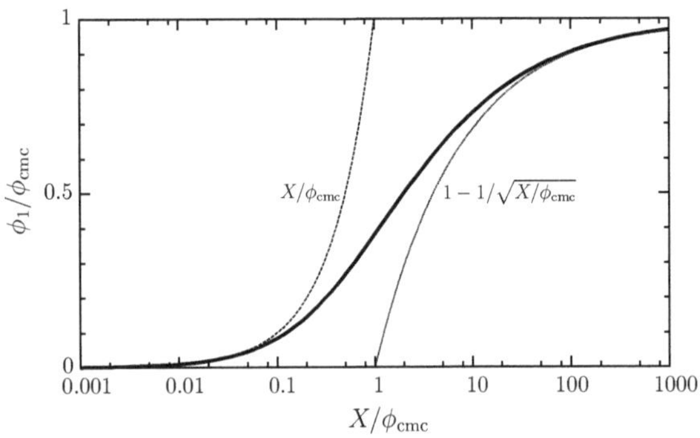

Fig. 6 Monomer concentration for the case of a cylindrical micelle aggregation scenario. The *dashed* and *dotted curves* indicate the small- and large-concentration limits from Eq. (38). The full solution shows a cross over at the cmc

The second moment of n is given by

$$\langle n^2 \rangle = \frac{\sum_{n=1}^{\infty} n^2 \tilde{X}_n}{\sum_{n=1}^{\infty} \tilde{X}_n} = \frac{\sum_{n=1}^{\infty} n^3 \tilde{\phi}_n}{\sum_{n=1}^{\infty} n \tilde{\phi}_n} \stackrel{*}{=} \frac{1 + \tilde{\phi}_1(4 + \tilde{\phi}_1)}{(1 - \tilde{\phi}_1)^2} , \tag{40}$$

where at $*$ we used Eqs. (36a) and (36c). Hence, the variance of n is

$$\sigma_n^2 = \langle n^2 \rangle - \langle n \rangle^2 = \frac{2\tilde{\phi}_1}{(1 - \tilde{\phi}_1)^2} \stackrel{\#}{=} 2\tilde{X} , \tag{41}$$

where at # we again used the solution (37). This answer is important, because it shows that the width of the distribution essentially scales with its mean, and hence

$$\frac{\sigma_n}{\langle n \rangle} = \sqrt{\frac{2\tilde{X}}{1 + 4\tilde{X}}} = \frac{1}{\sqrt{2}} - \mathcal{O}(\tilde{X}^{-1}). \tag{42}$$

Distributions of cylindrical micelles are hence "wide" no matter how large the micelles are; there is no "law of large micelles," or a $1/\sqrt{n}$ like asymptotics toward a sharp mean. Remarkable as this is, it is of course not unexpected, for that is what exponential distributions do.

Planar bilayers. Again, the first question to address is: what is ε_n for an aggregate that assembles in a planar fashion? To make headway, though, we need to make further assumptions about its geometry. We will assume that it stays flat, and that it will be circular. The latter follows because the amphiphiles at the bilayer disc's edge will have a higher free energy per molecule than the one in the flat region (for reasons analogous to the elevated free energy of monomers at the ends of cylindrical micelles). This excess free energy per unit length acts as a *line tension* (in this case usually called *edge tension*), and minimizing it at constant overall area of the aggregate means that the shape has to be a circle.

If the circular aggregate has area $A = \pi R^2$, its circumference is $C = 2\pi R = 2\sqrt{\pi A}$. The excess free energy of the edge is $E_{\text{edge}} = 2\pi R\gamma = 2\sqrt{\pi A}\gamma$, with γ being the edge tension—a material parameter. Since the number of lipids in the aggregate is approximately $n = 2A/a_\ell$, with a_ℓ being the area per lipid, we get $A = \frac{1}{2} n a_\ell$, and hence $E_{\text{edge}} = \sqrt{2\pi n a_\ell}\gamma$. The replacement for Eq. (31) is hence

$$\varepsilon_n = \varepsilon_\infty + \frac{E_{\text{edge}}}{n} = \varepsilon_\infty + \frac{\sqrt{2\pi n a_\ell}\gamma}{n} = \varepsilon_\infty + \frac{\alpha k_{\text{B}} T}{\sqrt{n}} , \tag{43}$$

where $\alpha = \sqrt{2\pi a_\ell}\beta\gamma$ is a dimensionless number that's again a fair bit larger than 1. To estimate it, let's take the DOPC values of $a_\ell \simeq 0.7\,\text{nm}^2$ (Kučerka et al. 2006) and $\gamma \simeq 20\,\text{pN}$ (Portet and Dimova 2010), from which we get $\alpha \approx 10$. Notice that the only difference between the cylindrical and the planar case is that in the latter the excess term is proportional to $1/\sqrt{n}$ instead of $1/n$. We will see that this changes the physics in a big way.

Inserting this expression for the energy per monomer into the general form of the aggregate distribution, Eq. (20), we get

$$\begin{aligned} X_n = n\,\phi_n &= n\left[\phi_1\,\mathrm{e}^{\beta(\varepsilon_1-\varepsilon_\infty-\alpha k_\mathrm{B}T/\sqrt{n})}\right]^n \\ &= n\left[\phi_1\,\mathrm{e}^{\beta(\varepsilon_1-\varepsilon_\infty)}\right]^n \mathrm{e}^{-\alpha\sqrt{n}} \\ &= n\left[\phi_1\,\mathrm{e}^{\alpha}\right]^n \mathrm{e}^{-\alpha\sqrt{n}}\,, \end{aligned} \tag{44}$$

where the last step again follows because this equation must also be true for $n = 1$. The normalization condition (17) then becomes

$$X = \sum_{n=1}^{N} X_n = \sum_{n=1}^{N} n\left[\phi_1\,\mathrm{e}^{\alpha}\right]^n \mathrm{e}^{-\alpha\sqrt{n}}\,. \tag{45}$$

The term $\mathrm{e}^{-\alpha\sqrt{n}}$ decreases with n, while for the term $[\phi_1\,\mathrm{e}^{\alpha}]^n$ the asymptotic behavior depends on whether $\phi_1\mathrm{e}^{\alpha}$ is bigger or smaller than 1. Assume it is bigger than 1. Then this term grows with n, and it asymptotically grows *faster* than $\mathrm{e}^{-\alpha\sqrt{n}}$ decreases. This might get us worried, for if we again replace $N \to \infty$ (because N will be macroscopically big), the sum in Eq. (45) would diverge. So let us assume that, instead, the expression $\phi_1\mathrm{e}^{\alpha}$ is smaller than 1. In that case, we can calculate

$$X = \sum_{n=1}^{\infty} n\left[\phi_1\,\mathrm{e}^{\alpha}\right]^n \mathrm{e}^{-\alpha\sqrt{n}} \le \sum_{n=1}^{\infty} n\,\mathrm{e}^{-\alpha\sqrt{n}} \approx \int_0^\infty \mathrm{d}n\; n\,\mathrm{e}^{-\alpha\sqrt{n}} = \frac{12}{\alpha^4}\,. \tag{46}$$

This is a pretty disastrous finding, though: apparently, the total amount of material we can add to the system is bounded from above. What if we wanted to add more material—who is going to stop us? (Not excluded volume—that was not part of the model!)

The solution to this conundrum is subtle: the assumption that N can be replaced by infinity is wrong—despite the fact that N could really be an Avogadro number of molecules. But large is not the same as infinite, and the normalization condition only enforces $\phi_1\mathrm{e}^{\alpha} \le 1$ if we really sum all the way up to infinity. If the sum is *finite*, there is no reason to demand that $\phi_1\mathrm{e}^{\alpha} \le 1$, because finite sums cannot diverge! More specifically, even if this term would ultimately outcompete $\mathrm{e}^{-\alpha\sqrt{n}}$, if $\phi_1\mathrm{e}^{\alpha}$ is only ever so slightly bigger than 1, this will only happen near the upper bound of the sum—showing us that the value of this sum will likely depend very critically on just how much $\phi_1\mathrm{e}^{\alpha}$ exceeds 1.

Unfortunately, it is quite tricky to see how this plays out analytically, because the normalization sum (45) turns out to be a very delicate interplay between very small and very large terms. To brace ourselves for what is actually happening here, we shall first look at a numerical example. Let us assume that $\alpha = 10$, that we

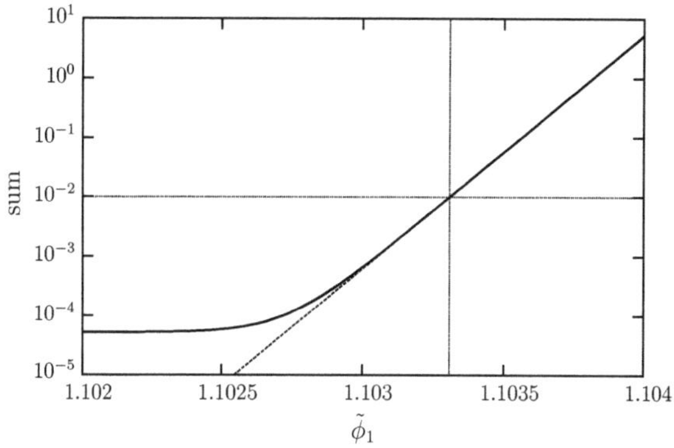

Fig. 7 The *solid curve* is the *right hand side* of Eq. (47) as a function of the parameter $\tilde{\phi}_1$, for $\alpha = 10$ and $N = 10,000$; the *dashed curve* is the large-n-approximation from Eq. (49)

have $N = 10,000$ molecules in the system (really an incredibly small number by experimental standards, but this might be a typical number to be used in a simulation), and let us demand that we want to ultimately gain a total concentration of $X = 10^{-2}$ (notice that this is larger than the erroneous upper bound of $12/\alpha^4 = 1.2 \times 10^{-3}$). If we abbreviate $\tilde{\phi}_1 = \phi_1 e^{\alpha}$, then we have to numerically solve the following equation for $\tilde{\phi}_1$:

$$10^{-2} = \sum_{n=1}^{10,000} n\, \tilde{\phi}_1^n\, e^{-10\sqrt{n}} . \tag{47}$$

Figure 7 plots the right hand side of this equation as a function of the parameter $\tilde{\phi}_1$ in the interesting range. Up to $\tilde{\phi}_1 \approx 1.1025$, the right hand side grows linearly (and extremely weakly) with $\tilde{\phi}_1$, but at around this point a big change happens, and the sum picks up extremely rapidly—becoming a power law with an exponent of about 10, 000. (This also shows why it is very hard to treat this problem numerically with even bigger values of N.) The value 10^{-2} is reached at $\tilde{\phi}_1 \approx 1.10330764$ and hence $X_1 \approx 5.009 \times 10^{-5}$.

Inserting this value for $\tilde{\phi}_1$ into the distribution function for X_n from Eq. (44), we can plot it over the entire range of permissible n values: from $n = 1$ to $n = 10,000$; this is done in Fig. 8. Initially, the distribution function drops precipitously: one finds $X_2 \approx 1.756 \times 10^{-6} \approx X_1/30$ and $X_3 \approx 1.211 \times 10^{-7} \approx X_1/400$. But at $n = 2566$ the function attains a minimum, after which it again begins to rapidly grow. At its largest n-value it becomes $X_{10,000} \approx 4.696 \times 10^{-4} \approx 10X_1$, showing it is about 10 times more likely to find a lipid in that aggregate than to find it in isolation! Another way of looking at this is the following: 99% of all monomers are found in aggregates with a size of at least 9, 890. And yet another illustration is the following: Look at the cumulative normalized distribution of X_n, namely, $f(m) = X^{-1} \sum_{n=1}^{m} X_n$. It rapidly rises from 0 to about 0.0052 when m rises from 1 to 10. However, after that it stays virtually constant, until about 9, 800, when it begins to rise again. In other

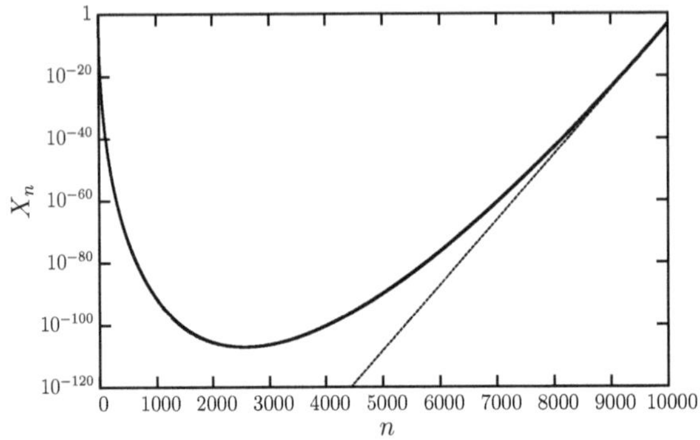

Fig. 8 The *solid curve* is the distribution function $X_n = n\,\phi_n$ from Eq. (44), using the numerical parameters $\alpha = 10$, $N = 10{,}000$, and $X = 0.01$, which implies the numerical solution $\tilde{\phi}_1 \approx 1.10330764$ and hence $X_1 \approx 5.009 \times 10^{-5}$. The *dashed curve* is the approximate distribution from Eq. (48), using the value for $\tilde{\phi}_1$ determined via the first-order approximation in Eq. (52), $\tilde{\phi}_1^{(1)} \approx 1.10330882$. Using $\tilde{\phi}_1^{(1)}$ in the full distribution (instead of the exact $\tilde{\phi}_1$) leads to a curve that is indistinguishable from the exact one on this plot, with a normalization that is about 1% off

words, with the exception of about half a percent of small oligomers, *virtually the whole system forms one giant aggregate.*

With these observations we are now in a better position to develop a decent approximate solution for the normalization condition (45). Notice that we need to analytically describe the region in that sum which strongly increases (the "uptake" in Fig. 7), and that this comes from the aggregates—meaning, the large-n part of the distribution function. Hence it is probably a good idea to expand the summands in Eq. (45) around the upper end, $n = N$, and preferably in such a fashion that we can perform the sum. But given the exponential variation of X_n, it is wise to do that expansion in the exponent:

$$
\begin{aligned}
X_n = n\,\tilde{\phi}_1^n \mathrm{e}^{-\alpha\sqrt{n}} &= \tilde{\phi}_1^n \exp\left\{-\alpha\sqrt{n} + \ln n\right\} \\
&= \tilde{\phi}_1^n \exp\Bigg\{ -\alpha\left[\sqrt{N} + \frac{1}{2\sqrt{N}}(n-N)\right] \\
&\qquad\qquad + \ln N + \frac{1}{N}(n-N) + \mathcal{O}\Big((n-N)^2\Big)\Bigg\} \\
&\approx N\mathrm{e}^{-\alpha\sqrt{N}/2}\left(\tilde{\phi}_1\,\mathrm{e}^{-\alpha/2\sqrt{N}}\right)^n .
\end{aligned}
\tag{48}
$$

This expansion permits us to do the sum, since it turns into a simple geometric series:

$$
\sum_{n=1}^{N} n\,\tilde{\phi}_1^n \mathrm{e}^{-\alpha\sqrt{n}} \approx N\mathrm{e}^{-\alpha\sqrt{N}/2}\,\frac{y^{N+1}-1}{y-1} \qquad \text{with} \quad y = \tilde{\phi}_1\,\mathrm{e}^{-\alpha/2\sqrt{N}} . \tag{49}
$$

Since y is slightly larger than 1, but N is huge, y^{N+1} will be very large compared to 1. (In our above numerical example we would find $y \approx 1.0494987$ and hence $y^{N+1} \approx 1.0494987^{10{,}001} \approx 6.92 \times 10^{209}$.) We can hence neglect the "-1" in the numerator, but of course not in the denominator.

The normalization condition now becomes

$$\Xi := \frac{X\,\mathrm{e}^{\alpha\sqrt{N}/2}}{N} = \frac{y^{N+1}}{y-1}\,, \tag{50}$$

but this is again impossible to solve analytically. However, we can get increasingly good approximations by iteration. First, recall that the right hand side really emerged as a geometric series, and so it is given by $y^N + y^{N-1} + y^{N-2} + \cdots$. Let us take the dominant term, y^N, and solve the equation. We then get

$$y = \Xi^{\frac{1}{N}}\,. \tag{51}$$

Even though only approximate, this already looks remarkably good, since it gives $\tilde{\phi}_1 = 1.103645$ for our numerical example, about 0.03% off. And yet, inserting this value into the normalization condition gives a value about 20 times too big. We need to do better. In fact, we can improve the solution by iterating the defining equation, à la $y^{(i+1)} = [\Xi(y^{(i)} - 1)]^{1/(N+1)}$, where $y^{(0)} = \Xi^{1/N}$ is our initial simple result. At first-order we get

$$\begin{aligned}
\tilde{\phi}_1^{(1)} &= \mathrm{e}^{\alpha/2\sqrt{N}}\; y^{(1)} \\
&= \mathrm{e}^{\alpha/2\sqrt{N}}\left[\Xi\left(\Xi^{1/N} - 1\right)\right]^{1/(N+1)} \\
&= \mathrm{e}^{\alpha/2\sqrt{N}}\left[\frac{X}{N}\,\mathrm{e}^{\alpha\sqrt{N}/2}\left(\left(\frac{X}{N}\right)^{1/N}\mathrm{e}^{\alpha/2\sqrt{N}} - 1\right)\right]^{1/(N+1)}\,.
\end{aligned} \tag{52}$$

With the numerical example from above ($X = 0.01$, $\alpha = 10$, and $N = 10{,}000$), this gives $\tilde{\phi}_1^{(1)} - 1.10330882$, which differs from the exact numerical solution only by 1 part in 10^6, and now the normalization condition is only 1% off. Unfortunately, further iterations do not gain us much anymore, because we are still solving an approximate equation, not the exact one.

There is more to be learned. First, even the simplest solution becomes exact in the thermodynamic limit $N \to \infty$. Performing it, we get

$$\phi_1 = \mathrm{e}^{-\alpha}\lim_{N\to\infty}\left\{\mathrm{e}^{\alpha/2\sqrt{N}}\left(\frac{X\,\mathrm{e}^{\alpha\sqrt{N}/2}}{N}\right)^{1/N}\right\} = \mathrm{e}^{-\alpha}\,, \tag{53}$$

showing that—again—we have a critical "micelle" concentration. Since bilayer patches are usually not viewed as "micelles," this is more commonly called the *critical aggregate concentration* and abbreviated as "cac": $\phi_{\mathrm{cac}} = \mathrm{e}^{-\alpha}$.

The scenario looks superficially similar to what we have seen in the spherical case: the normalization condition becomes a polynomial with a constant term, a linear term, and one term with a large power (compare Eqs. (29) and (50)), and the "largeness" of that power makes the transition. However, in the spherical micelle case that power was given by the micelle size, and hence it was *mesoscopic*—of order 10^2. In the bilayer case that power is *macroscopic*—the total number of molecules in the system, conceivably of the order of Avogadro's number, but more importantly: *extensive*. It will *by definition* diverge in the thermodynamic limit. It hence follows that *the aggregation transition for bilayers is a true phase transition*—at least in the model we have studied here.

Alas, our model is defective. The $1/\sqrt{n}$ correction to ε_n (see Eq. (43)), on which the whole scenario hinges, comes from the $\sqrt{n}$ divergence of the edge energy for increasingly large flat circular aggregates. But bilayer patches do not have to stay flat. Once they exceed a critical size, it is preferable for them to close up, make an edgeless spherical vesicle, and pay bending energy instead, because bending energy does not scale with size. This was first discussed by Helfrich (1974). Hence, vesiculation caps the edge energy, moving the correction term back to a $1/n$ form, for which we expect a wide exponential distribution function like in the case of cylindrical micelles. Unfortunately, in reality things are now a lot more complicated, because we can no longer ignore kinetics. In any case, we still encounter an aggregation transition once the amphiphile concentration in solution exceeds a critical aggregate concentration.

2 Fluid Elastic Sheets: From Three to Two Dimensions

The previous section has shown that there is something special about two-dimensional assemblies of amphiphiles. Spherical micelles are by construction microscopic, and cylindrical micelles are tenuous threads, constantly breaking and re-merging, with a corresponding wide length distribution. In contrast, two-dimensional amphiphilic sheets are endowed by thermodynamics with certain inalienable rights, among them extensivity, stability, and universal elasticity. They arise as macroscopic persistent entities, for which we therefore expect an effective large-scale theory to exist, whose key degrees of freedom are emergent and independent of the microscopic realization, and whose key physical parameters are functions of the underlying structure, but might as well be taken as fundamental at the emergent level.

This situation arises frequently in physics: a system is known to have an underlying structure, but we can describe it effectively (and very elegantly) at a level that completely ignores this structure. For instance, fluid dynamics need not know about atoms. Its laws follow from thermodynamics and symmetry, only its parameters (mass density and viscosity in the simplest case) reflect the details of the constituents. The same is true for elasticity theory, where we can see even more clearly how local microscopic symmetries leave traces in the macroscopic description (they dictate the number and type of elastic moduli).

In such a situation there are two ways for how to proceed, and they differ quite fundamentally in their "philosophy":

- Bottom-up approaches strive to reduce larger scale phenomenology to a microscopic description at a smaller scale that is considered more fundamental. In particular, they aim to elucidate the dependence of the larger scale parameters on the microscopic foundation.
- Top-down approaches ignore the underlying structure and postulate a macroscopic theory from scratch, constrained only by symmetry. The parameters of this theory are not themselves predictable, for they depend on the microscopic structure which this approach purposefully ignores. But one can always measure them at the macroscopic level, so the endeavor is self-contained.

Both approaches are perfectly valid and have their own advantages and drawbacks. The top-down approach, for instance, need not wrestle with underlying microscopic degrees of freedom—say, trying to eliminate them by performing partial traces in phase space or other scale-bridging procedures. But decorating all symmetry-permissible terms with phenomenological parameters might be dangerous, for they need not be independent: a relation between them, enforced by subtleties of the underlying microphysics, could be missed. The bottom-up approach, in contrast, necessarily captures such effects, which is probably its biggest strength. But given our poor ability to actually *do* the math needed to rigorously coarse-grain a Hamiltonian, approximations along the way might cloud the path of emergence. Moreover, often we do not know the underlying microscopic theory all that well, and so we instead start with what we perceive to be a good *model* of the microphysics. This often works flawlessly, in the sense of giving a perfectly acceptable macroscopic theory—but this is to be expected: after all, hardly any microscopic details survive the emergence process. The macroscopic theory only depends on very generic symmetry considerations and the microscopic details matter only inasmuch as they predict macroscopic coefficients or produce correlations between them. If we cannot measure both the microscopic and the macroscopic parameters, it is very difficult to test whether these predicted connections are fulfilled, and hence it is usually impossible to be sure that our microscopic model was correct. This, of course, is the well-known bane of scientists looking for The Fundamental Laws: there is more than one way to skin a cat.

In our experience, combining both approaches to elucidate the path of scale-bridging, being aware what powers and limits each modus operandi, and being skeptical of too freely floating phenomenology as well as suspiciously specific model-building—these are attitudes that will deepen one's understanding of the key physics. Indeed, one of the goals of the book you are holding is to explore this duality for lipid membranes, for which phenomenological geometric Hamiltonians can be written down, which in turn can also be motivated by underlying microphysics that considers the lipid constituents.

In this section we wish to discuss one particular connection between large-scale membrane theory and an underlying more microscopic model that is interesting because it is *itself* already coarse-grained. It is a description of a thin two-dimensional fluid elastic sheet, and the question is, how to bootstrap ourselves up to larger scales and one dimension lower: large-scale two-dimensional curvature elastic surfaces. This program has been proposed and worked through in an important and seminal paper by Hamm and Kozlov (2000). The goal of this section is to revisit their elegant derivation, but here and there keep a few higher order terms which Hamm and Kozlov have neglected, but for which one can make good arguments to keep them.

Before we now dive into membrane elasticity—a little heads-up: unlike the previous section, this one will start to use numerous tools from surface differential geometry. The notation follows a recent review one of us has written (Deserno 2015), which introduces the basic formalism, derives most of the key identities, and also provides several applications to membrane elasticity. But then, our view and usage of differential geometry in this context has been very heavily influenced by Jemal Guven, who also has a chapter in this book. We hence strongly recommend that the reader also consults the master, not merely his apprentices.

2.1 *The Starting Point: Thin Fluid Elastic Sheets*

It is well-known that if u_{ij} is the Cauchy strain tensor, the most general quadratic expression for the elastic energy density we can write down is

$$e_{3\mathrm{d}} = \frac{1}{2}\lambda_{ijkl}\, u_{ij}u_{kl} \;, \tag{54}$$

where λ_{ijkl} is the elastic modulus tensor (Landau and Lifshitz 1986). Without loss of generality, the exchange symmetries $i \leftrightarrow j$, $k \leftrightarrow l$, and $ij \leftrightarrow kl$ can be assumed, leaving at most 21 independent components. But we want to use this expression for the energy of a fluid lipid monolayer, and in that case additional symmetries reduce the number of components much further (Hamm and Kozlov 2000; Campelo et al. 2014).

Area strain. Assume the leaflet lies in the xy-plane. First note that the two reflection symmetries $(x, y, z) \to (-x, y, z)$ and $(x, y, z) \to (x, -y, z)$ imply that neither an x- nor a y index can occur in λ_{ijkl} an odd number of times. Curiously, this implies that the same must hold for the z-index, even though a monolayer does *not* have an up-down reflection symmetry that would enforce this all by itself. Furthermore, one consequence of in-plane isotropy is that the x- and the y-directions are indistinguishable, and so their λ-coefficients must be equal. This already massively reduces the permissible terms to the following six:

$$\begin{aligned} e_{3\mathrm{d}} =& \frac{1}{2}\lambda_{xxxx}\big(u_{xx}^2 + u_{yy}^2\big) + \lambda_{xxyy}u_{xx}u_{yy} + 2\lambda_{xyxy}u_{xy}^2 \\ &+ \lambda_{xxzz}\big(u_{xx} + u_{yy}\big)u_{zz} + 2\lambda_{xzxz}\big(u_{xz}^2 + u_{yz}^2\big) + \frac{1}{2}\lambda_{zzzz}u_{zz}^2 \;, \end{aligned} \tag{55}$$

where the prefactors account for obvious permutation multiplicities—such as $\lambda_{xyxy} = \lambda_{yxyx} = \lambda_{xyyx} = \lambda_{yxxy}$. It is now useful to rework the quadratic strain expressions in the following way:

$$\begin{aligned} e_{3\text{d}} =& \frac{1}{2}\lambda_{xxxx}\big(u_{xx} + u_{yy}\big)^2 + \big(\lambda_{xxyy} - \lambda_{xxxx}\big)\big(u_{xx}u_{yy} - u_{xy}^2\big) \\ &+ \big(2\lambda_{xyxy} + \lambda_{xxyy} - \lambda_{xxxx}\big)\, u_{xy}^2 \\ &+ \lambda_{xxzz}\big(u_{xx} + u_{yy}\big)u_{zz} + 2\lambda_{xzxz}\big(u_{xz}^2 + u_{yz}^2\big) + \frac{1}{2}\lambda_{zzzz}u_{zz}^2 \,. \end{aligned} \tag{56}$$

At this point we can exploit full in-plane rotational symmetry. The first two strain terms in Eq. (56) are quadratic invariants under in-plane rotation: they are (i) the square of the trace and (ii) the determinant of the strain tensor's xy-subspace, respectively. But the term in the second line is not an invariant, and there is no term left to combine it with to remedy this flaw; hence, *this term must vanish.*

Next, let us make use of in-plane fluidity, which implies that the energy cannot change under in-plane shear deformations—meaning, in-plane shear stresses must vanish. One such deformation is a simple shear, u_{xy}, and its associated shear stress is

$$0 \overset{!}{=} \sigma_{xy} = \frac{\partial e_{3\text{d}}}{\partial u_{xy}} = -2\big(\lambda_{xxyy} - \lambda_{xxxx}\big)\, u_{xy} \,. \tag{57}$$

Since this must hold for $u_{xy} \neq 0$, we must have $\lambda_{xxyy} = \lambda_{xxxx}$, and so the second term in Eq. (56) must vanish, too.

Finally, recall that we intend to describe a *thin* leaflet, which has the following consequence: the normal stress σ_{zz} at the leaflet's upper and lower surface vanishes if the surface is free, but since the leaflet is thin, σ_{zz} does not have much opportunity to considerably grow anywhere within the material. We will hence assume that it vanishes *throughout* the material, and this implies

$$0 \overset{!}{=} \sigma_{zz} = \frac{\partial e_{3\text{d}}}{\partial u_{zz}} = \lambda_{xxzz}\big(u_{xx} + u_{yy}\big) + \lambda_{zzzz}u_{zz} \,, \tag{58}$$

and this means that the in-plane and transverse strains are related by

$$u_{zz} = -\frac{\lambda_{xxzz}}{\lambda_{zzzz}}\big(u_{xx} + u_{yy}\big) =: -\tilde{\nu}\,\big(u_{xx} + u_{yy}\big) \,. \tag{59}$$

The dimensionless parameter $\tilde{\nu}$ is related to the usual Poisson ratio ν via $\tilde{\nu} = \nu/(1-\nu)$. Inserting this into Eq. (56), what remains is

$$e_{3\text{d}} = \frac{1}{2}\tilde{E}(u_{xx} + u_{yy})^2 + 2\lambda_{xzxz}(u_{xz}^2 + u_{yz}^2) \,, \tag{60}$$

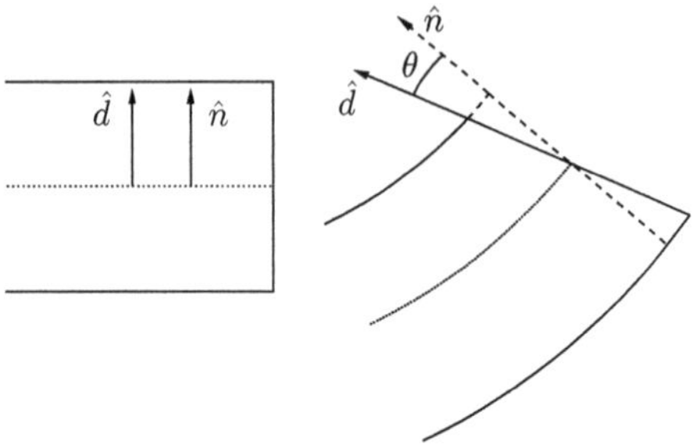

Fig. 9 The material director $\hat{\boldsymbol{d}}$ in a flat thin plate is by construction aligned with the local surface normal $\hat{\boldsymbol{n}}$. But upon bending, $\hat{\boldsymbol{d}}$ may deviate from $\hat{\boldsymbol{n}}$ by an angle θ due to transverse shear (Figure adapted from Reddy (2006))

where we defined the effective modulus

$$\tilde{E} = \lambda_{xxxx} - \frac{\lambda_{xxzz}^2}{\lambda_{zzzz}} = \lambda_{xxxx} - \tilde{\nu}^2 \lambda_{zzzz} \,. \tag{61}$$

The first part in the energy (60) has now been recast in terms of a local area strain, which we will soon relate to the extent of bending.

Lipid tilt. For the second term in Eq. (60), a connection to area strain is not possible, because the strains u_{xz} and u_{yz} correspond to a local *transverse shear*, i.e., a deformation related to the fact that the material director of a sheet need not coincide with the surface normal, even if it does so for the flat sheet—see Fig. 9. This term can instead be related to *lipid tilt*—if we decide that a lipid's orientation is the appropriate indicator for the local material director.[1] To do so quantitatively, it is useful to define a locally transverse tilt-field $\boldsymbol{T}$ that measures the deviation between material director and surface normal (Hamm and Kozlov 2000):

$$\boldsymbol{T} = T^l \boldsymbol{e}_l = \frac{\hat{\boldsymbol{d}}}{\hat{\boldsymbol{n}} \cdot \hat{\boldsymbol{d}}} - \hat{\boldsymbol{n}} \,. \tag{62}$$

This definition makes the transversality of $\boldsymbol{T}$ manifest, since $\boldsymbol{T} \cdot \hat{\boldsymbol{n}} = 0$ by construction (i.e., independent of any other conditions that would have to hold, such as $\boldsymbol{T}$ being the solution of some Euler–Lagrange equation). Also, $\boldsymbol{T}$ is not normalized; instead, its magnitude is $|\boldsymbol{T}| = \tan\theta$, where θ is the tilt angle (i.e., the angle between $\hat{\boldsymbol{d}}$ and $\hat{\boldsymbol{n}}$). Alternatively, we can write

$$\frac{1}{\cos\theta} = \frac{1}{\cos\arctan|\boldsymbol{T}|} = \sqrt{1+\boldsymbol{T}^2} = 1 + \frac{1}{2}\boldsymbol{T}^2 + \mathcal{O}\big((\boldsymbol{T}^2)^2\big) \,. \tag{63}$$

Since within first-order shear deformation plate theory $2u_{xz} = \boldsymbol{T} \cdot \boldsymbol{x}$ and $2u_{yz} = \boldsymbol{T} \cdot \boldsymbol{y}$ (Reddy 2006), this leads to

[1] Here we assume that a flat membrane is untilted—which is true for fluid phases, but *not* necessarily so for membranes in the gel phase.

$$u_{xz}^2 + u_{yz}^2 = \tfrac{1}{4} T_l T^l \,, \tag{64}$$

and that permits us to replace the second term in Eq. (60) in a way that involves tilt:

$$e_{3\mathrm{d}} = \frac{1}{2} \tilde{E} \, (u_{xx} + u_{yy})^2 + \frac{1}{2} \lambda_{xzxz} T_l T^l \,, \tag{65}$$

We would now have to relate the two deformations—especially the area strain—to the geometry of a curved membrane. But before we do that, it is important to realize that we are not yet done with the energy density: a very crucial term is missing, because its origin requires us to think beyond a thin sheet of local moduli—i.e., we must go beyond Eq. (54).

Lateral prestress. Consider again what we are trying to model: a thin self-assembled in-plane fluid leaflet made up of amphiphilic molecules. Now focus on the fact that, unlike a homogeneous thin sheet, a lipid monolayer has internal structure that underlies its very reason of existence: the strongly positionally varying solubility of lipids—which gives rise to the self-assembly process that shields the tails from the embedding solvent by placing the head groups in between. One important consequence of this assembly-driven cohesion is that it leaves the membrane under *internal pre-stresses*—meaning, stresses that do not locally vanish in the equilibrium state, only globally. As we will soon see, they contribute to the deformation energy.

Let us first explore, what kind of remaining stresses are permissible by symmetry. Evidently, for a flat membrane lying in the xy-plane the stress tensor $\mathbf{\Pi}$ is diagonal in the $\{x, y, z\}$ coordinate system. Due to translational symmetry, it can only depend on z, and due to rotational symmetry, the x- and y-components must be identical:

$$\mathbf{\Pi} = \mathrm{diag}\big(\Pi_{xx}(\boldsymbol{r}), \Pi_{yy}(\boldsymbol{r}), \Pi_{zz}(\boldsymbol{r})\big) = \mathrm{diag}\big(\Pi_{||}(z), \Pi_{||}(z), \Pi_{\perp}(z)\big) \,. \tag{66}$$

In mechanical equilibrium $\mathbf{\Pi}$ must be divergence free, $\partial_i \Pi_{ij} = 0$. This equation immediately implies that $\Pi_{\perp}(z) = \Pi_{\perp}$ is a constant, and so it must be equal to the isotropic ambient pressure acting on the membrane. The tangential component $\Pi_{||}(z)$, however, is not restricted by this argument and could be a pretty complicated function of z. As we will soon discover, it indeed is.

Now imagine that we place a small patch of membrane inside a cuboid box of area A and height z. How would the energy change if we (isothermally and reversibly) deform that box in a volume-preserving way such that $A \to A + \delta A$ and $z \to z - \delta z = z - (z/A)\delta A$? Following Rowlinson and Widom (2002 Chap. 2.5), the vertical compression requires the work

$$\delta W_{\perp} = A \, \delta z \, \Pi_{\perp} = z \, \delta A \, \Pi_{\perp} \,. \tag{67}$$

In contrast, the lateral expansion requires the work

$$\delta W_{||} = -\delta A \int_{-z/2}^{z/2} \mathrm{d}z \; \Pi_{||}(z) \,. \tag{68}$$

Hence, the total change in free energy is

$$\delta F = \delta W_{\perp} + \delta W_{\parallel} = \delta A \int_{-z/2}^{z/2} \mathrm{d}z \left\{ \Pi_{\perp} - \Pi_{\parallel}(z) \right\} . \tag{69}$$

The expression under the integral is the positionally resolved effective lateral mechanical tension acting in the membrane. It is often simply called the *lateral stress profile*:

$$\sigma_0(z) = \Pi_{\perp} - \Pi_{\parallel}(z) . \tag{70}$$

What does this function look like for a membrane?

First, consider that there is the equivalent of a hydrophilic–hydrophobic interface at the backbone of a lipid, and so roughly at that height in the monolayer we have a relatively large lateral tension. This is where the bilayer is being pulled together, where the effect is localized that gives rise to a membrane in the first place. As a consequence, both the tails and the heads of the lipids are now being compressed, leaving us with a positive pressure (or negative tension) in the tail and upper head region that strives to expand the leaflet. For a membrane that is not subject to a net lateral tension, these stresses must balance, such that the net total stress (the integral over $\sigma_0(z)$) vanishes, thereby setting the equilibrium area per lipid. Hence, we expect $\sigma_0(z)$ to be a function that features (positive) peaks near the two hydrophilic/hydrophobic transition regions in a lipid bilayer, while being negative both in the center and further out beyond the transition regions, such that the overall positive and negative areas balance.

Figure 10 shows the function $\sigma_0(z)$ as measured for a particular lipid membrane model (the MARTINI version of DMPC, at a temperature of 300 K). Our overall expectations are met, even though we could not have anticipated all the extra wiggles. What might look extremely surprising, though, is how very large the effective stresses are: hundreds of bars! However, upon second thought, this makes sense: a typical value for the oil-water surface tension is about 50 mN/m (Goebel and Lunkenheimer 1997). Chemistry and Fig. 10 suggest that the transition between the hydrophilic and hydrophobic environment occurs over a region of approximately 1 nm width, and

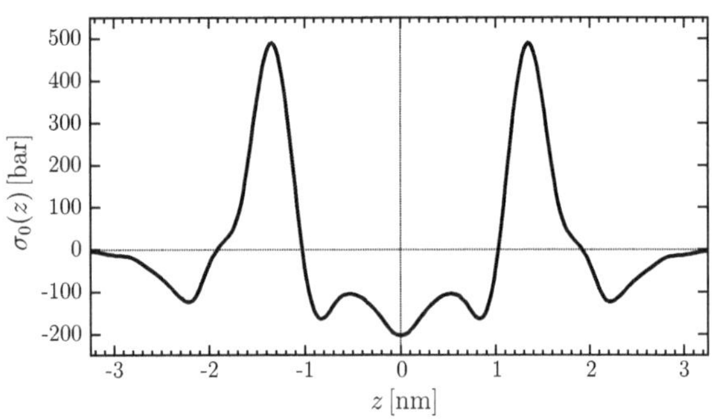

Fig. 10 Lateral stress profile $\sigma_0(z)$ of a lipid bilayer, using a coarse-grained model of the lipid DMPC (MARTINI force field) at 300 K. This profile is based on simulation results presented in (Wang and Deserno 2015)

hence the pressure we would expect at the peak is about

$$\sigma_0(z_{\rm peak}) \sim \frac{50\,{\rm mN/m}}{1\,{\rm nm}} = 500\,{\rm bar}\ . \tag{71}$$

Which is very close to what the simulation finds (fortuitously so, of course, but it is only the order of magnitude that counts).

Armed with the new insight that an in-plane lateral stress $\sigma_0(z)$ exists in a lipid membrane, we should amend the monolayer elastic energy from Eq. (65) with a term that penalizes stretching or compression against that pre-existing stress, which leads to a term that is *linear* in the area strain:

$$e_{3\rm d} = \sigma_0(z(\zeta))\varepsilon(\zeta) + \frac{1}{2}\tilde{E}\,\varepsilon(\zeta)^2 + \frac{1}{2}\lambda_{xzxz}T_l T^l\ . \tag{72}$$

Here we also defined two more concepts:

1. $z(\zeta)$ is the transverse coordinate z of a piece of material in the flat monolayer as a function of its transverse position ζ in the curved monolayer. Since curving leads to local lateral stretching or compression, this impacts the transverse coordinates, because the Poisson ratio generally does not vanish—see Eq. (58). We will soon exploit this to connect z with ζ.
2. $\varepsilon(\zeta)$ is the lateral area strain as a function of the curved transverse coordinate ζ. To first-order in ζ, it is equal to $u_{xx} + u_{yy}$, but at next order it differs. But since this difference takes the form of a lateral shear, which meets no resistance in fluid leaflets, we can ignore it—that's how Hamm and Kozlov (2000) argue. One could also state, though, that the true area strain should linearly couple to the true area stress, and that is why ε should naturally multiply the stress profile σ_0. Of course, the outcome is the same. Also, notice that the difference only matters in the linear (pre-stress) term, because it becomes higher than quadratic order in the already quadratic elastic term.

2.2 Decomposing the Membrane Deformation into Three Stages

It should now become quite evident that the reason curvature will enter our final expression for a surface energy density functional is that bending the leaflet will give rise to positionally varying strains. To describe them, we need to carefully distinguish coordinates in the flat and curved sheet. It turns out that a convenient way of doing this is to decompose the strain by defining an *intermediate state* between the original flat bilayer and the final curved one: a state where lipids have tilted, resulting in a change of thickness and area per lipid of the leaflet that is uniform throughout its width. From there, any further strain is now a function that depends at least linearly on the transverse position and thus describes higher order curvature-induced strains.

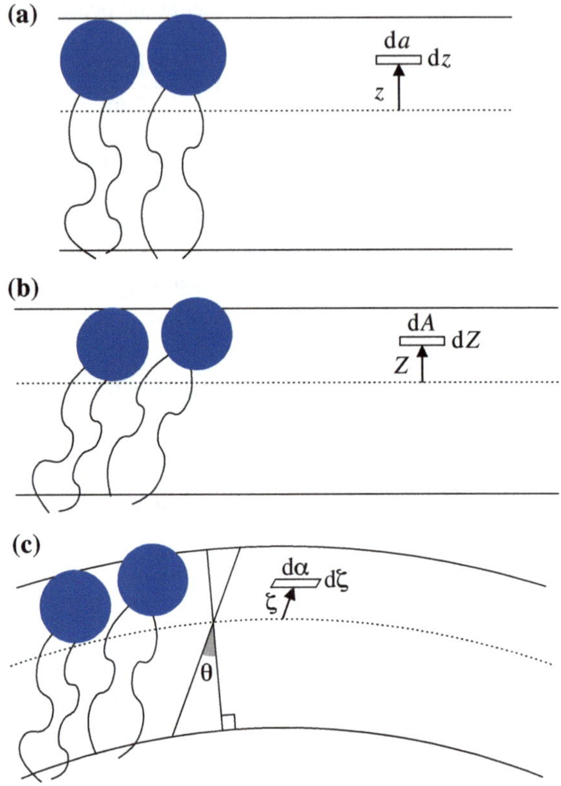

Fig. 11 The flat and untilted monolayer state in (**a**) is first transformed to a flat but tilted state (**b**) in which thickness and area per lipid have changed. From there a subsequent bending deformation, which leaves the area at the pivotal plane invariant, leads to the final curved state (**c**)

Figure 11 illustrates these stages, as well as the notation we will use to describe them: for coordinates or differentials in the initial, intermediate, and final state, we will use lower case roman, upper case roman, and lower case Greek letters, respectively. In particular, local area element and transverse height differential in these three states will be denoted as

$$\begin{aligned} \text{initial, (a):} \quad & \{\mathrm{d}a; \mathrm{d}z\} , \\ \text{intermediate, (b):} \quad & \{\mathrm{d}A; \mathrm{d}Z\} , \\ \text{final, (c):} \quad & \{\mathrm{d}\alpha; \mathrm{d}\zeta\} . \end{aligned}$$

While in the first two states the transverse coordinates z or Z are perpendicular to the area element, this is *not* the case in the final state, for which the coordinate ζ aligns with the local lipid direction. As a consequence, the volume element is *not* simply $\mathrm{d}\alpha\,\mathrm{d}\zeta$ but instead $\mathrm{d}\alpha\,\mathrm{d}\zeta\cos\theta$, where θ is the tilt angle.[2] This will become important below.

[2]During the workshop Jemal Guven pointed out that the additional required factor $\cos\theta$ is the equivalent of what in a $3+1$ foliation treatment of general relativity is called the "Lapse function" (Wheeler 1964).

The area element $d\alpha$ in the curved configuration (c) will generally be different from the area element dA of the intermediate configuration (b): further "outside" it will be stretched, while further "inside" it will be compressed. But there will exist one particular location in the leaflet, called the "pivotal plane," at which the area element is unchanged. We will use this specific location as the *reference surface* for the curved configuration, the transverse location from which ζ will be measured and to which all curvatures shall refer. Hence, we get $d\alpha(\zeta = 0) = dA$, while away from the pivotal plane the changed area element leads to the higher order lateral strain

$$\epsilon_\zeta = \frac{d\alpha - dA}{dA} . \tag{73}$$

Since ϵ_ζ is local, quantifying it requires not only the lateral location on the leaflet, but also the transverse coordinate ζ. In contrast, the strain leading from the initial state (a) to the intermediate state (b) is by construction independent of ζ. We will hence refer to it as the zeroth order strain, which we can express as

$$\epsilon_0 = \frac{dA - da}{da} . \tag{74}$$

Obviously, the *total* area strain upon transitioning from state (a) to state (c) can be expressed through ϵ_0 and ϵ_ζ:

$$\epsilon(\zeta) = \frac{d\alpha - da}{da} = (1 + \epsilon_0)(1 + \epsilon_\zeta) - 1 . \tag{75}$$

Observe that the decomposition through some intermediate state is not unique. Other states could have been chosen, and more than one intermediate state is possible. But since the final state of deformation and its associated elastic energy is indeed a *thermodynamic state*, it does not matter by what specific path it is reached. The particular sequence of strains we have chosen to get from the initial to the final state is motivated by convenience, but our final answer will not depend on it.

2.3 *The Link Between Curvature and Local Area Strain*

Let us henceforth assume that we describe the shape of a curved monolayer via the location of its pivotal plane.[3] Any other surface, displaced from the pivotal plane by some amount, will generally *not* have a vanishing local area strain, and so there will be a local contribution to the elastic energy coming from (i) the local stress–strain

[3]Notice that if the monolayer leaflet were not fluid, a deformation that starts from a flat leaflet and ends up with one that has a non-vanishing Gaussian curvature cannot be isometric by virtue of the *Theorema Egregium*. Hence, this approach of writing the elastic energy by looking at the stretching away from a pivotal plane relies by construction on fluidity.

work when stretching or compressing against the pre-existing stress σ_0 and (ii) the elastic contribution quadratic in the local strain. We need to calculate how big this strain is.

More precisely, if $\boldsymbol{X}$ is a point on the pivotal plane, we arrive at the new shifted point by displacing it a constant distance ζ along the local material direction $\hat{\boldsymbol{d}}$:

$$\boldsymbol{X}' = \boldsymbol{X} + \zeta\,\hat{\boldsymbol{d}}\,. \tag{76}$$

We need to know the area strain on the shifted surface $\boldsymbol{X}'$, which in turn depends on the local displacement direction $\hat{\boldsymbol{d}}$.

Area strain for parallel surfaces. The easiest situation is if $\hat{\boldsymbol{d}} = \hat{\boldsymbol{n}}$, in which case the shifted surface is called a *parallel surface*. To calculate the resulting area strain, we must compare an area element $\mathrm{d}A'$ on the parallel surface with its corresponding area element $\mathrm{d}A$ on the original parent surface. Recall that the tangent vectors on the parent surface are given by $\boldsymbol{e}_i = \nabla_i \boldsymbol{X}$, where ∇_i is the metric-compatible covariant derivative. The area element on the parallel surface is hence

$$\boldsymbol{e}'_i = \nabla_i(\boldsymbol{X} + \zeta\hat{\boldsymbol{n}}) = \boldsymbol{e}_i + \zeta K_i^{\,j}\boldsymbol{e}_j\,, \tag{77}$$

where K_{ij} is the curvature tensor and where we used the Weingarten equation $\nabla_i\hat{\boldsymbol{n}} = K_i^{\,j}\boldsymbol{e}_j$. To get the area element, we need the metric determinant g' on the parallel surface, and that we get from the cross products of the two tangent vectors:

$$\begin{aligned}
\sqrt{g'} = |\boldsymbol{e}'_1 \times \boldsymbol{e}'_2| &= \Big|(\boldsymbol{e}_1 + \zeta K_1^{\,j}\boldsymbol{e}_j) \times (\boldsymbol{e}_2 + \zeta K_2^{\,k}\boldsymbol{e}_k)\Big| \\
&= \Big|\boldsymbol{e}_1 \times \boldsymbol{e}_2 + \zeta(K_1^1 + K_2^2)\,\boldsymbol{e}_1 \times \boldsymbol{e}_2 \\
&\qquad + \zeta^2(K_1^1K_2^2 - K_1^2K_2^1)\,\boldsymbol{e}_1 \times \boldsymbol{e}_2\Big| \\
&= \Big|(1 + K\zeta + K_{\mathrm{G}}\zeta^2)\,(\sqrt{g}\,\hat{\boldsymbol{n}})\Big| \\
&= \sqrt{g}\,(1 + K\zeta + K_{\mathrm{G}}\zeta^2)\,,
\end{aligned} \tag{78}$$

where K and K_{G} are trace and determinant of the curvature tensor K_{ij}.

The maybe slightly unorthodox use of individual components can be avoided by proceeding a little bit more formally. The calculation is a bit longer, but it will turn out to be quite useful when we go beyond this simple case. Define the Levi–Civita symbol $\epsilon_{ij} = \epsilon^{ij}$ such that $\epsilon_{11} = \epsilon_{22} = 0$ and $\epsilon_{12} = -\epsilon_{21} = 1$. Furthermore, define the Levi–Civita *tensor density* $\varepsilon_{ij} = \sqrt{g}\,\epsilon_{ij}$, which also implies $\varepsilon^{ij} = \epsilon^{ij}/\sqrt{g}$. Now, the cross product between the two tangent vectors can be written as

$$
\begin{aligned}
e'_1 \times e'_2 &= \frac{1}{2}\sqrt{g}\,\varepsilon^{ij} e'_i \times e'_j \\
&= \frac{1}{2}\sqrt{g}\,\varepsilon^{ij} (e_i + \zeta K_i^k e_k) \times (e_j + K_j^l e_l) \\
&\overset{*}{=} \frac{1}{2}\sqrt{g}\,\varepsilon^{ij} \left[\varepsilon_{ij} + \zeta(\varepsilon_{il} K_j^l + \varepsilon_{kl} K_i^k) + \zeta^2 \varepsilon_{kl} K_i^k K_j^l)\right] \hat{n} \\
&= \frac{1}{2}\sqrt{g}\left[\varepsilon^{ij}\varepsilon_{ij} + \zeta(\varepsilon^{ij}\varepsilon_{il} K_j^l + \varepsilon^{ij}\varepsilon_{kj} K_i^k) + \zeta^2 \varepsilon^{ij}\varepsilon_{kl} K_i^k K_j^l\right] \hat{n} \,, \qquad (79)
\end{aligned}
$$

where at $*$ we used $e_i \times e_j = \varepsilon_{ij}\hat{n}$. If we now apply the identities $\varepsilon^{ij}\varepsilon_{ij} = 2$, $\varepsilon^{ij}\varepsilon_{ik} = g_k^j = \delta_k^j$ (i.e., the Kronecker-δ), and the definition of the determinant, $\det(K_{ij}) = \frac{1}{2}\varepsilon^{ij}\varepsilon^{kl} K_{ik} K_{jl}$, the last line immediately reproduces Eq. (78) when taking the modulus.

Since $\mathrm{d}A = \sqrt{g}\,\mathrm{d}u^1\mathrm{d}u^2$ and $\mathrm{d}A' = \sqrt{g'}\,\mathrm{d}u^1\mathrm{d}u^2$, we find the area strain

$$
\epsilon_\zeta = \frac{\mathrm{d}A' - \mathrm{d}A}{\mathrm{d}A} = \frac{\sqrt{g'}}{\sqrt{g}} - 1 = K\zeta + K_{\mathrm{G}}\zeta^2 \,. \qquad (80)
$$

This is quite remarkable because it is exact: No corrections beyond quadratic order in ζ occur.

Area strain for more general lipid-shifted surfaces. If the direction of shift, $\hat{d}$, is not along the surface normal but along the lipid orientation, we instead have

$$
\hat{d} = \frac{T^j e_j + \hat{n}}{\sqrt{1 + T_j T^j}} = T^j e_j + \left(1 - \tfrac{1}{2} T_j T^j\right)\hat{n} + \mathcal{O}(|T|^3) \,. \qquad (81)
$$

It is worthwhile to note that we deviate here from Hamm and Kozlov (2000), for these authors do not normalize the orientation vector. Clearly, this only matters at higher order, but the difference does have a physical interpretation. Recall that we want ζ to measure a given distance along a lipid. If we do *not* normalize the lipid director, the displacement $|\zeta d|$ along a tilted lipid is *longer* than for an untilted one, while this distance remains unchanged if we use the normalized director. Which one is correct hence depends on whether lipids stretch upon tilting. Hamm and Kozlov assumed that lipids stretch by laterally shearing, which exactly corresponds to *not* normalizing the orientation vector in the numerator of Eq. (81). However, more recently Kopelevich and Nagle (2015) showed in a simulation study that there is virtually no correlation between a lipid's length and its orientation, suggesting that lipids *rotate* upon tilting. In that case the normalized orientation vector is the more appropriate choice, which leads to the lipid-shifted surface

$$
X' = X + \zeta\left[T^j e_j + \left(1 - \tfrac{1}{2} T_j T^j\right)\hat{n}\right] + \mathcal{O}(T^3) \,. \qquad (82)
$$

The remainder of the calculation follows the one for parallel surfaces, except that the form of $\boldsymbol{X}'$ results in more complex expressions. To begin with, the tangent vectors are

$$\begin{aligned} \boldsymbol{e}'_j &= \nabla_j \left\{ \boldsymbol{X} + \zeta \left[T^j \boldsymbol{e}_j + \left(1 - \tfrac{1}{2} T_j T^j\right) \hat{\boldsymbol{n}} \right] \right\} \\ &= \boldsymbol{e}_j + \zeta \left[\left(\tilde{K}_j{}^k - \tfrac{1}{2} K_j^k T_l T^l \right) \boldsymbol{e}_k - \tilde{K}_{jl} T^l \hat{\boldsymbol{n}} \right] , \end{aligned} \tag{83}$$

where we defined the *effective curvature tensor*

$$\tilde{K}_{ij} := K_{ij} + \nabla_i T_j \ . \tag{84}$$

Warning: $\tilde{K}_{ij}$ is generally *not* a symmetric tensor (unlike K_{ij}), because $\nabla_i T_j \neq \nabla_j T_i$. This means we must be careful when contracting indices, or when raising one of them: $\tilde{K}_i{}^j$ is *not* the same as $\tilde{K}^j{}_i$.

Calculating the cross product of the tangent vectors is now a bit more tedious, but still straightforward. First,

$$\begin{aligned} \boldsymbol{e}'_1 \times \boldsymbol{e}'_2 =& \frac{1}{2} \sqrt{g}\, \varepsilon^{ij} \boldsymbol{e}'_i \times \boldsymbol{e}'_j \\ =& \frac{1}{2} \sqrt{g}\, \varepsilon^{ij} \left\{ \boldsymbol{e}_i + \zeta \left[\left(\tilde{K}_i{}^k - \tfrac{1}{2} K_i^k \boldsymbol{T}^2 \right) \boldsymbol{e}_k - \tilde{K}_{im} T^m \hat{\boldsymbol{n}} \right] \right\} \\ & \times \left\{ \boldsymbol{e}_j + \zeta \left[\left(\tilde{K}_j{}^l - \tfrac{1}{2} K_j^l \boldsymbol{T}^2 \right) \boldsymbol{e}_l - \tilde{K}_{jn} T^n \hat{\boldsymbol{n}} \right] \right\} . \end{aligned} \tag{85}$$

Making use of $\boldsymbol{e}_i \times \boldsymbol{e}_j = \varepsilon_{ij} \, \hat{\boldsymbol{n}}$ as well as $\hat{\boldsymbol{n}} \times \boldsymbol{e}_i = \varepsilon_{ij} \, \boldsymbol{e}^j$, all cross products can again be expressed as Levi–Civita tensor densities. Two of them contract either into a metric or create a determinant. The one case where that does not happen, they form an expression that will not matter up to order ζ^2. We then find

$$\begin{aligned} \boldsymbol{e}'_1 \times \boldsymbol{e}'_2 =& \sqrt{g} \left\{ \hat{\boldsymbol{n}} \left[1 + \zeta (\tilde{K} - \tfrac{1}{2} K \boldsymbol{T}^2) + \zeta^2 (\tilde{K}_{\mathrm{G}} - K_{\mathrm{G}} \boldsymbol{T}^2) \right] \right. \\ & \left. + \boldsymbol{e}^i \left[\zeta \tilde{K}_{im} T^m + \zeta^2 (\text{irrelevant stuff}) \right] \right\} , \end{aligned} \tag{86}$$

where the trace and determinant of the effective curvature tensor are

$$\tilde{K} = \mathrm{Tr}(\tilde{K}_{ij}) = g^{ij} \tilde{K}_{ij} \quad , \quad \tilde{K}_{\mathrm{G}} = \det(\tilde{K}_{ij}) = \tfrac{1}{2} \varepsilon^{ij} \varepsilon^{kl} \tilde{K}_{ik} \tilde{K}_{jl} \ . \tag{87}$$

Moreover, Eq. (86) already exploits the fact that this expansion is only supposed to be accurate up to maximally order $K^2 T^2$. This for instance means that terms like $\tilde{K}^2 T^2$ can be replaced by $K^2 T^2$, since the "extra T" in $\tilde{K}$ would contribute at higher order.

Up to order ζ^2, the square of Eq. (86) is hence given by

$$\frac{|\boldsymbol{e}_1' \times \boldsymbol{e}_2'|^2}{g} = \Big[1 + \zeta(\tilde{K} - \tfrac{1}{2}K\boldsymbol{T}^2) + \zeta^2(\tilde{K}_{\rm G} - K_{\rm G}\boldsymbol{T}^2)\Big]^2 + \zeta^2 K_{im}K^i_n T^m T^n , \quad (88)$$

and so the ratio of metric determinants is

$$\frac{\sqrt{g'}}{\sqrt{g}} = 1 + (\tilde{K} - \tfrac{1}{2}K\boldsymbol{T}^2)\zeta + (\tilde{K}_{\rm G} - K_{\rm G}\boldsymbol{T}^2 + \tfrac{1}{2}K_{im}K^i_n T^m T^n)\zeta^2 . \quad (89)$$

We hence find the following area strain:

$$\epsilon_\zeta = \frac{{\rm d}\alpha - {\rm d}A}{{\rm d}A} = \frac{\sqrt{g'}}{\sqrt{g}} - 1 = \epsilon_1\,\zeta + \epsilon_2\,\zeta^2 , \quad (90)$$

with the first- and second-order contribution

$$\epsilon_1 = \tilde{K} - \tfrac{1}{2}K T_i T^i , \quad (91a)$$
$$\epsilon_2 = \tilde{K}_{\rm G} - \big(K_{\rm G}\, g_{mn} - \tfrac{1}{2}K_{im}K^i_n\big)T^m T^n . \quad (91b)$$

The transverse dimension: Poisson ratio effects. We have just seen how the *lateral area element* changes due to curvature—both with and without accounting for tilt. However, the *transverse length element* will change, too, and the extent to which this happens is dictated by the Poisson ratio. It will affect the zeroth order strain ϵ_0 as well as the connection between the differentials dz, dZ, and dζ, which we have been careful to distinguish.

Let us begin with the zeroth order strain ϵ_0. Following the finding by Kopelevich and Nagle (2015) that lipids *rotate* upon tilting, we must have dZ = dz cos θ, and so the zeroth order transverse strain is

$$u^0_{zz} = \frac{{\rm d}Z - {\rm d}z}{{\rm d}z} = \cos\theta - 1 \overset{*}{=} \frac{1}{\sqrt{1+|\boldsymbol{T}|^2}} - 1 = -\frac{1}{2}T_l T^l + \mathcal{O}(|\boldsymbol{T}|^4) , \quad (92)$$

where at "*" we used $|\boldsymbol{T}| = \tan\theta$.

Furthermore, recall that the normal stresses vanish at the top and bottom of the leaflet. If it is sufficiently thin, this implies that the normal stress vanishes *throughout* the leaflet, since they has not much opportunity to grow appreciably. This relates the transverse and lateral strains (Landau and Lifshitz 1986)

$$-\frac{\nu}{1-\nu}(u^0_{xx} + u^0_{yy}) =: -\tilde{\nu}(u^0_{xx} + u^0_{yy}) = u^0_{zz} \overset{(92)}{=} -\frac{1}{2}T_l T^l , \quad (93)$$

where ν is Poisson's ratio for the present anisotropic material, which in terms of the elastic tensor λ_{ijkl} is

$$\nu = \frac{\lambda_{xxzz}}{\lambda_{xxzz} + \lambda_{zzzz}} \quad \text{or} \quad \tilde{\nu} := \frac{\nu}{1-\nu} = \frac{\lambda_{xxzz}}{\lambda_{zzzz}} . \tag{94}$$

These elastic coefficients could in general depend on their transverse position through the leaflet, but notice that *the area strain cannot*, because we assumed it to arise from a rigid rotation of the lipids. Hence, in Eq. (94) $\tilde{\nu}$ must denote the *average* value across the leaflet.

Finally, since at $\mathcal{O}(\zeta)$ the zeroth order area strain equals the sum of the in-plane diagonal components of the strain tensor, it is found to be

$$\epsilon_0 = \frac{1}{2\tilde{\nu}} T_l T^l . \tag{95}$$

Notice that ν can in principle be zero, in which case $\tilde{\nu}$ would also vanish, seemingly leading to a divergent strain. However, at a vanishing Poisson ratio a lateral surface stress would not lead to a reduction of thickness, and hence lipids cannot in fact tilt. Indeed, Eq. (93) shows that if $\tilde{\nu} = 0$ then the tilt vanishes as well, and hence the area strain remains finite. At any rate, the physically relevant situation for soft fluid leaflets is $\tilde{\nu} \approx 1$, not a vanishing Poisson ratio.

Next, let us look at the connection between the transverse area elements in the initial and final configuration. To begin with, note that the difference in alignment between the Z- and ζ-coordinate again implies that $\mathrm{d}Z = \mathrm{d}\zeta \cos\theta$. Combining this with the usual Poisson relation between lateral and transverse strain *normal to the lateral direction*, we get

$$\frac{\mathrm{d}\zeta \cos\theta - \mathrm{d}Z}{\mathrm{d}Z} = -\tilde{\nu}\epsilon_\zeta . \tag{96}$$

Inserting $\mathrm{d}Z = \cos\theta\,\mathrm{d}z$ from Eq. (92), $\cos\theta$ cancels in the relation between the transverse differentials z and ζ:

$$\mathrm{d}\zeta = \mathrm{d}z\left[1 - \tilde{\nu}\left(\epsilon_1\zeta + \epsilon_2\zeta^2\right)\right] . \tag{97}$$

The expansion coefficients ϵ_1 and ϵ_2 are those given in Eq. (91a) and (91b). This connection constitutes a differential equation for $\zeta(z)$, and it can be solved by a straightforward quadrature. Fortunately, though, we will only need the solution up to order z^2:

$$\zeta(z) = z - \tfrac{1}{2}\tilde{\nu}\epsilon_1 z^2 + \mathcal{O}(z^3) . \tag{98}$$

Just as in the case of the zeroth order area strain, $\tilde{\nu}$ in principle depends on the position within the leaflet, but since we do not know the functional form, we could not in general integrate the differential equation. However, at the order in z that we strive for, all we could and need account for is a linear deviation away from its average value. Since $\tilde{\nu}$ is anyways most likely very close to 1, this extra work seems hardly justified, and in order to keep things simple, we will again just take the average value of the Poisson ratio in Eq. (98).

The volume element Having calculated the lateral and transverse coordinate differentials in the deformed configuration, we can now calculate the volume element in the coordinates we need—which are the transverse position z in the flat untilted state and the area element $\mathrm{d}A$ of the flat tilted state, which by definition is identical to the area element $\mathrm{d}\alpha(\zeta = 0)$ of the curved leaflet at its pivotal plane. To calculate the volume element, we also must recall that the new volume element is generally not orthogonal, since the ζ-direction has an angle θ with respect to the membrane normal, and so we get a projection factor $\cos\theta$. Putting everything together, we find

$$\mathrm{d}V = \mathrm{d}A\,\mathrm{d}z\big[1 - \tfrac{1}{2}T_l T^l\big]\big[1 + (1-\tilde{\nu})\epsilon_\zeta \underline{- \tilde{\nu}\epsilon_\zeta^2}\big]\,, \tag{99}$$

where we use Eq. (73) and the Poisson ratio relation Eq. (96).

Notice that Eq. (99) has one disconcerting feature: in the incompressible limit, $\tilde{\nu} = 1$, the only contribution to area strain should come from tilt (namely, ϵ_0), but here we get another contribution from geometry—the underlined term. This trouble is not specific to our particular problem but more generally reflects the fact that the Poisson ratio is a first-order concept. To see this, consider an area strain ϵ_A and a transverse strain ϵ_z. Together, they result in a volume strain $\epsilon_V = (1+\epsilon_A)(1+\epsilon_z) - 1 = \epsilon_A + \epsilon_z + \epsilon_A\epsilon_z$. And with the usual Poisson ratio connection $\epsilon_z = -\tilde{\nu}\epsilon_A$, we get $\epsilon_V = (1-\tilde{\nu})\epsilon_A - \tilde{\nu}\epsilon_A^2$. The last term does not vanish in the incompressible limit $\tilde{\nu} = 1$, and it is exactly the source of the underlined term in Eq. (99). To avoid this inconsistency, we will drop the underlined quadratic term.

2.4 *From Three Dimensions to Two*

Putting everything together, we then arrive at the following overall elastic energy, which is correct up to order ζ^2, squared curvature, squared tilt, and biquadratic terms:

$$\begin{aligned}
\mathcal{H}_\mathrm{m} = \int \mathrm{d}A\,\mathrm{d}z \Big[1 - \tfrac{1}{2}T_l T^l + (1-\tilde{\nu})\Big(\tilde{K} - K T_l T^l\Big)\zeta\Big] \times \\
\Big\{\sigma_0(z(\zeta))\Big[\ \tfrac{1}{2\tilde{\nu}}T_l T^l + \Big(\tilde{K} + \tfrac{1-\tilde{\nu}}{2\tilde{\nu}}K\ T_l T^l\Big)\zeta \\
+ \Big(\tilde{K}_\mathrm{G} - \tfrac{1}{2}\Big(K_\mathrm{G} g_{ij} - K_{ki}K^k_j\Big)T^i T^j\Big)\zeta^2\Big] \\
+ \tfrac{1}{2}\tilde{E}\Big(\tilde{K}^2 + \tfrac{1}{\tilde{\nu}}K_\mathrm{G} T_l T^l\Big)\zeta^2 \\
+ \tfrac{1}{2\tilde{\nu}}\tilde{E}K\zeta T_l T^l + \tfrac{1}{2}\lambda_{xzxz}T_l T^l\Big\}\,.
\end{aligned} \tag{100}$$

Here we made one further approximations: we dropped biquadratic terms which exhibit an additional factor of $1-\tilde{\nu}$. Since we will invariably be close to the incompressible limit, this multiplies the small biquadratics by yet another smallness parameter, which we will ignore for simplicity.

From this expression we get the *elastic surface energy density* by performing the integral over z, which also requires us to insert the functional dependence $\zeta(z)$ from Eq. (98). Doing this integral, we arrive at the surface energy density

$$e_{2d} = \frac{1}{2}\kappa_{\rm m}(\tilde{K} - K_{0,\rm m})^2 + \overline{\kappa}_{\rm m}\tilde{K}_{\rm G} + \frac{1}{2}\kappa_{\rm t,m}M'_{ij}T^iT^j \ . \tag{101}$$

This expression now features numerous new elastic constants, but all of them have expressions in terms of the underlying elastic model:

$$\kappa_{\rm m} = \int {\rm d}z \left[\tilde{E}(z) - \tilde{\nu}\sigma_0(z)\right] z^2 \ , \tag{102a}$$

$$\overline{\kappa}_{\rm m} = \int {\rm d}z \, \sigma_0(z) \, z^2 \ , \tag{102b}$$

$$\kappa_{\rm t,m} = \int {\rm d}z \, \lambda_{xzxz}(z) \ , \tag{102c}$$

$$\kappa_{{\rm m},\nu} = \int {\rm d}z \, \frac{1}{\tilde{\nu}} \left[\tilde{E}(z) - \tilde{\nu}\sigma_0(z)\right] z^2 \ , \tag{102d}$$

$$-\kappa_{\rm m} K_{0,\rm m} = \int {\rm d}z \, \sigma_0(z) \, z \ , \tag{102e}$$

$$-\kappa_{\rm m} K_{0,\rm t} = \int {\rm d}z \, \lambda_{xzxz}(z) \, (1 - \tilde{\nu}) \, z \ , \tag{102f}$$

$$\kappa_{\rm m} K'_{0,\rm m} = \int {\rm d}z \, \frac{1}{\tilde{\nu}} \left[\tilde{E}(z) + (2 - 3\tilde{\nu})\sigma_0(z)\right] z \ , \tag{102g}$$

The quadratic tilt term in Eq. (101) is not merely characterized by a scalar modulus but instead by a full *tensor*, which has the form

$$M'_{ij} = \left[1 + \ell^2 K \left(K'_{0,\rm m} - K_{0,\rm t}\right) - \ell^2 K^2 + \left(\ell_\nu^2 - r_{\rm m}\ell^2\right) K_{\rm G}\right] g_{ij} + r_{\rm m}\ell^2 K_{ki} K^k_j \ . \tag{103}$$

Here, ℓ is a characteristic length defined from bending and tilt moduli, while the other length scale ℓ_ν is defined via the new modulus $\kappa_{{\rm m},\nu}$:

$$\ell^2 = \frac{\kappa_{\rm m}}{\kappa_{\rm t,m}} \quad , \quad \ell_\nu^2 = \frac{\kappa_{{\rm m},\nu}}{\kappa_{\rm t,m}} \ . \tag{104}$$

Moreover, the dimensionless number $r_{\rm m}$ is given by

$$r_{\rm m} = \frac{\overline{\kappa}_{\rm m}}{\kappa_{\rm m}} \ . \tag{105}$$

In the absence of tilt, the stability of the quadratic curvature expression Eq. (101) requires $-2 \leq r_{\rm m} \leq 0$ (Deserno 2015), and so $r_{\rm m} < 0$.

Observe that in the absence of tilt Eq. (101) simplifies to the Helfrich Hamiltonian. Moreover, if the curvature radii are large compared to the characteristic scales ℓ and ℓ_ν), the tilt tensor approaches the metrix, $M'_{ij} \to g_{ij}$, and the expression for the surface energy density reduces to the original expression by Hamm and Kozlov (2000).

Disentangling Tilt and Curvature in $\tilde{K}_{\rm G}$. The effective curvature $\tilde{K} = K + \nabla_l T^l$ is the sum of the total curvature and the divergence of the tilt. This separates tilt and curvature quite nicely, and shows for instance that the divergence of tilt can be viewed as a position-dependent dynamic spontaneous curvature. Unfortunately, it is not quite so easy to see how we can wrest the tilde from $\tilde{K}_{\rm G}$. But it is possible. To do so, recall the definition

$$\begin{aligned}
\tilde{K}_{\rm G} &= \frac{1}{2}\varepsilon^{ij}\varepsilon^{kl}\tilde{K}_{ik}\tilde{K}_{jl} \qquad (\text{now use } \varepsilon^{ij}\varepsilon^{kl} = g^{ik}g^{jl} - g^{il}g^{jk}) \\
&= \frac{1}{2}\Big(\tilde{K}^2 - \tilde{K}_i{}^k\tilde{K}_k{}^i\Big) , \\
&= \frac{1}{2}\Big[\Big(K_i^i + \nabla_i T^i\Big)\Big(K_j^j + \nabla_j T^j\Big) - \Big(K_i^j + \nabla_i T^j\Big)\Big(K_j^i + \nabla_j T^i\Big)\Big] \\
&= \frac{1}{2}\Big[K^2 - K_i^j K_j^i + 2\Big(K\nabla_i T^i - K_i^j \nabla_j T^i\Big) + \nabla_i T^i \nabla_j T^j - \nabla_i T^j \nabla_j T^i\Big] \\
&= K_{\rm G} + \Big(K\nabla_i T^i - K_i^j\nabla_j T^i\Big) + \frac{1}{2}\Big(\nabla_i T^i\nabla_j T^j - \nabla_i T^j\nabla_j T^i\Big) . \qquad (106)
\end{aligned}$$

As the next step, recall that the above expression occurs under an integral. We aim to integrate the second and third parenthesis by parts, which means "swapping one derivative and one sign," as well as getting one boundary term. Doing so, we find

$$\tilde{K}_{\rm G} = K_{\rm G} + \Big(\nabla_i K_j^i - \nabla_j K\Big)T^j + \frac{1}{2}T^i\Big(\nabla_j\nabla_i - \nabla_i\nabla_j\Big)T^j + \nabla_i B^i , \qquad (107)$$

where the last term is the total divergence of

$$B^i = KT^i - K_j^i T^j + \frac{1}{2}\Big(T^i\nabla_j T^j - T^j\nabla_j T^i\Big) . \qquad (108)$$

Now notice that the expression in the first parenthesis of Eq. (107) vanishes due to the contracted Codazzi–Mainardi equation. The expression in the second parenthesis is more interesting: this is the commutator of covariant derivatives, and as is well know, *it does not vanish in curved geometries*. Instead, we have

$$[\nabla_a, \nabla_b]V_c = R_{abcd}V^d , \qquad (109)$$

where R_{abcd} is the Riemann tensor. In the present case we hence find

$$T^i[\nabla_j, \nabla_i]T^j = g^{jk}T^i[\nabla_j, \nabla_i]T_k = g^{jk}T^i R_{jikl}T^l = R_{il}T^iT^l = K_G \boldsymbol{T}^2 , \quad (110)$$

where $R_{il} = g^{jk}R_{jikl}$ is the Ricci tensor, which in two dimensions is simply given by $R_{il} = K_G\, g_{il}$. This shows that—up to a boundary term—we can disentangle the tilt from the effective Gaussian curvature, finding

$$\tilde{K}_G = K_G + \frac{1}{2}K_G\boldsymbol{T}^2 . \quad (111)$$

As it turns out, the boundary term is in many cases irrelevant. Notice that we are writing down a theory for a monolayer. If this monolayer is part of a closed vesicle, it has no boundary. But even if we have a bilayer membrane with an open edge or a pore, the *monolayer* is continuous and boundary-free, since it wraps around the edges. A case where we cannot ignore the edge hence needs to actually provide an edge. One way in which this could happen is if a membrane contains transmembrane proteins, which locally provide an end to the monolayer. Now the boundary term will matter, but we will not look at this case here.

Observe what the disentanglement (111) does to our energy density from Eq. (101): removing the tilde from $\tilde{K}_G$ creates the new term $\frac{1}{2}\overline{\kappa}_m K_G \boldsymbol{T}^2$, which we can incorporate into the effective tilt modulus tensor of Eq. (103), where it cancels the K_G part in its isotropic contribution.

The elastic parameters. The two-dimensional elastic functional (101) contains seven new parameters, and the set of Eq. (102) shows how they depend on the underlying elastic tensor $\lambda_{ijkl}(z)$ and the pre-stress $\sigma_0(z)$. Of these parameters, κ_m, $\overline{\kappa}_m$, $\kappa_{t,m}$, and $K_{0,m}$ already appear in the treatment by Hamm and Kozlov (2000), and in fact are given by the same microscopic expressions (if we specialize to the incompressible limit $\tilde{\nu} = 1$). On the other hand, the three parameters $\kappa_{m,\nu}$, $K_{0,t}$, and $K'_{0,m}$ are new. They are related to the novel biquadratic terms, and in order to judge their relevance, we need to estimate their magnitude. To keep things simple, we will assume that the elastic tensor λ_{ijkl} is in fact constant throughout the leaflet, for this allows us to evaluate the moment-integrals analytically.

Let us start with the inverse length $K_{0,t}$ from Eq. (102f), the form of which mimics the spontaneous curvature term $K_{0,m}$. In contrast to the latter, however, $K_{0,t}$ is usually negligible. To see this, consider the following:

$$\begin{aligned} -\kappa_m K_{0,t} &= \int_0^{d_m} dz\, \lambda_{xzxz}(z)\,(1-\tilde{\nu})\,(z-z_0) \\ &\approx \lambda_{xzxz}\,(1-\tilde{\nu})\int_0^{d_m} dz\,(z-z_0) \\ &= \lambda_{xzxz}\,(1-\tilde{\nu})\frac{d_m^2 - 2d_m z_0}{2} \\ &\approx \frac{1}{2}\kappa_{t,m}(1-\tilde{\nu})(d_m - 2z_0)\,, \end{aligned} \quad (112)$$

where $d_{\rm m}$ is the monolayer thickness and where we explicitly centered the transbilayer integral around z_0. We also used Eq. (102c) to rewrite $\kappa_{\rm t,m} = \lambda_{xzxz} d_{\rm m}$ in the constant-λ-approximation. Dividing out $\kappa_{\rm m}$ and using Eq. (104), we get

$$K_{0,\rm t} \approx \frac{1-\tilde{\nu}}{2\ell^2}(2z_0 - d_{\rm m}) \ . \tag{113}$$

This is only nonzero if (i) the monolayer is compressible and (ii) its pivotal plane is not in the center of the leaflet. In a recent simulation study Wang and Deserno (2016) found $z_0 = 1.32\,\rm nm$ for a united-atom model of the lipid DMPC (Berger et al. 1997; Lindahl and Edholm 2000). Taking the monolayer thickness to be the distance between the bilayer midplane and the position of the phosphate atom (as a proxy for the Luzzati plane), the authors find $d_{\rm m} = 1.80\,\rm nm$. Also using the value $\ell \approx 1.61\,\rm nm$ determined in the same paper, we arrive at $K_{0,\rm t} \approx 0.16(1-\tilde{\nu})\,\rm nm^{-1}$, or a corresponding curvature radius of $1/K_{0,\rm t} \approx 6/(1-\tilde{\nu})\,\rm nm$. In practice we do not expect any strong deviation from incompressibility, and even if we assume $\nu \approx 0.45$, we still find $1/K_{0,\rm t} \approx 33\,\rm nm$, *much* larger than any of the other microscopic length scales (such as $d_{\rm m}$, z_0, or ℓ). It is hence a very good approximation to neglect the $K_{0,\rm t}$ term altogether.

The other two new terms contain the Poisson ratio in a way that leaves their incompressible limit finite, and for the sake of estimating magnitudes, we will hence set $\tilde{\nu} = 1$. This immediately shows that $\kappa_{\rm m,\nu} = \kappa_{\rm m}$ and hence also $\ell_\nu = \ell$. The final expression, $\kappa_{\rm m} K'_{0,\rm m}$ is then found to be the first moment of $\tilde{E}(z) - \sigma_0(z)$. With the approximation $\tilde{E}(z) = \tilde{E} = $ const., we then find

$$\kappa_{\rm m} K'_{0,\rm m} \approx \int {\rm d}z\,\left[\tilde{E}(z) - \sigma_0(z)\right] z \approx \tilde{E}\int_0^{d_{\rm m}} {\rm d}z\,(z - z_0) + \kappa_{\rm m} K_{0,\rm m} \ ,$$

where we again explicitly centered the z-integral. We hence find

$$K'_{0,\rm m} = K_{0,\rm m} - \frac{\tilde{E} d_{\rm m}}{2\kappa_{\rm m}}(2z_0 - d_{\rm m}) \ . \tag{114}$$

If the pivotal plane is in the middle of the leaflet, then $K'_{0,\rm m} = K_{0,\rm m}$. However, usually the pivotal plane of a lipid monolayer is located closer to the headgroup region, often about $\frac{2}{3}$ up along the lipid. Using this rule of thumb, we get

$$K'_{0,\rm m} \approx K_{0,\rm m} - \frac{\tilde{E} d_{\rm m}^2}{6\kappa_{\rm m}} \qquad (\text{if } z_0 = \tfrac{2}{3} d_{\rm m}) \ . \tag{115}$$

If we now apply the constant-$\tilde{E}$-approximation also to Eq. (102a), we get

$$\kappa_{\rm m} \approx \tilde{E}\int_0^{d_{\rm m}} {\rm d}z (z - z_0)^2 - \overline{\kappa}_{\rm m} = \frac{1}{3}\tilde{E} d_{\rm m}(d_{\rm m}^2 - 3 d_{\rm m} z_0 + 3z_0^2) - \overline{\kappa}_{\rm m} \ . \tag{116}$$

And if we again specialize to the good guess $z_0 = \frac{2}{3}d_{\text{m}}$, we find

$$\frac{\tilde{E}d_{\text{m}}^3}{9} = \kappa_{\text{m}} + \overline{\kappa}_{\text{m}} \qquad (\text{if } z_0 = \tfrac{2}{3}d_{\text{m}}) \,, \tag{117}$$

which together with Eq. (115) leads to

$$K'_{0,\text{m}} \approx K_{0,\text{m}} - \frac{1 + r_{\text{m}}}{z_0} \qquad (\text{if } z_0 = \tfrac{2}{3}d_{\text{m}}) \,. \tag{118}$$

This expression is quite curious, because the "correction" part $(1 + r_{\text{m}})/z_0$ can be anything between zero and very large. It vanishes for $r_{\text{m}} = -1$, which is a perfectly permissible value for the Gaussian elastic ratio. On the other hand, it is equally possible that r_{m} is somewhere between -1 and 0, say $-\frac{1}{2}$, in which case the additional term is $-1/2z_0$, and this is a very strong spontaneous curvature. Recall that Wang and Deserno (2016) found $z_0 = 1.32$ nm for a united-atom model of DMPC, which gives $-1/2z_0 \approx -0.38\,\text{nm}^{-1}$, much larger (in magnitude) than typical lipid spontaneous curvatures. For comparison, the conventional spontaneous curvature $K_{0,\text{m}}$ for DMPC is about $0.025\,\text{nm}^{-1}$ (Venable et al. 2015), and lysophosphatidylcholine, one of the most strongly positively curved lipids, has a spontaneous curvature radius of about $0.26\,\text{nm}^{-1}$ (Kooijman et al. 2005). The reason why such a potentially large $K'_{0,\text{m}}$ does not majorly affect bilayer stability and morphology is that it does not directly enter the bending term—only the ordinary spontaneous curvature $K_{0,\text{m}}$ does.

Putting things together. We can finally write down a (slightly approximated) version of the surface energy functional, in which we ignore $K_{0,\text{t}}$, identify $\kappa_{\text{m},\nu} = \kappa_{\text{m}}$, wrest the tilde from the Gaussian curvature, and also disentangle the term $K_{ki}K^k_j$ by virtue of the once-contracted Gauss equation $K_{ki}K^k_j = KK_{ij} - K_{\text{G}}g_{ij}$:

$$e_{\text{2d}} = \frac{1}{2}\kappa_{\text{m}}(K + \nabla_i T^i - K_{0,\text{m}})^2 + \overline{\kappa}_{\text{m}}K_{\text{G}} + \frac{1}{2}\kappa_{\text{t,m}}M_{ij}T^iT^j \tag{119}$$

with

$$M_{ij} = \Big[1 + \ell^2\big(KK'_{0,\text{m}} - K^2 + (1 - r_{\text{m}})K_{\text{G}}\big)\Big]g_{ij} + r_{\text{m}}\ell^2 KK_{ij} \,. \tag{120}$$

2.5 *Some Consequences of the Curvature-Tilt Functional*

As stated before, the theory presented here follows the lead of Hamm and Kozlov (2000), but it retains some of the higher order terms which they have neglected—specifically biquadratic terms such as $K_{\text{G}}T^2$, which is quadratic in both curvature and tilt. Hamm and Kozlov eliminate such terms in their treatment whenever they explicitly occur, on account of them being higher order than the usual quadratic terms. And yet, the tilde over K_G, which they do not ignore, is effectively a biquadratic term.

To be consistent, two paths are possible. The simple one is to eliminate all biquadratics, *including* the tilde over K_G. The perhaps more interesting one is to keep them all, because they are responsible for some fascinating new physics. However, one could object against this on the ground that if we keep biquadratic terms, we should also keep quartic ones, such as K^4, K_G^2, or $T^2(\nabla_k T^k)^2$. This is, in principle, a valid concern. However, there are good *pragmatic* reasons for working with a theory that drops these terms, despite the issue of a consistent order termination: the biquadratic terms create *qualitative* changes in the curvature-tilt theory, because they introduce a *new mode of coupling* between curvature and tilt that is absent on the quadratic level. In consequence, they spawn "new physics"—as we will soon see. The same cannot be said for the quartic terms, which (at least initially) only *quantitatively* change the physics, for instance by affecting the curvature energy and hence changing equilibrium shapes, while only *indirectly* affecting the partnering field. Of course, ultimately we would need all terms for truly quantitative predictions, but it is easier to investigate how a novel curvature-tilt coupling affects the basic physics without simultaneously having to deal with all other conceivable nonlinearities on the non-coupled side of the energy functional. We hence learn, what new physics is in store, and so we can create hypotheses worthy of testing with more refined approaches. Incidentally, it is of note that the geometric transformations we have discussed above indeed create terms quartic in curvature, but they do not create purely quartic tilt terms.

As anticipated when we started, the new two-dimensional surface functional comes with a number of coupling coefficients in front of terms that are permitted by symmetry, but *the underlying elastic theory predicts their values in terms of the underlying parameters*, such as λ_{ijkl} or $\sigma_0(z)$. Crucially, this is not only true for the "classical" parameters which Hamm and Kozlov (2000) already wrote down, but also for all higher order terms. This means that any ad hoc extension of their original functional by terms such as $K_G T^2$ would likely miss the fact that the corresponding prefactors are not new coefficients but related to the existing ones, such as $\overline{\kappa}_{\mathrm{m}}$.

An important general finding is that all biquadratic terms act as position-dependent contributions to the tilt modulus. This is mathematically obvious, but then, it would be equally conceivable to have them enter as position-dependent contributions to the bending modulus. After all, the following (simplified toy) expressions are perfectly equivalent:

$$\frac{1}{2}\kappa_{\mathrm{m}}K^2 + \frac{1}{2}\kappa_{\mathrm{t,m}}\left[1 + \frac{A}{\kappa_{\mathrm{t,m}}}K^2\right]T^2 = \frac{1}{2}\kappa_{\mathrm{m}}\left[1 + \frac{A}{\kappa_{\mathrm{m}}}T^2\right]K^2 + \frac{1}{2}\kappa_{\mathrm{t,m}}T^2 \,. \quad (121)$$

But while equivalent, from a *practical* point of view the notion that the tilt modulus gets modified is more useful. To begin with, at sufficiently large scale tilt becomes irrelevant,[4] And hence, bending is all there is. It then makes sense to solve the problem iteratively by starting with the shape solution in the absence of tilt, and then take this to calculate the tilt field *at a given shape background*. Moreover, there are interesting cases where the shape is given and need not really be solved for, such as when we ask what the tilt field is at the edge of a membrane or within a small pore, where a monolayer tightly curves around to connect the two individual leaflets. In this case, again, it makes sense to solve for the tilt field in the presence of a shape, but not the other way around. Of course, should there ever be a situation where the opposite point of view is more useful, it is trivial to rewrite our equations to reflect this shift in philosophy.

Observe that the biquadratic terms do not merely amend the tilt modulus in a local curvature-dependent way; they amend it in an *anisotropic* way, because M_{ij} is not merely proportional to g_{ij}: the last term in Eq. (120) involves the curvature *tensor* K_{ij}. As a consequence, the eigenvectors of M_{ij} coincide with those of K_{ij}, and so the principal curvature directions of the surface also play a special role for *tilting*. To make this more explicit, assume that $\boldsymbol{p} = p^i \boldsymbol{e}_i$ and $\boldsymbol{q} = q^i \boldsymbol{e}_i$ are the two principal directions of K_{ij} (at some local point), so that we can write it as $K_{ij} = K_p p_i p_j + K_q q_i q_j$, where K_p and K_q are the principal curvatures. The anisotropic term in the tilt energy density can hence be written as

$$\begin{aligned} \frac{1}{2}\overline{\kappa}_{\mathrm{m}} K K_{ij} T^i T^j &= \frac{1}{2}\overline{\kappa}_{\mathrm{m}} K \Big[K_p\, p_i p_j + K_q\, q_i q_j \Big] T^i T^j \\ &= \frac{1}{2}\overline{\kappa}_{\mathrm{m}} K \Big[K_p\, T_p^2 + K_q\, T_q^2 \Big] , \end{aligned} \tag{122}$$

where $T_p = T^i p_i = \boldsymbol{T} \cdot \boldsymbol{p}$ is the p-component of the tilt field, and T_q is the q-component. For instance, imagine a straight membrane edge, where the p-direction points "around" the edge, and the q-direction points along the edge. In that case, $K_q = 0$ and $K_p \approx 1/z_0$, giving the contribution

$$\text{straight edge:} \quad \frac{1}{2}\overline{\kappa}_{\mathrm{m}} K K_{ij} T^i T^j = \frac{1}{2}\overline{\kappa}_{\mathrm{m}} \frac{1}{z_0^2} T_p^2 \tag{123}$$

This term leaves any tilt *along* the edge unaffected, but it *lowers* the cost for tilting *around* the edge—since $\overline{\kappa}_{\mathrm{m}} < 0$. In fact, it is easy to see that the full edge tilt energy density is given by

$$e_{\text{2d,edge}} = \frac{1}{2}\kappa_{\text{t,m}} \left[1 + \frac{\ell^2}{z_0^2} \Big(K'_{0,\mathrm{m}} z_0 - 1 \Big) \right] T^2 + \frac{1}{2}\overline{\kappa}_{\mathrm{m}} \frac{1}{z_0^2} T_p^2 \,. \tag{124}$$

---

[4]This is merely a consequence of the fact that the length ℓ, which pits curvature against tilt, is microscopic. On scales larger than ℓ, tilt therefore only enters as a minor correction to the overall bending physics of then problem.

However, there is something quite disconcerting about this expression: the ratio ℓ^2/z_0^2 can be bigger than 1. In fact, taking the numbers which Wang and Deserno (2016) found for DMPC ($\ell = 1.61$ nm and $z_0 = 1.32$ nm) we get $\ell^2/z_0^2 \approx 1.5$. Now, $r_{\rm m} < 0$, and $K'_{0,\rm m}z_0$ is generally very negative—see Eq. (118). We hence must conclude that for curvatures as large as the ones we encounter at an open edge, the effective tilt energy density is negative, and this could in principle drive the tilt to grow beyond all bounds. For the tilt *around* the edge this cannot happen in practice over the short region of the edge, since the tilt divergence term in Eq. (119) prevents the tilt from changing too rapidly. But notice that Eq. (124) shows that the effective tilt modulus *along* the edge can also become negative,[5] and in that case the finite-region-argument does not save us. Hence, it truly *is* worrisome that the functional can cease to be bounded below. This, of course, is a direct consequence of us having neglected quartic terms, which would have to stabilize it (since the microscopic theory we started out with is clearly bounded below). We thereby have encountered a case where we are pushing our theory to its limits. But we also discover remarkable physics that is hidden at that border, for even if we catch the divergence by a quartic term, we have now run into a phase transition, and so it is conceivable that *strongly curved regions create spontaneous tilt*. A more refined theory is necessary to probe this, but even without such a better theory, the "circumstantial evidence" that exciting things can happen in highly curved regions might motivate us to look for them in experiments or simulations.

Clearly, the anisotropic term in M_{ij} vanishes if $K = 0$, meaning that *on minimal surfaces the tilt modulus is always isotropic*. This is curious, because minimal surfaces are anything but isotropic. The other possibility for M_{ij} being isotropic is if the curvature tensor is locally proportional to the metric, $K_{ij} = c\, g_{ij}$ with some (possible position dependent) function $c(u^1, u^2)$. What do such surfaces look like? Inserting this special form of K_{ij} in the contracted Gauss–Codazzi equation, we find

$$0 = \nabla_j K - \nabla_i K^i_j = \nabla_j(2\,c) - \nabla_i(c\, g^i_j) = 2\nabla_j c - \nabla_j c = \nabla_j c\,, \qquad (125)$$

and hence $c = \text{const}$. We then have $K_{ij} = c\, g_{ij}$ with a *constant* prefactor c. Such surfaces are spheres (do Carmo 1976), and so the resulting isotropy of M_{ij} is much less mysterious.

3 Measuring the Bending Modulus

In the previous section we have derived a curvature-tilt functional, following the original treatment of Hamm and Kozlov (2000). The functional form of many terms in that theory is often highly intuitive, in the sense that we could have confidently predicted that these terms would show up; but there is of course nothing intuitive

[5] Since $T^2 = T_p^2 + T_q^2$, the tilt modulus along the edge (the q-direction) is just the prefactor of T^2 in Eq. (124).

about their *prefactors*. Revisiting the bottom-up and top-down philosophies discussed at the beginning of Sect. 2, we now have two choices: either we derive the resulting moduli from the lower level theory, or we need to determine them on the level of the larger scale theory.

In the present case, the lower level theory was built on the notion of a pre-stressed thin fluid elastic sheet, quantified by the elastic modulus tensor λ_{ijkl} and the stress profile $\sigma_0(z)$, and we know from Eqs. (102) how the parameters of the curvature-tilt functional relate to the lower level input. However, we have not yet addressed the question where we would get λ_{ijkl} and $\sigma_0(z)$ from. Again, we have two choices here. One of them is that there could be an even lower level theory that predicts these objects, based on even more fundamental parameters. And yet, the reader might be wondering whether we are merely begging the question, for where would *these* parameters come from? An *even* lower level model? And where would *its* parameters come from? What saves us from an infinite regress? The answer is, usually, that at some point we *declare* that we know the theory and the parameters. We state that this is the most fundamental level we care about, and that on this level we happen to have a theory that we trust. For instance, we could state that the lowest level we care about is atomistic chemistry (meaning, we ignore nuclei, quarks, strings, ...), and that we are maybe even willing to trust the force fields of classical molecular dynamics to be applicable to this problem. Being poor calculators of such complex systems, we then most likely think hard what type of simulation would give us, say, a modulus tensor, and then we run such a simulation and "measure" that tensor.

The other choice is to forgo the hope of predicting the elastic modulus tensor λ_{ijkl} and the stress profile $\sigma_0(z)$ from some underlying theory and instead measure them in experiment. Once we have them, we can then plug the results into Eqs. (102) and derive the curvature-tilt parameters, such as the bending modulus.

Thinking about the second option, the following question might stir: why not measure the parameters of the curvature-tilt theory *directly*? Why should we even take the detour over the lower level theory? Why not cut out the middle man?

The question is serious. After all, we have just noted that the *form* of the terms in the higher level theory is often very clear: symmetry principles usually go a long way in telling us which terms can or cannot appear. Hence, their presence in a theory rests on something stronger and more fundamental than the particular lower level model we have chosen to construct. Stated differently: if the higher level theory can be phrased completely in terms of observables that *emerge* on that higher level, the specific details for *how* that emergence happens need not concern us in order to have a perfectly workable theory on that level. We do not *have* to dig down into the details. But if we *do* care about emergence, then a key worry might be whether we got the underlying model right. It is then a good idea to measure things on *both* levels, followed by a series of tests that scrutinize the putative connections between the two theories. Or we might at least test whether connections predicted entirely on the level of emergent quantities, which are consequences of the model, turn out to be satisfied. If they hold up, the underlying model is promising. If they fail, it is (likely) wrong.

Notice that we are merely retracing the thoughts of the beginning of Sect. 2, with the specific issue of model parameters in mind. We hope the reader will not consider them trite, because in this interplay between tiers of modeling lies a core element of science and epistemology.

The purpose, then, of this last section is to discuss, how some of the parameters entering the theories discussed so far can be measured. As we have argued above, determining parameters on two different levels, and then checking whether they connect according to some proposed model of emergence, is the probably most thorough way to probe nature. However, this is a rather extensive endeavor, and it would warrant a book on its own, just dealing with the special case of lipid membranes. We will hence restrict to make some comments about measuring *one* parameter, one which happens to live at the emergent level, and illustrate the maybe unexpected richness of problems and opportunities that arise even in this narrow corner. Specifically, we will discuss using *simulations* to find a parameter. Purists do not call this "measurement," and they are strictly right: we don't query nature, we merely query a theoretical model invented to represent nature. In some sense, we use computers to solve a problem we are yet incapable to tackle analytically, but whose answer follows inevitably from that model. And there it is again: tier-bridging.

Let us hence ask: how can we find the value of a membrane's bending modulus κ in a simulation?

3.1 Active Versus Passive Strategies

When measuring spring constants, there are two conceptually different things one could do. First, one could simply deform the spring and monitor, how much force is required for a given deformation. But if the spring is very soft, this requires measuring very small forces. In fact, the spring could be *so* soft that thermal fluctuations all by themselves already deform the spring. In that case we not only have to measure a presumably *very* tiny deformation force; we would also have to figure out how to correct for the effects of thermal noise. However, there is an opportunity here: if fluctuations *alone* deform the spring, maybe this suffices as a deformation? After all, we know the strength of thermal fluctuations, and if we can measure the spring's stochastic response, we ought to be able to back out its stiffness.

This second approach—measuring fluctuations to infer rigidities—is very popular in many fields of soft matter physics. The reason is that soft matter has (almost by definition) small spring constants (read now: moduli), which can be inferred by the way they pit themselves against the thermal breeze. Lipid membranes are a good example for this, and we will begin with a discussion for how this connection works—before concluding that we can do better by actively deforming a membrane.

Membrane undulation spectrum. Consider a flat membrane patch of area $L \times L$, and imagine it being subject to periodic boundary conditions. This is not only theoretically convenient; it is the most natural choice in simulations. Even if the membrane is *on average* flat, thermal fluctuations will roughen it up by adding stochastic undulations of the shape. However, these will be small, and so we can likely get away with a parametrization of the membrane that describes the geometry as a quadratic-level deviation from flatness: linear Monge gauge.

As a brief reminder: in Monge gauge, a membrane's shape is described by a height function $h(\boldsymbol{r})$ above a flat (horizontal) reference plane, with $\boldsymbol{r}$ being the position within that plane. Let ∇ be the gradient operator in that base plane. If $|\nabla h| \ll 1$, the expressions for area element and curvatures simplify significantly, and can be written as (Deserno 2015)

$$\mathrm{d}A = \sqrt{1+(\nabla h)^2}\,\mathrm{d}^2r \;\approx\; \left(1+\tfrac{1}{2}(\nabla h)^2\right)\mathrm{d}^2r\;, \tag{126a}$$

$$K = -\nabla\cdot\left(\frac{\nabla h}{\sqrt{1+(\nabla h)^2}}\right) \;\approx\; -\nabla^2 h\;, \tag{126b}$$

$$K_{\mathrm{G}} = \frac{\det(\partial_i\partial_j h)}{(1+(\nabla h)^2)^2} \approx \det(\partial_i\partial_j h)\;. \tag{126c}$$

Hence, the membrane Hamiltonian (including bending, but ignoring both spontaneous curvature and tilt for now, and adding a membrane tension σ) can be written as

$$E = \int \mathrm{d}A\left\{\frac{1}{2}\kappa K^2 + \overline{\kappa}K_{\mathrm{G}} + \sigma\right\} \tag{127a}$$

$$\approx \int \mathrm{d}^2r\left\{\frac{1}{2}\kappa(\nabla^2 h)^2 + \frac{1}{2}\sigma(\nabla h)^2\right\} + \text{const.} \tag{127b}$$

where we eliminated the Gaussian term, because it vanishes under periodic boundary conditions—courtesy of the Gauss–Bonnet theorem.

The resulting Hamiltonian in Eq. (127b) is quadratic, but it contains gradients and Laplacians. These can be removed by going into Fourier space (since Fourier modes are the eigenfunctions of the gradient operator). Hence, let us Fourier expand the shape $h(\boldsymbol{r})$ according to

$$h(\boldsymbol{r}) = \sum_{\boldsymbol{q}} \tilde{h}_{\boldsymbol{q}} \mathrm{e}^{\mathrm{i}\boldsymbol{q}\cdot\boldsymbol{r}} \quad \text{with } \boldsymbol{q} = \frac{2\pi}{L}\begin{pmatrix} n_x \\ n_y \end{pmatrix} \quad \text{and } n_x, n_y \in \mathbb{N}\,. \tag{128}$$

Since we want this expansion to be real, we must require of the Fourier coefficients that $\tilde{h}_{-\boldsymbol{q}} = \tilde{h}^*_{\boldsymbol{q}}$. Inserting this expansion into the quadratic Hamiltonian (127b), we find

$$E = \frac{1}{2}\int \mathrm{d}^2r \left\{ \sum_{\boldsymbol{q},\boldsymbol{q}'} \tilde{h}_{\boldsymbol{q}}\tilde{h}_{\boldsymbol{q}'}\Big[\kappa(-q^2)(-q'^2) + \sigma(\mathrm{i}q)(\mathrm{i}q')\Big]\mathrm{e}^{\mathrm{i}(\boldsymbol{q}+\boldsymbol{q}')\cdot\boldsymbol{r}}\right\}$$

$$= \frac{1}{2} \sum_{q,q'} \tilde{h}_q \tilde{h}_{q'} \left(\kappa q^2 q'^2 - \sigma q q'\right) \underbrace{\int \mathrm{d}^2 r \; \mathrm{e}^{\mathrm{i}(q+q')\cdot r}}_{=L^2 \delta_{q,-q'}}$$

$$= \frac{1}{2} L^2 \sum_q \left|\tilde{h}_q\right|^2 \left(\kappa q^4 + \sigma q^2\right) . \tag{129}$$

This final form shows that if the membrane shape is expressed using the Fourier components $\tilde{h}_q$ as degrees of freedom, then the Hamiltonian is not merely quadratic but *diagonal*—all degrees of freedom are independent. From the equipartition theorem we then immediately find that the mean squared amplitude of every Fourier mode is given by

$$\left\langle \left|\tilde{h}_q\right|^2 \right\rangle = \frac{k_B T}{L^2(\kappa q^4 + \sigma q^2)} . \tag{130}$$

This formula, and variants of it, underly a vast number of methods and papers for measuring the bending modulus κ—both in simulation and, in fact, experiment. The basic idea is that if we can access the fluctuation spectrum, we can fit to this equation and extract κ.

But let's now investigate how much of a membrane deformation we are talking about. First, notice that the bending rigidity will of course *reduce* the fluctuations—as will the tension. To get the biggest effect, let us imagine that we set the tension to zero.[6] Since the (root mean square) curvature will scale like $q^2|\tilde{h}_q|$, the typical (root mean square) radius of curvature $\overline{R}_q$ of any given Fourier mode q is going to be

$$\overline{R}_q \sim \frac{1}{\sqrt{\langle K^2 \rangle}} \sim \frac{1}{\sqrt{\langle (q^2|\tilde{h}_q|)^2 \rangle}} \stackrel{\sigma=0}{=} L\sqrt{\frac{\kappa}{k_B T}} . \tag{131}$$

Since for a typical bilayer membrane we have $\kappa \sim 10\ldots50\,k_B T$, we find $\overline{R}_q \sim 3\ldots7\,L$, showing that—independently of mode—the radius of curvature is several times bigger than the size of the bilayer. These are very weak curvatures! Not only are they hard to pick up in a simulation,[7] they are also much smaller than many curvatures we are likely to later impose on membranes (say, when we simulate vesicles), raising the question whether at much larger curvatures the quadratic theory assumed in Eq. (127a) actually holds.

The reason this happens is that the rigidity κ is actually not really small compared to thermal energy $k_B T$. It is comfortably larger than thermal energy, ensuring that membranes do not fluctuate themselves into bits and pieces, and so while flickering of membranes is readily observed, it is still a small effect.

---

[6] In simulations that can be achieved relatively easily by suitable boundary conditions.

[7] Recall that we must measure the undulation spectrum $\langle|\tilde{h}_q|^2\rangle$ over a sufficiently wide q-range to plausibly fit a spectrum, and that this spectrum decays rapidly with q.

Force along a cylindrical membrane tube. The observations from the previous section suggest that we could instead look at an actively imposed deformation a membrane and measure the force required to impose it. Several years ago, Harmandaris and Deserno (2006) have proposed to study a cylindrical membrane tube (connected through periodic boundary conditions into one "infinitely long" cylinder) and measure the axial force along it. It is easy to see that such a force should exist: the fixed number of lipids in the simulation box will give rise to a membrane of some given overall area $A = 2\pi RL$, where R and L are cylinder radius and length, respectively. If we change the length of the cylinder, we change R (since A must stay constant), and so we change the bending energy E. This results in a force F, given by

$$F = \frac{\partial E}{\partial L}\bigg|_A = \frac{\partial}{\partial L}\bigg|_A \left[\frac{1}{2}\kappa\left(\frac{1}{R}\right)^2 \times A\right] = \frac{\partial}{\partial L}\bigg|_A \left[\frac{1}{2}\kappa\left(\frac{2\pi L}{A}\right)^2 \times A\right]$$
$$= \kappa\left(\frac{2\pi L}{A}\right)\left(\frac{2\pi}{A}\right) \times A = \frac{2\pi\kappa}{R}. \qquad (132)$$

Hence, measuring the force and the radius gives the rigidity: $\kappa = FR/2\pi$. Moreover, we can impose *much* larger curvatures than would ever happen under passive undulation conditions, and so we can test how far the quadratic curvature Hamiltonian (127a) can be trusted. Harmandaris and Deserno (2006) found that—for the coarse grained model they studied (Cooke et al. 2005)—it worked with remarkable accuracy down to curvature radii equal to a few times the membrane thickness—much better than one would probably have any right to hope! Also, the measured rigidity was compatible with what was previously measured from monitoring membrane shape undulations (i.e., exploiting Eq. (130)), but it could be measured more precisely with the same simulation overhead.

There is a big snag, though: as nice and intuitive as this method appears, it fundamentally relies on two conditions that are hardly ever met in a realistic simulation context, both of which are related to the equilibration of a chemical potential. First, the simulation setup divides the simulation box into a region inside the tube, and a region outside. These do not easily communicate, because the solvent (water, or a coarse grained version of it) usually does not diffuse fast enough through a bilayer (on the time scales relevant for the simulation). While in reality the chemical potential of water is equilibrated across the two sides, in a simulation it generally is not (we do not know ahead of time how much water we really need to put into the two environments), and it will not automatically equilibrate. Second, the chemical potential of the lipids in the two bilayer leaflets also has to be the same, since lipids can flip-flop between leaflets. But again, this typically is much too slow a process to significantly happen during the course of a simulation, so unless we set up the system already in equilibrium (and we cannot easily do that, because we do not know how many more lipids we would have to place in the outside leaflet), we have no chance of instead converging to it. Harmandaris and Deserno did not have these difficulties, since the highly coarse grained lipid model which they used (Cooke et al. 2005) (a) has no solvent and (b) has a sufficiently high flip–flop rate. But for any more highly resolved and not necessarily solvent free model, the "pulling-a-tube" method does not readily work.

And yet, this idea of an active deformation remains enticing—we just need to find a way to circumvent the unfortunate equilibration troubles. The path to glory exists, and it involves looking at a different deformation.

3.2 *Buckling for Fluid Membranes*

In a very important paper, Noguchi (2011) presented a method that solves this problem (without actually needing it for the model he used): if we place a membrane into a box that is too small for that membrane, it will buckle. Choosing a large aspect ratio, we end up with a very well-defined one-dimensional deformation, an example of which is shown in Fig. 12. Clearly, maintaining that shape requires a force, which ought to encode the stiffness of the membrane—buckling a more rigid membrane ought to be harder. In fact, it seems clear that this force ought to be proportional to the bending rigidity κ. In the following we provide a solution to this problem, following Hu et al. (2013), which pushes the analytical treatment slightly farther than Noguchi did.

The shape of a one-dimensional buckle. If we parametrize the membrane in the angle-arclength parametrization $\psi(s)$ indicated in Fig. 12, the relevant curvature along the buckle is given by $-\dot{\psi}$. Since the curvature in the perpendicular direction vanishes, we get $K = -\dot{\psi}$ and $K_{\mathrm{G}} = 0$. The curvature elastic Hamiltonian (again,

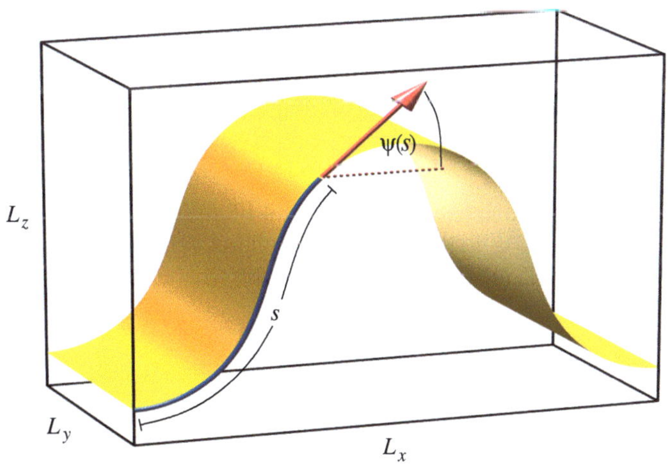

Fig. 12 Geometry of a buckled membrane, and illustration of the angle-arclength parametrization that can be used to describe it: it gives the angle $\psi(s)$ of the local profile with respect to the horizontal as a function of the arclength measured along the buckle. Reprinted from Hu et al. (2013), with the permission of AIP Publishing

without tilt) is then given by

$$E = L_y \int_0^L \mathrm{d}s \left\{ \frac{1}{2}\kappa\dot{\psi}^2 + f_x \left[\cos\psi - \frac{L_x}{L} \right] \right\} . \tag{133}$$

The second term in the integrand enforces the constraint that the membrane fits into the box—meaning, that the total distance traversed horizontally equals L_x. Physically, the associated Lagrange multiplier f_x is nothing but the force (per unit length) required to ensure that this constraint is satisfied.

A simple functional variation gives the Euler–Lagrange equation that $\psi(s)$ needs to satisfy in oder to minimize this energy:

$$\ddot{\psi} + \lambda^{-2}\sin\psi = 0 \qquad \text{with } \lambda = \sqrt{\frac{\kappa}{f_x}} , \tag{134}$$

where we encounter a new characteristic length λ. If we multiply this equation with $\dot{\psi}$, we find

$$0 = \dot{\psi}\ddot{\psi} + \lambda^{-2}\dot{\psi}\sin\psi = \frac{\mathrm{d}}{\mathrm{d}s}\left[\frac{1}{2}\dot{\psi}^2 - \lambda^{-2}\cos\psi \right] , \tag{135}$$

showing that the expression in square brackets is conserved and hence a first integral. We can make this constant more explicit by evaluating the expression at an inflection point of the buckle, where $\dot{\psi} = 0$. Calling the value of the angle at that point ψ_i, we get

$$\frac{1}{2}\dot{\psi}^2 - \lambda^{-2}\cos\psi = -\lambda^{-2}\cos\psi_\mathrm{i} , \tag{136}$$

a first-order differential equation whose quadrature can be found by separation of variables:

$$\frac{s}{\lambda} = \int_0^s \frac{\mathrm{d}s'}{\lambda} = \int_0^{\psi} \frac{\mathrm{d}\psi'}{\sqrt{2(\cos\psi' - \cos\psi_\mathrm{i})}} = \mathrm{F}\left[\arcsin\frac{\sin(\psi/2)}{\sin(\psi_\mathrm{i}/2)} \,\middle|\, \sin^2\frac{\psi_\mathrm{i}}{2} \right] . \tag{137}$$

Here, $\mathrm{F}[z|m]$ is the incomplete elliptic integral of the first kind. (For all subsequent special functions—a veritable panoply of elliptic functions and integrals—see Abramowitz and Stegun (1970)). After defining the elliptic parameter

$$m = \sin^2\frac{\psi_\mathrm{i}}{2} , \tag{138}$$

inverting Eq. (137) leads to the angle $\psi(s)$

$$\psi(s) = 2\,\arcsin\left\{ \sqrt{m}\,\mathrm{sn}\left[s/\lambda \,\middle|\, m \right] \right\} , \tag{139}$$

and integrating the cosine and sine of that expression gives a parametric representation of the buckle:

$$x(s) = 2\lambda\,\mathrm{E}\big[\mathrm{am}[s/\lambda\,|\,m]\,\big|\,m\big]\,, \tag{140a}$$

$$z(s) = 2\lambda\,\sqrt{m}\,\big(1 - \mathrm{cs}[s/\lambda\,|\,m]\big)\,. \tag{140b}$$

For instance, the second equation (140b) shows that the buckle amplitude is $z_{\mathrm{a}} = z(L/4) = 2\lambda\sqrt{m}$.

Fixing the constraints. The solutions (139) or (140) to the buckle's differential equation contain two integration constants: first, λ—which really stands in for the unknown Lagrange multiplier f_x; and second, m—which encodes the angle which the buckle makes at its inflection point. The first one is of great interest to us, the second one not so much—but it is the one that causes technical troubles, because in a simulation we do not fix the angle but the extent of a buckle's compression—essentially, L_x. Of course, we could always *measure* the inflection angle in our simulation, but this is laborious, for it would require us to explicitly determine the membrane shape. Instead, it is much more convenient to do a bit more work and re-express the constant m in terms of a more natural one, namely the compressional strain γ, defined as

$$\gamma = \frac{L - L_x}{L}\,. \tag{141}$$

To do so, recall that the two constants are fixed by the two boundary conditions of the problem, which are

$$\psi(L/4) = \psi_{\mathrm{i}} \qquad \text{and} \qquad x(L/4) = L_x/4\,. \tag{142}$$

Using Eq. (137), the first condition implies

$$\frac{L}{4\lambda} = \mathrm{F}\Big[\frac{\pi}{2}\,\Big|\,m\Big] = \mathrm{K}[m]\,. \tag{143}$$

And using Eq. (140a), the second one yields

$$L_x = 8\lambda\,\mathrm{E}[m] - L\,. \tag{144}$$

Between these two equations, the length λ can be eliminated, leading to the tran scendental equation

$$\gamma(m) = 2\left(1 - \frac{\mathrm{E}[m]}{\mathrm{K}[m]}\right)\,, \tag{145}$$

which we now "merely" have to invert for $m(\gamma)$ in order to make the strain γ the independent variable. Unfortunately, this cannot be done in closed form. But it is quite easy to find an accurate series expansion solution, by making the ansatz

$$m(\gamma) = \sum_{i=1}^{\infty} a_i\,\gamma^i\,, \tag{146}$$

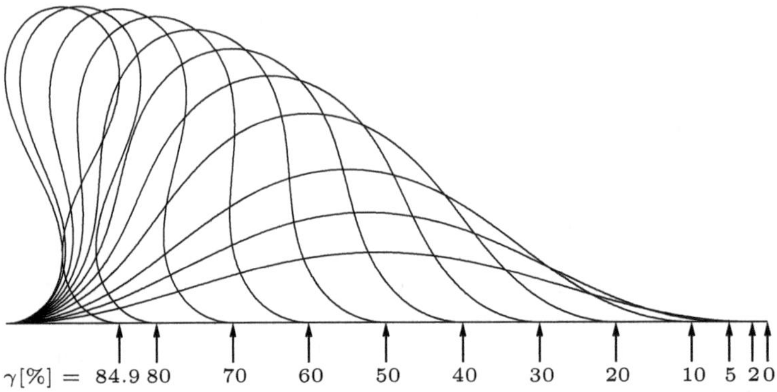

Fig. 13 Sequence of buckles, with the buckling strain γ (in percent) given below the arrow at the right end of the buckle. The buckle self-touches at $\gamma \approx 84.87\%$; notice also that a strain of merely 10% already reaches about half the transverse amplitude of that final touching-state

inserting this into Eq. (145), *again* expanding the right hand side in a Taylor series in γ, comparing equal powers of γ on both sides, and thus obtain a set of equations that will determine the coefficients a_i. Most symbolic algebra packages do this in seconds, and one finds

$$m(\gamma) = \gamma - \frac{1}{8}\gamma^2 - \frac{1}{32}\gamma^3 - \frac{11}{1024}\gamma^4 - \frac{17}{4096}\gamma^5 - \frac{55}{32\,768}\gamma^6 - \cdots \tag{147}$$

Hu et al. (2013) tabulate the coefficients up to order γ^{10} and show that the accuracy (compared to an "exact" numerical solution, and restricted to relevant values of $\gamma \lesssim 0.5$) is always better than 2×10^{-9}. In other words: we now have to all intents and purposes an *analytical* solution of the buckling problem. As an illustration, Fig. 13 shows a sequence of buckles for increasing strain γ

Stress–strain relation. The stress f_x required to compress the buckle enters in the length scale λ, and now that we know $m(\gamma)$, Eq. (143) can be solved for the stress strain relation:

$$f_x(\gamma) = \kappa \left(\frac{4}{L}\mathrm{K}\big[m(\gamma)\big]\right)^2 \tag{148a}$$

$$= \kappa \left(\frac{2\pi}{L}\right)^2 \left[1 + \frac{1}{2}\gamma + \frac{9}{32}\gamma^2 + \frac{21}{128}\gamma^3 + \frac{795}{8192}\gamma^4 + \cdots\right]. \tag{148b}$$

Notice that the stress is directly proportional to the rigidity (as expected) and inversely proportional to the square of the buckle's contour length. Also, the limit $\gamma \to 0$ is discontinuous, showing that a finite stress is required to induce even an infinitesimal strain—the hallmark of a buckling transition. After onset of buckling, the stress continues to grow monotonically. The initial post-buckling slope is $\frac{1}{2}$ (independent in fact of details of the boundary conditions), and the remaining terms only provide a small correction to them—about 7% at $\gamma = 50\%$.

Of course, for compressible materials the initial rise cannot be discontinuous. Since a lipid membrane has a finite area compressibility K_A, we would hence expect the initial rise to be linear, but with a much bigger slope. The crossover strain γ^* occurs, roughly, where compression and buckling have equal stresses, leading to the condition

$$K_A\gamma^* = \kappa\left(\frac{2\pi}{L}\right)^2 . \tag{149}$$

Using microscopic theories (such as the ones from Sect. 2), we can relate the area compressibility and the bending modulus. In our special case this is difficult, because the bending modulus also involves the stress profile. But mere scaling already suggests a relation $\kappa \propto K_A d^2$, where d is the membrane thickness. Imagining lipid bilayers as two thin homogeneous slidable plates *without* internal prestress gives a constant of proportionality of $\frac{1}{36}$ (for a Poisson ratio of $\frac{1}{2}$) (Deserno 2015), leading to

$$\gamma^* = \frac{\kappa}{K_A}\left(\frac{2\pi}{L}\right)^2 \sim \frac{\pi^2}{9}\left(\frac{d}{L}\right)^2 \approx \left(\frac{d}{L}\right)^2 . \tag{150}$$

For the systems studied by Hu et al. (2013), this is always smaller than about 1%. Notice, however, that a finite compressibility also changes the buckling problem itself. The corrections are small if the area compressibility is small (in the sense that $\sqrt{\kappa/K_A}$ is microscopic), but the resulting theory is extremely fascinating, as Oshri and Diamant (2016) show. For instance, while there is a well-known analogy between the one-dimensional Euler elastic studied here and the mathematical pendulum (observe that Eq. (134) is nothing but the pendulum equation), the *compressible* elastic can be exactly mapped to the *relativistic* pendulum.

Evidently, the idea is now to simulate buckles at various different strains (bigger at least than the crossover strain γ^*) and fit the measured stress–strain relation to Eq. (148)—using κ as the sole fitting parameter. As Hu et al. (2013) demonstrate, this works very well for models all the way from strongly coarse grained to virtually fully atomistic.

The stress tensor for membrane buckles. We can learn more about the stress distribution in a buckle, and in particular the isotropic tension σ within it, by looking at the *membrane stress tensor* $\boldsymbol{f}^a$ (Capovilla and Guven 2002, 2004; Guven 2004). Guven and Vázquez–Montejo provide a pedagogical introduction in this volume to the necessary mathematics, and it is also covered in a recent review by one of us (Deserno 2015). Briefly, if we draw a curve on a membrane surface with tangent vector $\boldsymbol{t} = t^a\boldsymbol{e}_a$, tangential co-normal $\boldsymbol{l} = l^a\boldsymbol{e}_a$, and membrane normal $\boldsymbol{n} = \boldsymbol{l}\times\boldsymbol{t}$, the traction $\boldsymbol{f}$ acting onto the membrane side into which $\boldsymbol{l}$ points is given by

$$\boldsymbol{f} = l_a\boldsymbol{f}^a = \left[\frac{1}{2}\kappa\left(K_\perp^2 - K_\parallel^2\right) - \sigma\right]\boldsymbol{l} + \kappa K K_{\perp\parallel}\,\boldsymbol{t} - \kappa(\nabla_\perp K)\,\boldsymbol{n} . \tag{151}$$

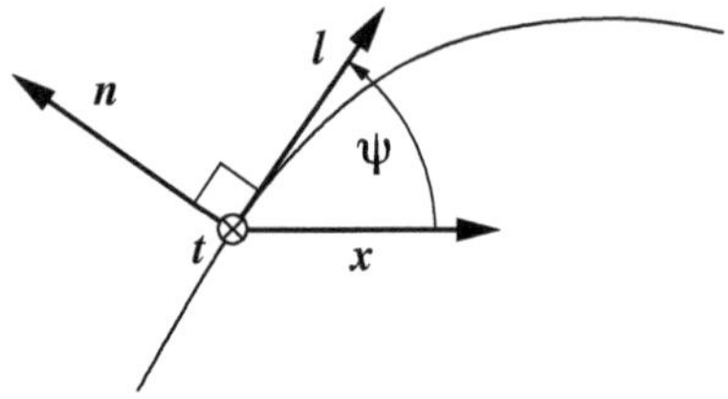

Fig. 14 Cross cut through part of a buckle, defining the local $(\boldsymbol{l}, \boldsymbol{t}, \boldsymbol{n})$ coordinate system, and the angle ψ which the buckle makes with the horizontal $\boldsymbol{x}$. Notice that $\boldsymbol{l} \cdot \boldsymbol{x} = \cos\psi$ and $\boldsymbol{t} = \boldsymbol{y}$

Here, $K_\perp = l^a l^b K_{ab}$ and $K_{||} = t^a t^b K_{ab}$ are the normal curvatures into $\boldsymbol{l}$ and $\boldsymbol{t}$ direction, respectively, while $K_{\perp||}$ is the off-diagonal element of the curvature tensor in the $(\boldsymbol{l}, \boldsymbol{t})$ basis; $\nabla_\perp K = l^a \nabla_a K$ is the gradient of K along $\boldsymbol{l}$. Let's check the sign: if $\kappa = 0$ and we merely have surface tension (this would correspond for instance to a soap film), we have $\boldsymbol{f} = -\sigma \boldsymbol{l}$, showing that a surface tension of magnitude σ *pulls* (minus sign!) tangentially onto the side into which $\boldsymbol{l}$ points.

Let us now specialize this to the case of a straight line which runs in the flat direction of the buckle (the y-direction in Fig. 12). The local geometry is sketched in Fig. 14. Since this line is straight, $K_{||} = 0$, and since it is also a line of curvature, $K_{\perp||} = 0$. Hence, the traction $\boldsymbol{f}$ is given by

$$\boldsymbol{f} = \left[\frac{1}{2}\kappa K_\perp^2 - \sigma\right]\boldsymbol{l} - \kappa(\nabla_\perp K_\perp)\,\boldsymbol{n} = \left[\frac{1}{2}\kappa\dot{\psi}^2 - \sigma\right]\boldsymbol{l} + \kappa\,\ddot{\psi}\,\boldsymbol{n}\,, \tag{152}$$

where in the second step we used $\nabla_\perp = \frac{\mathrm{d}}{\mathrm{d}s}$ and $K_\perp = -\dot{\psi}$.

Now, a crucial thing to realize is that $\boldsymbol{f}$ must be constant and horizontal. Constant, because the stress tensor is divergence free, $\nabla_a \boldsymbol{f}^a = 0$, or in our one-dimensional case, $\mathrm{d}\boldsymbol{f}/\mathrm{d}s = 0$, and since there are no sources of stress along the buckle, the traction is constant. There *are* sources at the ends, and they push the buckle horizontally; hence $\boldsymbol{f} \propto \boldsymbol{x}$. This means that there are two ways for how to get the magnitude of $\boldsymbol{f}$: you could either project it onto $\boldsymbol{x}$, or you could square it. This leads to the two equations

$$f_x = \boldsymbol{f} \cdot \boldsymbol{x} = \left[\frac{1}{2}\kappa\dot{\psi}^2 - \sigma\right]\cos\psi - \kappa\,\ddot{\psi}\,\sin\psi\,, \tag{153a}$$

$$f_x^2 = \boldsymbol{f} \cdot \boldsymbol{f} = \left[\frac{1}{2}\kappa\dot{\psi}^2 - \sigma\right]^2 + \kappa^2\,\ddot{\psi}^2\,. \tag{153b}$$

Between these two equations, we can eliminate the higher derivative $\ddot{\psi}$ and thereby arrive at a differential equation that is one order lower:

$$\frac{1}{2}\kappa\dot{\psi}^2 - \sigma = f_x\,\cos\psi\,. \tag{154}$$

In other words, *stress conservation has given us a first integral of the shape equation*—and we did not even have to *write down* the shape equation. Observe that Eq. (154) is the analog of Eq. (136), but in this case we also get a *mechanical* interpretation of the constant of integration, not just a geometrical one. Picking the position such that we are at an inflection point—just as we had done in Eq. (136)—we find

$$\sigma = -f_x \cos\psi_i \ . \tag{155}$$

Hence, the isotropic tension (which couples to the area per lipid) is not equal to the (negative of the) buckling stress, but equal to that stress times the cosine of the inflection angle. In particular, it vanishes if $\psi_i = \frac{\pi}{2}$, which happens at $m = \frac{1}{2}$ or $\gamma \approx 0.543$.

Advantages and drawbacks. Now that we have seen how buckling a membrane gives rise to an observable, f_x, that will encode the bending modulus, κ, let us briefly stop and ponder the benefits and limitations that come with this particular method of determining a membrane's rigidity, especially in comparison with more traditional fluctuation approaches.

Advantages:

- The signal we measure, f_x, is directly proportional to the observable we care about, κ. In the fluctuation case it was *inversely* proportional: $|\tilde{h}_q|^2 \propto \kappa^{-1}$. Hence, the buckling method should become better if membranes get stiffer, and worse if they get softer. Since κ is on the order of a few tens of $k_B T$, we already are in the limit where fluctuations are visible but weak. Moreover, fitting the q^{-4} dependence predicted in Eq. (130) requires a range of q-values, and if we want just one order of magnitude in q, we encounter a drop of four orders of magnitude in $|\tilde{h}_q|^2$. Indeed, we are looking at very weak signals then.
- In fluctuation methods, the fluctuations are the *signal* from which the observable κ is deduced, and hence we need to sample them adequately. In contrast, in the buckling protocol fluctuations are *noise*—an unwanted *perturbation*. Not sampling them properly affects the error of our result much less than in a fluctuation method. To see that fluctuations are indeed subdominant, consider the persistence length ℓ_p of the equivalent one-dimensional "polymer," which is given by $\ell_p = \kappa L_y / k_B T$. This is typically several tens times L_y. For common situations this makes the persistence length substantially larger than the buckle's length, and so its deformations are dominated by the ground state energy.[8]
- The method makes no strong assumptions about the microphysics that gives rise to a bending rigidity in the first place. It measures the emergent macroscopic modulus, not a microscopic object that is predicted to coincide with it within the framework of a particular scale-bridging theory. Hence, the bending modulus derived from buckling can serve as a reference which microscopic predictions must meet. It has this property in common with the classical fluctuation method based on Eq. (130), but not with *every* fluctuation method. For instance Watson et al. (2012) propose a method that extracts the bending rigidity from the *orientation fluctuations* of

[8]Hu et al. (2013) discuss buckle fluctuations in a little bit more detail. They conclude that systematic fluctuation corrections exist, but that they are small for f_x. They are *not* necessarily small for the force f_y acting along the buckle's ridges, though.

lipids; it works with significantly smaller membrane sizes than what Eq. (130) tends to need (and is hence *much* more efficient), but it relies on an underlying microscopic theory for how curvature and tilt couple (which, incidentally, is of similar nature as the one discussed in Sect. 2).

- Fluctuation methods involve relatively weak curvatures for typical values of the bending rigidity, as Eq. (131) shows. In order to test whether quadratic curvature elasticity holds for curvatures beyond the weak fluctuation-induced ones, we have to impose them actively. For instance, by simulating tethers, Harmandaris and Deserno (2006) showed that within the statistics available at that time, Cooke-model membranes (panel (b) in Fig. 1) can be bent into curvature radii approaching the thickness of the membrane without significant deviations from quadratic curvature elasticity. The buckling method opens this possibility to membrane models for which tether pulling does not work (because, as discussed above, it is hard to equilibrate the chemical potential of solvent and lipids).

Drawbacks:

- Studying buckles is technically more involved than studying a flat membrane. First, they must be created[9]; and second, they require bigger simulation boxes in the z-direction, hence necessitating more solvent.
- Buckled membranes are not stress free. This does not merely refer to the externally applied buckling stress f_x, but the resulting tension $\sigma = -f_x \cos\psi_{\rm i}$—see Eq. (155). Since σ couples to the area per lipid, buckled membranes usually have their lipids under a compression, and so they are not, strictly speaking, in the thermodynamically relaxed state that is probed with the fluctuation formula from Eq. (130). This matters in particular if the membrane is close to a phase transition for which the area per lipid could change. For instance, if a fluid membrane is close to its main phase transition temperature (below which it goes into a gel phase with a smaller area per lipid), the additional imposed compressive stress can drive (parts of) the membrane—via Le Châtelier's principle—into a gel phase, thus obviating the applicability of the buckling protocol. An exception is the strain leading to $\psi_{\rm i} = \frac{\pi}{2}$, at which point $\sigma = 0$.
- The buckling protocol cannot be applied to mixtures without some substantial extensions. The reason is that the buckle's local geometry changes with position, and different lipid species could prefer different regions—for instance regions where the local monolayer curvature better matches their own spontaneous curvature. Hence, the nontrivial geometry constitutes a driving force for a nonuniform lipid distribution (and even trigger demixing, in the most extreme case). One can of course account for these effects, and most likely even learn more about the mixture in that way, but this requires additional modeling.

---

[9]One efficient construction method proceeds via the analytical expressions for position and angle of a buckle—Eqs. (139) and (140)—for mapping a flat membrane (leaflet-wise) into a buckled one. If a lipid's center of mass in the flat configuration has coordinates (x_0, y_0, z_0), map it to $(x(x_0), y_0, z(x_0))$ and rotate it around the y-axis by the angle $\psi(x_0)$.

Thermodynamics of the membrane bending modulus. Before we move on to some striking deviations from Euler buckling, let us conclude this section with a little detour through the thermodynamics of membrane bending. The buckling force f_x arises because the curved membrane has a higher energy than the flat one. Or to be more precise—and now we have to be—because it has a higher *free energy*: we compress the lipid bilayer at constant temperature. It is crucial to realize that even if we ignore *large wavelength* thermal undulations, we by no means study a system that *microscopically* sits in an energy ground state. The lipid constituents have a considerable number of degrees of freedom (translation, rotation, bond length and angle vibrations, dihedral rotations) which explore their permissible phase space and whose non-sharp distribution functions "store" a substantial amount of entropy. Of course, none of this is explicitly accounted for in the Helfrich Hamiltonian—so where did it go? The answer is that it went into the *parameters*—for instance the moduli. The microscopic wiggling of the molecular constituents is captured by effective parameters on the macroscale. If so, the Helfrich Hamiltonian really describes a free energy, and since the curvatures K and $K_{\rm G}$ merely capture the geometry, the subdivision into energetic and entropic contributions to the free energy happens at the level of the moduli. One might hence ask: is there a way to disentangle them?

If we integrate the stress strain relation $f_x(\gamma)$ over γ, we get back the free energy $\mathcal{E}(\gamma)$, and it is easy to see that per unit area it is given by

$$\frac{\mathcal{E}(\gamma)}{LL_y} = \int_0^\gamma {\rm d}\gamma' \; f_x(\gamma') \stackrel{(148)}{=} \kappa \left(\frac{2\pi}{L}\right)^2 \left[\gamma + \frac{1}{4}\gamma^2 + \frac{3}{32}\gamma^3 + \cdots\right] . \tag{156}$$

Now, in a simulation we can also measure the plain *energy*—simply by evaluating the total microscopic Hamiltonian of the system. Even at zero strain it will have some nonzero value, but if we buckle the membrane, this energy changes. Let us define $E(\gamma) = E_{\rm sim}(\gamma) - E_{\rm sim}(0)$, the *excess energy* which the buckled membrane has relative to the stress-free flat state. How does it compare to $\mathcal{E}(\gamma)$?

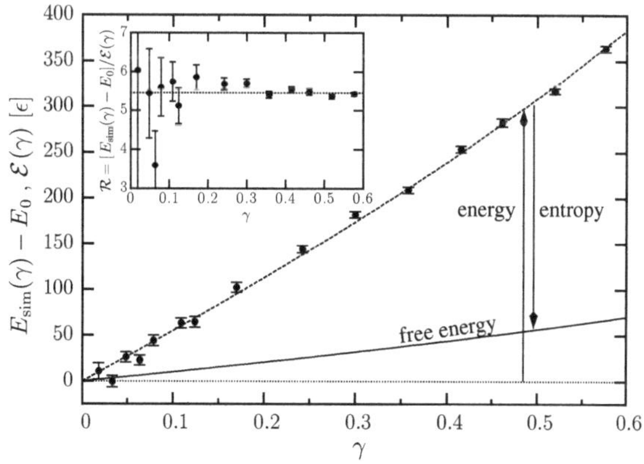

Fig. 15 Free energy $\mathcal{E}(\gamma)$ (*solid*) and energy $E(\gamma)$ (*dashed*) in units of $\epsilon = 1.1\,k_{\rm B}T$ as a function of strain γ. The *inset* shows the ratio $\mathcal{R} = E/\mathcal{E}$, which is largely independent of γ and hence a property of the modulus. Reprinted from Hu et al. (2013), with the permission of AIP Publishing

Figure 15 shows a plot of $E(\gamma)$ and $\mathcal{E}(\gamma)$ versus the strain γ, for the Cooke model at standard conditions (see Hu et al. (2013) for details). The energy increases much more rapidly than the free energy, indicating that the entropic contribution will bring down the true cost of bending—or, in other words, entropy *favors* bending. The inset in Fig. 15 shows the *ratio* $\mathcal{R} = E/\mathcal{E}$ of these two quantities. Notice that $\mathcal{R}$ is remarkably constant, indicating that geometry "cancels" and all we see is the ratio between energy and free energy as captured in the bending modulus. It hence makes sense to talk of the energetic and entropic contribution of the bending modulus, and thus to "take it apart" as we do with any ordinary free energy:

$$\kappa = \kappa_E - T\kappa_S \,. \tag{157}$$

Moreover, using well-known thermodynamic identities, we can write

$$\kappa_E = \kappa + T\kappa_S = \kappa - T\frac{\partial \kappa}{\partial T} = \kappa\left(1 - \frac{T}{\kappa}\frac{\partial \kappa}{\partial T}\right) = \kappa\left(1 - \frac{\partial \log \kappa}{\partial \log T}\right) , \tag{158}$$

and hence

$$\mathcal{R} = \frac{\kappa_E}{\kappa} = 1 - \frac{\partial \log \kappa}{\partial \log T} \,. \tag{159}$$

This is a differential equation for the temperature dependence of the bending modulus which we can integrate—provided we know $\mathcal{R}(T)$. Assuming we can expand it as a series in the smallness parameter $\log(T/T_0)$,

$$\mathcal{R}(T) = \sum_{n=0}^{\infty} \frac{\mathcal{R}_n}{n!} \log^n \frac{T}{T_0} \tag{160a}$$

$$= \mathcal{R}_0 + \mathcal{R}_1 \frac{T - T_0}{T_0} + \frac{\mathcal{R}_2 - \mathcal{R}_1}{2}\left(\frac{T - T_0}{T_0}\right)^2 + \cdots , \tag{160b}$$

the integration is trivially done, leading to

$$\log \frac{\kappa(T)}{\kappa_0} = (1 - \mathcal{R}_0) \log \frac{T}{T_0} - \sum_{n=2}^{\infty} \frac{\mathcal{R}_{n-1}}{n!} \log^n \frac{T}{T_0} \,. \tag{161}$$

This expresses the functional form of $\kappa(T)$ in a log-log fashion. Notice that for T close to T_0 this boils down to a simple power law, with corrections only at the *quadratic* level:

$$\kappa(T) \approx \kappa_0 \left(\frac{T_0}{T}\right)^{\mathcal{R}_0 - 1} \left[1 - \frac{\mathcal{R}_1}{2}\left(\frac{T - T_0}{T_0}\right)^2 + \cdots\right] . \tag{162}$$

By explicitly calculating $\kappa(T)$ over the range $0.95 \leq \frac{T}{T_0} \leq 1.11$, Hu et al. (2013) have shown (using the standard Cooke model) that a simple power law relation indeed describes the data very well. This is quite advantageous, because it means that by also measuring $\mathcal{R}$ from the buckling simulations (essentially at no extra cost), one can predict the bending modulus κ in the *vicinity* of the simulation temperature, not just *at* it.

Notice that if $\mathcal{R}_0 > 1$, heating softens the membrane. We would probably have expected this to be true no matter what, but we now see that this occurs if and only if buckling increases the energy more rapidly than it increases the free energy. Interestingly, this need not always be true: $\mathcal{R}(T) < 1$ is thermodynamically possible and does in fact occur. Its hallmark is "anomalous swelling," the phenomenon that the spacing in a multilamellar stack of membranes unexpectedly increases upon cooling, which happens for some lipids a few degrees above their main phase transition. Chu et al. (2005) have argued that this swelling is due to an increased Helfrich fluctuation repulsion between the lamellae, which indeed points toward a softening of the modulus.

3.3 *Buckling for Gel-Phase Membranes*

When fluid membranes are cooled, they ultimately reach a temperature at which they change into a new phase that is both more ordered and more rigid—the so-called "gel phase"; this is called the "main transition" of a membrane (see Nagle (1980) for a review of the theory). Many subtleties exist about this transition, and some membranes even change first into an intriguing corrugated phase (the so-called "ripple phase"), but none of this will concern us here. For now we are very modest and merely want to know, how much stiffer a gel phase is, and how we can measure that.

Experiments indicate that gel-phase membranes are at least an order of magnitude stiffer than fluid-phase membranes (Lee et al. 2001; Dimova et al. 2000; Steltenkamp et al. 2006). Hence, the observable signal from fluctuation methods drops by at least an order of magnitude, while the signal from active methods increases by the same factor. Relatively speaking, active methods should therefore be about two orders of magnitude more sensitive for measuring the membrane bending modulus. Since, furthermore, the buckling method works even if the membrane becomes less fluid (the deformation is *isometric* and can be realized even for solid sheets, such as paper), applying the technique discussed in this chapter seems ideally suited to study the rigidity of gel phases. Indeed, Diggins et al. (2015) have done just that. What they found, though, was highly surprising: the theory developed so far, in particular the stress strain relation from Eq. (148), does not describe their simulation data *at all*—not even *qualitatively*. Figure 16 shows the stress–strain relation extracted from simulations of a gel-phase membrane. In contrast to the prediction from Eq. (148),

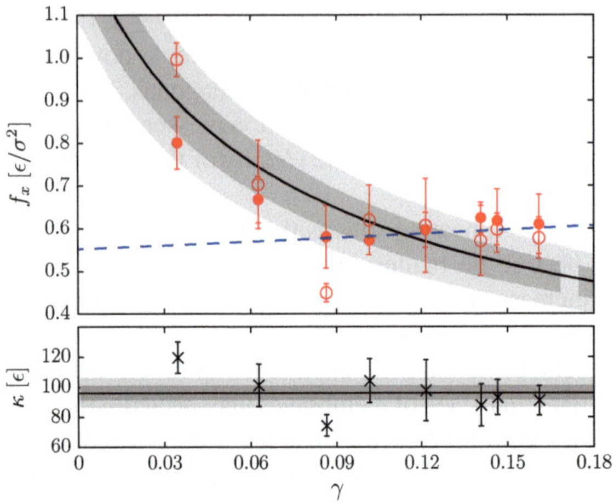

Fig. 16 Stress–strain relation for a Cooke buckle at $w_c/\sigma = 1.6$ and $k_B T/\epsilon = 0.85$. The *open circles* are the directly measured stress, the *filled circles* use additional information from the shape. The *blue dashed line* is a poor fit to Eq. (148), the *solid line* is the prediction from Eq. (164a) (surrounded by the 68 and 95% confidence bands). The *bottom panel* shows the inferred value of the bending rigidity. Reprinted with permission from Diggins et al. (2015); copyright 2015 American Chemical Society

which is clearly a monotonically *increasing* function, the opposite is true for the measured data: higher strains lead to smaller stresses, and thus *the compressibility is negative*.

Curvature softening. Based on a careful analysis of the resulting buckle shapes, which on average appear more "pointy" than classical Euler buckles, Diggins et al. (2015) conjecture that the reason for the discrepancy is a failure of quadratic curvature elasticity: assume that membranes soften upon bending, in the sense that their elastic energy does not keep growing quadratically with curvature but instead lags behind as one continues to increase the curvature. If so, it would be energetically advantageous to localize bending in small regions, rather than distributing it more evenly. This would explain the more "pointy" buckle shapes, but what would it predict for the stress–strain relation?

In order to be quantitative about the stresses, we first need a quantitative theory of curvature softening. The probably easiest phenomenological approach would be to amend the quadratic curvature energy density $e(K)$ by a quartic term that reduces the energy—in the spirit of adding a next order correction to Helfrich theory:

$$e(K) = \frac{1}{2}\kappa K^2 - \frac{1}{4}\kappa_4 K^4 + \mathcal{O}(K^6) . \tag{163}$$

Unfortunately, this is an awkward theory to work with: for $K > K^\star = \sqrt{\kappa/\kappa_4}$ this energy density *decreases* with curvature—all the way to minus infinity; the energy is not convex, not even bounded below. This will invariably create numerous artifacts and is hence ill-suited as an explanatory model for our findings. To fix this, Diggins et al. (2015) propose an alternative energy density which is both bounded below and in fact convex, but which *up to quartic order* coincides with the first guess from Eq. (163):

$$e(K) = \frac{\kappa}{\ell^2}\Big[\sqrt{1+K^2\ell^2}-1\Big] \tag{164a}$$

$$= \begin{cases} \frac{1}{2}\kappa K^2 - \frac{1}{8}(\kappa\ell^2)K^4 + \mathcal{O}(K^6)\ , & K \ll \ell^{-1} \\ \frac{\kappa}{\ell^2}\Big(|K\ell|-1\Big) + \mathcal{O}(K^{-1}) & , K \gg \ell^{-1} \end{cases} \tag{164b}$$

where ℓ is a new characteristic length scale, telling us where softening starts to set in. Notice that for sufficiently small K this looks like the curvature-softened first guess from Eq. (163), with $\kappa_4 = \frac{1}{2}\kappa\ell^2$. But beyond $K \sim \ell^{-1}$ the initial quadratic increase turns into a mere linear one. Stronger bending still always costs more energy, but at large curvature the differential price is much less than at small curvature. What does this imply for the stress–strain relation?

A new stress–strain relation. As it turns out, the Euler–Lagrange equation associated with this new energy density can still be turned into a first integral:

$$\frac{s}{\ell} = \int_{\psi_i}^{\psi(s)} \mathrm{d}s \, \frac{1}{\sqrt{\Big\{1-\tilde{f}_x\big[\cos(\psi(s)) - \cos\psi_i\big]\Big\}^{-2} - 1}}\ , \tag{165}$$

where $\tilde{f}_x = f_x\ell^2/\kappa = \ell^2/\lambda^2$ is the scaled buckling force.

Using the same series-inversion techniques as in the ordinary case, Diggins et al. (2015) arrive at a revised stress strain relationship:

$$f_x(\gamma,\delta) = \kappa\left(\frac{2\pi}{L}\right)^2\left[1+\frac{1}{2}\Big(1-3\delta^2\Big)\gamma + \frac{9}{32}\Big(1-\frac{14}{3}\delta^2+\frac{31}{3}\delta^4\Big)\gamma^2 + \cdots\right]\ , \tag{166}$$

which features the new parameter $\delta = \frac{2\pi\ell}{L}$ as a convenient dimensionless measure for exactly how strongly the situation deviates from the plain Euler case. Notice that in the limit $\delta \to 0$ Eq. (166) reduces to the first terms of the Eulerian stress–strain relation (148), and that for any nonzero δ the initial post-buckling slope of $\frac{1}{2}$ is reduced. In fact, at $\delta = \delta_c = 1/\sqrt{3} \approx 0.577$ that slope vanishes, and for $\delta > \delta_c$ the stress–strain relation starts with a negative slope. For the Cooke-model data in Fig. 16 Diggins et al. (2015) indeed find that δ is *much* bigger than that critical value: $\delta \approx 2.9$. Unfortunately, at these large values the series expansion from Eq. (166) no longer converges for all strains of interest, so a numerical solution needs to be sought. But this solution very nicely fits the measured data, hence supporting the contention that gel-phase membranes appear to soften upon bending, in a way that is captured reasonably well by the empirical energy density (164a).

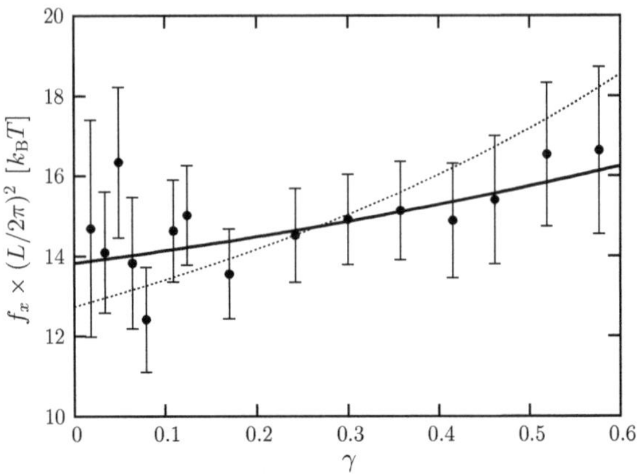

Fig. 17 Stress–strain relation for a buckle consisting of Cooke lipids—the data are from Hu et al. (2013). The *dashed curve* is a fit to the classical Euler stress–strain relation from Eq. (148), the *solid curve* is a fit to the revised stress–strain relation from Eq. (166) that allows for curvature softening—using the bending rigidity κ and the new variable δ as fitting parameters. The scaling of the vertical axis is such that the intercept will give the bending rigidity. Including curvature softening leads to a prediction for the value of κ that is about 10% bigger than what the classical fit yields

Given that gel-phase membranes soften, it is fair to ask whether this is also true for fluid-phase membranes. If they do, the effect cannot be very large, for otherwise it would have been observed in many earlier studies. But if the effect is small, finding it requires both good statistics and a quantitative model capable of identifying the softening is needed. Hence, one way to answer the question of fluid-phase curvature softening is to revisit the original buckling data from Hu et al. (2013) and fit them with the revised curvature softened theory (164a). The result is shown in Fig. 17. While the classical Euler fit is not truly poor, it does seem to have a slight overall bias—in the sense that the fit is too large at high strains and too small at low strains. Given that softening will reduce the slope of the stress–strain relation, we can expect that this deficiency is resolved by the new theory. Indeed, the solid curve in Fig. 17 shows the fit to Eq. (166), which is overall a better description of the data. Notice that this implies the bending rigidity (which can be read off at the intercept) to be larger than what the classical Euler fit would predict. Indeed, the latter would give $\kappa/k_BT = 12.7 \pm 0.3$, while the curvature softened theory yields the larger rigidity $\kappa/k_BT = 13.8 \pm 0.4$, with a value for the softening parameter of $\delta = 0.44 \pm 0.08$, or a characteristics length of $\ell/\sigma = 4.7 \pm 0.8$ (which is about the bilayer thickness).

Acknowledgements Many people have contributed in numerous ways to the results presented in this chapter. We would especially like to acknowledge Luca Deseri, Patrick Diggins IV, Jemal Guven, Mingyang Hu, Zach McDargh, and Pablo Vázquez-Montejo. MD would also like to thank David Steigmann for putting together this exciting summer school. Financial support from the National Science Foundation via the grants CMMI-0941690 and CHE-1464926 is also gratefully acknowledged.

References

M. Abramowitz, I.A. Stegun, *Handbook of Mathematical Functions* (Dover, New York, 1970)

O. Berger, O. Edholm, F. Jähnig, Molecular dynamics simulations of a fluid bilayer of dipalmitoylphosphatidylcholine at full hydration, constant pressure, and constant temperature. Biophys. J. **72**(5), 2002–2013 (1997)

F. Campelo, C. Arnarez, S.J. Marrink, M.M. Kozlov, Helfrich model of membrane bending: From Gibbs theory of liquid interfaces to membranes as thick anisotropic elastic layers. Adv. Colloid Interface Sci. **208**, 25–33 (2014)

R. Capovilla, J. Guven, Stresses in lipid membranes. J. Phys. A: Math. Gen. **35**(30), 6233 (2002)

R. Capovilla, J. Guven, Stress and geometry of lipid vesicles. J. Phys.: Condens. Matt. **16**(22), S2187 (2004)

N. Chu, N. Kučerka, Y. Liu, S. Tristram-Nagle, J.F. Nagle, Anomalous swelling of lipid bilayer stacks is caused by softening of the bending modulus. Phys. Rev. E **71**(4), 041904 (2005)

I.R. Cooke, M. Deserno, Coupling between lipid shape and membrane curvature. Biophys. J. **91**(2), 487–495 (2006)

I.R. Cooke, K. Kremer, M. Deserno, Tunable generic model for fluid bilayer membranes. Phys. Rev. E **72**(1), 011506 (2005)

M. Deserno, Fluid lipid membranes: from differential geometry to curvature stresses. Chem. Phys. Lipids. **185**, 11–45 (2015)

P. Diggins, Z.A. McDargh, M. Deserno, Curvature softening and negative compressibility of gel-phase lipid membranes. J. Am. Chem. Soc. **137**, 12752–12755 (2015)

R. Dimova, B. Pouligny, C. Dietrich, Pretransitional effects in dimyristoylphosphatidylcholine vesicle membranes: optical dynamometry study. Biophys. J. **79**(1), 340–356 (2000)

M. do Carmo, *Differential Geometry of Curves and Surfaces* (Prentice Hall, Englewood Cliffs, 1976)

A. Goebel, K. Lunkenheimer, Interfacial tension of the water/n-alkane interface. Langmuir **13**(2), 369–372 (1997)

J. Guven, Membrane geometry with auxiliary variables and quadratic constraints. J. Phys. A: Math. Gen. **37**(28), L313 (2004)

M. Hamm, M.M. Kozlov, Elastic energy of tilt and bending of fluid membranes. Eur. Phys. J. E **3**(4), 323–335 (2000)

V.A. Harmandaris, M. Deserno, A novel method for measuring the bending rigidity of model lipid membranes by simulating tethers. J. Chem. Phys. **125**(20), 204905 (2006)

W. Helfrich, The size of bilayer vesicles generated by sonication. Phys. Lett. A **50**(2), 115–116 (1974)

M. Hu, P. Diggins, M. Deserno, Determining the bending modulus of a lipid membrane by simulating buckling. J. Chem. Phys. **138**(21), 214110 (2013)

J.N. Israelachvili, D.J. Mitchell, B.W. Ninham, Theory of self-assembly of hydrocarbon amphiphiles into micelles and bilayers. J. Chem. Soc., Farad. Transact. 2: Mol. Chem. Phys. **72**, 1525–1568 (1976)

R.A.L. Jones, *Soft Condensed Matter* (Oxford University Press, Oxford, 2002)

E.E. Kooijman, V. Chupin, N.L. Fuller, M.M. Kozlov, B. de Kruijff, K.N.J. Burger, P.R. Rand, Spontaneous curvature of phosphatidic acid and lysophosphatidic acid. Biochemistry **44**(6), 2097–2102 (2005)

D.I. Kopelevich, J.F. Nagle, Correlation between length and tilt of lipid tails. J. Chem. Phys. **143**(15), 154702 (2015)

N. Kučerka, S. Tristram-Nagle, J.F. Nagle, Structure of fully hydrated fluid phase lipid bilayers with monounsaturated chains. J. Mem. Biol. **208**(3), 193–202 (2006)

L.D. Landau, E.M. Lifshitz, *Theory of Elasticity*, vol. 7, 3rd edn., Course of Theoretical Physics (Pergamon Press, Oxford, 1986)

C.-H. Lee, W.-C. Lin, J. Wang, All-optical measurements of the bending rigidity of lipid-vesicle membranes across structural phase transitions. Phys. Rev. E **64**, 020901 (2001)

E. Lindahl, O. Edholm, Mesoscopic undulations and thickness fluctuations in lipid bilayers from molecular dynamics simulations. Biophys. J. **79**(1), 426–433 (2000)

S.J. Marrink, H.J. Risselada, S. Yefimov, D.P. Tieleman, A.H. De Vries, The MARTINI force field: coarse grained model for biomolecular simulations. J. Phys. Chem. B **111**(27), 7812–7824 (2007)

J.F. Nagle, Theory of the main lipid bilayer phase transition. Ann. Rev. Phys. Chem. **31**(1), 157–196 (1980)

H. Noguchi, Anisotropic surface tension of buckled fluid membranes. Phys. Rev. E **83**(6), 061919 (2011)

O. Oshri, H. Diamant, Properties of compressible elastica from relativistic analogy. Soft Matter **12**(3), 664–668 (2016)

T. Portet, R. Dimova, A new method for measuring edge tensions and stability of lipid bilayers: effect of membrane composition. Biophys. J. **99**(10), 3264–3273 (2010)

J.N. Reddy, *Theory and Analysis of Elastic Plates and Shells* (CRC, London, 2006)

J.S. Rowlinson, B. Widom, *Molecular Theory of Capillarity* (Dover, New York, 2002)

S. Steltenkamp, M.M. Müller, M. Deserno, C. Hennesthal, C. Steinem, A. Janshoff, Mechanical properties of pore-spanning lipid bilayers probed by atomic force microscopy. Biophys. J. **91**(1), 217–226 (2006)

N.J. Turro, A. Yekta, Luminescent probes for detergent solutions. a simple procedure for determination of the mean aggregation number of micelles. J. Am. Chem. Soc. **100**(18), 5951–5952 (1978)

R.M. Venable, F.L.H. Brown, R.W. Pastor, Mechanical properties of lipid bilayers from molecular dynamics simulation. Chem. Phys. Lipids **192**, 60–74 (2015)

X. Wang, M. Deserno, Determining the pivotal plane of fluid lipid membranes in simulations. J. Chem. Phys. **143**(16), 164109 (2015)

X. Wang, M. Deserno, Determining the lipid tilt modulus by simulating membrane buckles. J. Phys. Chem. B **120**(26), 6061–6073 (2016)

M.C. Watson, E.G. Brandt, P.M. Welch, F.L.H. Brown, Determining biomembrane bending rigidities from simulations of modest size. Phys. Rev. Lett. **109**(2), 028102 (2012)

J.A. Wheeler, Geometrodynamics and the Issue of the Final State, in *Relativity, Groups and Topology*, ed. by C. DeWitt, B.S. DeWitt (Gordon and Breach, New York, 1964), p. 316

T.A. Witten, P.A. Pincus, *Structured Fluids: Polymers, Colloids, Surfactants* (Oxford University Press, Oxford, 2004)

The Geometry of Fluid Membranes: Variational Principles, Symmetries and Conservation Laws

Jemal Guven and Pablo Vázquez-Montejo

Abstract The behavior of a lipid membrane on mesoscopic scales is captured unusually accurately by its geometrical degrees of freedom. Indeed, the membrane geometry is, very often, a direct reflection of the physical state of the membrane. In this chapter we will examine the intimate connection between the geometry and the physics of fluid membranes from a number of points of view. We begin with a review of the description of the surface geometry in terms of the metric and the extrinsic curvature, examining surface deformations in terms of the behavior of these two tensors. The shape equation describing membrane equilibrium is derived and the qualitative behavior of solutions described. We next look at the conservation laws implied by the Euclidean invariance of the energy, describing the remarkably simple relationship between the stress distributed in the membrane and its geometry. This relationship is used to examine membrane-mediated interactions. We show how this geometrical framework can be extended to accommodate constraints—both global and local—as well as additional material degrees of freedom coupling to the geometry. The conservation laws are applied to examine the response of an axially symmetric membrane to localized external forces and to characterize topologically nontrivial states. We wrap up by looking at the conformal invariance of the symmetric two-dimensional bending energy, and examine some of its consequences.

J. Guven (✉)
Instituto de Ciencias Nucleares, Universidad Nacional Autónoma de México,
Apdo. Postal 70-543, 04510 Mexico City, Mexico
e-mail: jemal@nucleares.unam.mx; jemalguven@gmail.com

P. Vázquez-Montejo
Facultad de Matemáticas, Universidad Autónoma de Yucatán, Periférico Norte,
Tablaje 13615, 97110 Mérida, Yucatán, Mexico

D.J. Steigmann (ed.), *The Role of Mechanics in the Study of Lipid Bilayers*,
CISM International Centre for Mechanical Sciences 577,
DOI 10.1007/978-3-319-56348-0_4

1 Introduction

Bilayers of amphiphilic molecules form the essential component of all cellular membranes, not only the plasma membrane enclosing the cell but also every membrane contained within it. Under physiological conditions, this bilayer is a fluid along the membrane; stretching is very costly, but it shears freely. By the 70s, it was beginning to become clear that on mesoscopic scales, which are often the scales that are most relevant physiologically and certainly the scales on which the global architecture comes into focus, the biochemical details of the membrane composition get telescoped into a small number of material parameters. Indeed, the membrane morphology itself is described surprisingly well by the geometrical degrees of freedom of a two-dimensional smooth surface. The membrane radius of curvature ($\sim$20 nm and up) is *large* compared to the bilayer thickness $\sim$5 nm. This is better than one could have hoped. While these surfaces may be smooth, they are rarely simple, reflecting the complex functions they play. To understand this morphological variety, it is necessary to possess the appropriate geometrical language. For the most part, this was understood by the mid-nineteenth century, to wit that two tensor fields characterize the surface geometry: the metric marking distances along the surface and the extrinsic curvature quantifying how it bends along different tangent directions. The two are not unrelated.

The physical behavior on mesoscopic scales is largely controlled by the bending energy of this surface, proportional to a quadratic in the membrane extrinsic curvature, Canham (1970); Helfrich (1973). For reviews, see Seifert and Lipowsky (1995); Bassereau et al. (2014); Tu and Ou-Yang (2014). Significantly, this energy depends only on the two fundamental tensors. What is more, modulo topology and boundary conditions, to an unusually good first approximation it is also unique.

Because the energy is determined completely by the geometry, the distribution of stress established along the membrane must *in turn* depend only on the geometry. This contrasts with the transverse distribution of stress across the membrane which does depend sensitively on the molecular details of the bilayer, as well as its interaction with water. But when we zoom out, these details contribute only to the constant of proportionality in the bending energy. The role they play is to set the rigidity. In this chapter, we review various aspects of the connection between the membrane geometry and the physics shaping it. Of course, almost always, additional agents need to be accommodated, and this direct connection gets modulated by the fields describing them.

We first examine how the bending energy, not necessarily in equilibrium, responds to deformations of the surface geometry (Sects. 2, 4). This turns on the behavior of the two fundamental tensors under deformation. The shape equation describing the surface geometry depends on the geometry through scalars constructed using these two tensors. We will develop in parallel the description of the surface in terms of its height above a plane. This representation is most useful when gradients in this function are small. This allows us to understand the local behavior consistent with membrane equilibrium.

The energy possesses symmetries: even if we disguise the fact by parametrizing the surface in terms of a height function, it is invariant under reparametrization—as any meaningful description of a physical theory should be; however, the effect of a tangential deformation of the surface is equivalent to a reparametrization of the surface except—and this is important—where the surface terminates; this apparently innocuous identification distinguishes the geometrical degrees of freedom from any additional material degrees of freedom overlaying the surface. And it has its physical consequences. Understanding how the metric and curvature behave under normal deformations of the surface permits us to describe not only its approach to equilibrium but also its behavior out of equilibrium.

The surface energy is obviously invariant under the Euclidean motions of its three-dimensional environment: translations and rotations. The bending energy of a symmetric fluid membrane is also scale invariant; this property of the two-dimensional bending energy distinguishes it from its one-dimensional counterpart, or for that matter from higher dimensional generalizations. If it were not for a constraint fixing the length, bending energy would tend to stretch a loop; a hypothetical three-dimensional surface would collapse. In two dimensions, the bending energy is independent of size. If this energy is isotropic, it turns out to be even invariant under the (angle preserving) conformal transformations of three-dimensional space, a property of two-dimensional surfaces that one could be forgiven for failing to anticipate (Sect. 7). In particular, it is invariant under inversion in spheres. Two different, indeed very different, geometries may possess identical bending energies. The energies may be identical but the local stresses supporting these geometries will generally be different.

Each symmetry implies a conservation law. In particular, the translational (rotational) invariance implies the existence of a conserved stress (torque) tensor, Capovilla and Guven (2002a) (see Sect. 5). It would not be an exaggeration to claim that the recognition of the role of symmetry has been central in the development of physics since the beginning of the 20th century. Improbable as it may appear from a traditional biophysical point of view, fluid membranes on mesoscopic scales are no exception; it could be argued that they provide a physical system par excellence supporting this claim.

In an unadorned fluid membrane, the identification of tangential deformations of the surface with reparametrizations implies constraints on the stresses associated with each term in the Hamiltonian, whether or not the membrane happens to be in equilibrium. We show how the stress tensor can be used to quantify the forces mediated by the membrane geometry. We also show how the conservation of the stress tensor permits direct access to the physics of axially symmetric shapes. Conformal invariance, happy accident or not, provides unexpected insight into the behavior of membranes well outside the regime where perturbation theory is reliable; even when the full symmetry is broken by constraints (Sect. 7).

Section 5 exploits extensively an adaptation of the calculus of variations introduced by one of the authors, Guven (2004). This approach, exploiting the structure equations describing the surface, provides a very direct construction of the stress tensor. Surprisingly, as we show, one does not need to know how the metric or the curvature respond to surface deformations to determine the approach to equilibrium. We show how this approach can be tweaked to accommodate local constraints on the geometry, and specifically a constraint on the metric. It also provides a natural framework in which to examine interactions mediated by the membrane as well as the boundary conditions on free edges. At the end of this section we present a complementary approach to the variational problem in which the equilibrium surface itself is treated as an emergent quantity from the two fundamental tensors. This approach exploits not the structure equations but the integrability conditions on these two tensors which follow from these equations. Unexpectedly, this approach provides a criterion for assessing membrane stability, Guven and Vázquez-Montejo (2013a). As we will attempt to communicate, each of the different approaches provides valuable insight into the underlying physics.

If the membrane consisted only of lipids, and they all responded in the same way, there would not be a lot to say. Intracellular membranes, however, display striking morphological diversity: contrast the spherical nuclear envelope punctured by pores with the endoplasmic reticulum (ER); or indeed the laminar stacks forming the rough ER with the tubular network forming its smooth counterpart. Then there are the flattened cisternae of the Golgi apparatus, as well as the convoluted inner membrane of a mitochondrion. And each membrane morphology is exquisitely adapted to its function. The downside is that life-threatening physiological malfunctions can all too often be traced to flaws in assembly.

What lends a membrane its specific morphology is the modulation of the behavior of the fluid bilayer by local or global constraints or biases, localized external forces, its composition and, as increasing recognized in recent years, its interactions with proteins or other macromolecules, themselves very often assembled into one or two-dimensional semi-flexible structures (for example, see Sens et al. 2008; Amoasii et al. 2013; Terasaki et al. 2013). Remarkably, on the scales that interest us, this additional structure is captured by fields or by some effective one or two-dimensional elastic structure interacting with the surface geometry. To understand how these interactions shape a membrane, it is invaluable to think in terms of the stresses they induce in the membrane. We will illustrate this point using a number of physiologically relevant but simple examples. We examine the forces constricting a membrane. We also examine the forces and torques, topological in origin, that sculpt the morphology of a toroidal vesicle, as well as the distribution of stress associated with them. Here nontrivial topology arises without the intervention of any exterior agent. These stresses are contrasted with their counterparts in a topologically spherical vesicle whose poles are pushed together by an external agency.

Before we begin, we need to introduce a few geometrical ideas on which the framework is built.

2 Surface Geometry: Intrinsic Versus Extrinsic Elements

We begin with a quick summary of a few essential geometrical concepts. This is not the place to present all of the details which would require a monograph in its own right. The interested reader can consult, Do Carmo (1976); Spivak (1999) or, for a less technical treatment, Kreyszig (1991). Some of this material is also covered nicely by Deserno in one of his excellent reviews (most recently, Deserno 2015). The reader may wish to glide over this section. It will, however, serve to establish notation and conventions.

We describe the surface parametrically in terms of a mapping into three-dimensional Euclidean space, $\Sigma : (u^1, u^2) \rightarrow \mathbf{X}(u^1, u^2)$, where u^1 and u^2 are *any* two local coordinates. Our description of Euclidean space will be Cartesian. The tangents to the coordinate curves associated with this parametrization form two surface tangent vectors at every point[1]: $\mathbf{e}_a = \partial_a \mathbf{X}$, $a = 1, 2$. These two vectors, in turn, define the surface (unit) normal vector $\mathbf{n}$. Now any vector field on the surface can be decomposed with respect to the basis vectors $\{\mathbf{e}_1, \mathbf{e}_2, \mathbf{n}\}$.

The two fundamental tensors: By the mid 19th century, it was already recognized that two surface tensor fields—constructed using derivatives of $\mathbf{X}$—describe the surface geometry completely. The first of these is the metric tensor induced on the surface from its Euclidean environment, whose components with respect to the parametrization are given by the Euclidean scalar product[2]

$$g_{ab} = \mathbf{e}_a \cdot \mathbf{e}_b \,. \tag{1}$$

If $ds^2 = d\mathbf{x} \cdot d\mathbf{x}$ is the line element in three-dimensional Euclidean space, its *pullback* to the surface is given by $ds^2\Big|_\Sigma = \mathbf{e}_a \cdot \mathbf{e}_b \, du^a du^b$. Thus g_{ab} quantifies distances between points on the surface and thus characterizes what we think of as its intrinsic geometry. In particular, the area element on the surface is given by $dA = \sqrt{g}\, du^1 du^2$, where $g = \det g_{ab} = |\mathbf{e}_1 \times \mathbf{e}_2|^2$.[3] If $g \neq 0$, g_{ab} has an inverse g^{ab}: $g^{ac} g_{cb} = \delta^a{}_b$. If Ψ is a scalar function on the surface, then $g^{ab}\partial_a \Psi \partial_b \Psi$ is another scalar. Indices are raised (lowered) using g^{ab} (g_{ab}). Let $\mathbf{V} = V^a \mathbf{e}_a$ be a surface vector field[4]; the metric allows us to associate a covector field V_a with V^a through the relationship, $V_a = g_{ab} V^b$.

---

[1]We abbreviate $\partial_a = \partial_a \mathbf{X}/\partial u^a$.

[2]We are interested specifically in surface tensors and the scalars constructed out of them. Consider a surface reparametrization $(u^1, u^2) \rightarrow (\bar{u}^1(u^1, u^2), \bar{u}^2(u^1, u^2))$. Define $J^{\bar{a}}{}_b = \partial \bar{u}^a / \partial u^b$, with inverse $J_{\bar{a}}{}^c : J^{\bar{a}}{}_c J_{\bar{b}}{}^c = \delta^{\bar{a}}{}_{\bar{b}}$. Tensor fields transform under reparametrization by matrix multiplication on each index with the Jacobian matrix of the reparametrization or its inverse. In particular, the metric transforms by $\bar{g}_{\bar{a}\bar{b}} = J_{\bar{a}}{}^c J_{\bar{b}}{}^d g_{cd}$. Note that the three Cartesian embedding functions $\mathbf{X} = (X^1, X^2, X^3)$ are each scalars under reparametrization: $\bar{X}^1(\bar{u}^1, \bar{u}^2) = X^1(u^1, u^2)$, etc.

[3]Under reparametrization, $\sqrt{\bar{g}} = \det J^{-1} \sqrt{g}$.

[4]It is simple to show that V^a transforms like a vector under reparametrization.

The covariant derivative ∇_a compatible with g_{ab} coincides with the projection onto the tangent directions of the derivative along the tangent curves, Do Carmo (1992). If $\mathbf{V} = V^a \mathbf{e}_a$, then its covariant derivative, a surface tensor, is given by

$$\nabla_a V^b = \mathbf{e}^b \cdot \partial_a \mathbf{V} = \partial_a V^b + \Gamma^b_{ac} V^c \,, \tag{2}$$

where $\Gamma^b_{ac} = \mathbf{e}^b \cdot \partial_a \mathbf{e}_c$ is symmetric in its lower indices by construction. It is straightforward to show that Γ^c_{ab} can be expressed entirely in terms of g_{ab} and its derivatives:

$$\Gamma^b_{ac} = \frac{1}{2} g^{bd} (\partial_a g_{cd} + \partial_c g_{ad} - \partial_d g_{ac}) \,, \tag{3}$$

identifying it, not coincidentally, with the Christoffel connection. The action of ∇_a on covectors as well as higher order tensor fields follow from the identification $\nabla_a \Psi = \partial_a \Psi$ on a scalar and the product (Leibnitz) formula for differentiation. Note that the identity $\nabla_a g_{bc} = 0$ follows.

An intrinsic measure of curvature is provided by the Riemann tensor, which quantifies the failure of covariant derivatives to commute. For a covector field V_a, the Ricci identity

$$(\nabla_a \nabla_b - \nabla_b \nabla_a) V_c = \mathcal{R}_{abc}{}^d \, V_d \,, \tag{4}$$

defines the Riemann tensor, $\mathcal{R}_{abc}{}^d$. It is constructed out of Γ^a_{bc} and its first derivatives, Do Carmo (1976).

In the familiar Monge representation, the surface is described in terms of its height $h(\mathbf{r})$ above a plane. If the plane is itself parametrized by Cartesian coordinates, $\mathbf{r} = x\,\mathbf{i} + y\,\mathbf{j}$, one has $\mathbf{e}_j = (\delta^1{}_j, \delta^2{}_j, \partial_j h)$, so that $g_{ij} = \delta_{ij} + \partial_i h \, \partial_j h$. Its inverse is given by

$$g^{ij} = \delta_{ij} - \frac{\partial_i h \partial_j h}{1 + |\nabla_0 h|^2} \,, \tag{5}$$

where $|\nabla_0 h|^2 = \partial_i h \partial_i h$. While $|\nabla h_0|^2$ is a scalar on the plane, it is not a surface scalar. The relevant scalar is $|\nabla h|^2 = g^{ab} \partial_a h \partial_b h$.

This representation of a surface has proven very useful when gradients are small (see, for example, Fournier 2007). With respect to an appropriate plane, it is always valid locally; even on complex geometries. However, it has its limitations, most notably if we attempt to access global information when the geometry is closed or its topology is nontrivial so that gradients necessarily not only become large but also diverge in places.

In this representation, $\sqrt{g} = \sqrt{1 + |\nabla_0 h|^2}$, so that the area is given by

$$A = \int d\mathbf{r} \, \sqrt{1 + |\nabla_0 h|^2} \,, \tag{6}$$

where we use the abbreviation $d\mathbf{r}$ to denote the area element on the base plane. The easiest way to see this is to note that the matrix g_{ij}, defined by Eq. (5) has eigenvalues

$1 + |\nabla_0 h|^2$ (corresponding to the eigenvector $\partial_i h$) and 1 (corresponding to a vector orthogonal on the plane to $\partial_i h$).

The second fundamental surface tensor is the extrinsic curvature, with components given by

$$K_{ab} = \mathbf{e}_a \cdot \partial_b \mathbf{n}\,; \tag{7}$$

it quantifies how fast the unit normal vector rotates into one tangent direction as one moves it along another. It thus captures the way that the surface curves in the Euclidean environment. This tensor is symmetric; as such, the linear map on tangent vectors $K^a{}_b = g^{ac} K_{cb}$ has two real eigenvalues, C_1 and C_2 (the principal curvatures). The surface curvature along a given direction $\mathbf{V}$ is given by $C(V) = V^a K_{ab} V^b$; it is extremal along the two corresponding orthogonal (principal) directions, along which it assumes the values C_1 and C_2. To see this, construct the constrained quadratic, $\mathcal{C}(V) = C(V) - \Lambda(V^a g_{ab} V^b - 1)$, where Λ is a Lagrange multiplier. Now $\mathcal{C}(V)$ is stationary when $K_{ab} V^b = \Lambda V_a$, or, equivalently, V^a is an eigenvector of $K^a{}_b$, with eigenvalue Λ. $C(V)$ is completely determined once we know the angle $\mathbf{V}$ makes with (one of) the principle two directions. This may be simple linear algebra but the geometrical consequences are far-reaching.

If the geometry is simple, it is possible to get along fine without knowing that curvature is a tensor; but good luck if it is not. The tensorial nature of K_{ab} will play an essential role in teasing out the relationship between stress and geometry. The two principal directions turn out to possess physical significance in the interpretation of the surface stress tensor.

An elementary calculation using height functions is useful to ground the definition of K_{ab}. If the base plane is tilted so as to coincide with the tangent plane to the surface at a given point, the curvature there is given by the Hessian of the height function: $K_{ij} = -\partial_i \partial_j h$. Its trace at this point (which we denote K) is given by $K = -\nabla_0^2\, h$, where ∇_0^2 is the Laplacian defined on the base plane. Its determinant, correct to quadratic order,

$$\mathcal{K}_G = 2\det K_{ij} \approx (\nabla_0^2 h)^2 - (\nabla_i \nabla_j h)(\nabla_i \nabla_j h) = \nabla_i \left[\nabla_i h \nabla^2 h - \nabla_j h\, \nabla_i \nabla_j h\right] \tag{8}$$

is a divergence.

3 The Bending Energy

The bending energy of a homogeneous and isotropic fluid membrane is given by the Canham–Helfrich (CH) Hamiltonian, quadratic in the symmetric curvature invariants.These invariants can be constructed in terms of the trace $K = C_1 + C_2$, and the determinant $\mathcal{K}_G = C_1 C_2$ of $K^a{}_b = g^{ac} K_{cb}$,[5] so it reads, Canham (1970); Helfrich (1973); Evans (1974)

$$H_{CH}[\mathbf{X}] = \frac{1}{2}\kappa \int dA\,(K - C_0)^2 + \bar{\kappa}\int dA\,\mathcal{K}_G + \sigma A\,. \tag{9}$$

It involves two rigidity moduli, $\kappa \approx 20k_BT$ and $\bar{\kappa}$. The constant spontaneous curvature C_0 reflects an asymmetry between the two sides of the bilayer; the parameter σ is interpreted either as a chemical potential or as a surface tension controlling the area A.[6] It is important to emphasize, despite the persistence of claims to the contrary, that σ is rarely the complete mechanical tension in the membrane. This issue will be addressed below. As Gauss famously was first to observe, the scalar $\mathcal{K}_G$ (known as Gaussian curvature) is invariant under isometry, depending only on the metric tensor. What is more, according to the Gauss–Bonnet theorem, the corresponding integrated energy is topological, modulo a boundary addition, Do Carmo (1976); Spivak (1999); on a closed single component membrane, it is irrelevant as far as determining the shape is concerned. We will see that it does not contribute explicitly to the stress. It does, however, play a role in determining the equilibrium geometry though its contribution to boundary conditions, or if the membrane is inhomogeneous. It is a mistake to ignore it.

The quadratic approximation: In terms of the height function, the quadratic approximation for the energy (9)—good when height gradients are small ($|\nabla h| \ll 1$)—reads

$$H \approx \frac{1}{2}\kappa \int d\mathbf{r}\,(\nabla_0^2 h)^2 + \frac{1}{2}\Sigma \int d\mathbf{r}\,(\nabla_0 h)^2 + \text{constant}\,, \tag{10}$$

where $\Sigma = \sigma + \kappa C_0^2/2$ is what we provisionally called surface tension augmented by spontaneous curvature, Lipowsky (2013). We saw that the term linear in K appearing in Eq. (9) is a divergence in this approximation, so it does not contribute to the local energy at this order. The constant $= \Sigma A_0$, proportional to the projected area, is usually ignored. It does, however, contribute to the stress in the membrane. Indeed, in the familiar textbook demonstration that σ is tension in a soap film, using a square frame with an adjustable edge, this is the only term appearing in H.

If one expands the energy in powers of gradients, the quartic term is negative. Note that

$$\sqrt{1 + |\nabla_0 h|^2} \approx 1 + \frac{1}{2}\,|\nabla_0 h|^2 - \frac{1}{8}|\nabla_0 h|^4 + \cdots\,. \tag{11}$$

[5] It is straightforward to confirm that the two remaining symmetric quadratics, $C_1^2 + C_2^2$ and $(C_1 - C_2)^2$, can be expressed as linear combinations of K^2 and $\mathcal{K}_G$.

[6] As we will show below, controlling area locally is equivalent, in equilibrium, to controlling it globally.

To our knowledge, all attempts to improve perturbatively on the quadratic approximation have failed. As we will show sometimes nature is obliging and small gradients can tell us a lot about the system. But more often, they are too restrictive and one needs to approach the problem non-perturbatively. For the moment, let us examine where they take us.

Linearized shape equation: The Euler–Lagrange (EL) or shape equation, describing the stationary shapes of the energy (10), is given to linear order by

$$(\nabla_0^2 - \lambda^2)\, \nabla_0^2 h = 0\,, \tag{12}$$

with a length scale $l = \lambda^{-1} = \sqrt{\kappa/\Sigma}$, indicating the scale below which bending rigidity dominates capillary forces: $l \approx 100\,\text{nm}$ for a typical fluid membrane. The derivation of Eq. (12) is a straightforward exercise in the calculus of variations.[7]

It is useful to slow down a moment and examine the elementary solutions of the linearized shape equation (12) in some detail. This will provide a guide as to what geometries we can expect to observe locally on a free membrane, as well as priming us to recognize behavior suggesting that some additional agent is involved.

Harmonic functions: Solutions of the linearized shape equation include minimal surfaces satisfying Laplace's equation on the plane, $\nabla_0^2 h = 0$. By the definition of curvature, these are symmetric saddles almost everywhere, with $K = 0$ or $C_1 = -C_2$. Typically, however, minimal surfaces are inconsistent with the boundary conditions—there are also no closed minimal surfaces—but they do feature prominently as local approximations of the physical geometry; sometimes they do even better, as we describe in the next paragraph. The Helmholtz equation, $(\nabla_0^2 - \lambda^2)\, h = 0$, also clearly provides solutions of the EL equation, with the same caveat. Any equilibrium surface can be described locally as a linear combination of solutions to the Laplace and Helmholtz equations.

Let us parametrize the plane in terms of the complex coordinate $Z = re^{i\varphi}$, where r and φ are polar coordinates; solutions of the two-dimensional Laplace equation are given by the real and imaginary parts of analytic functions $f(Z)$. Particular solutions are generated by $f(Z) = Z^n$, $n = 0, \pm 1, \pm 2, \ldots$; the dipole $n = 1$ describes a tilt; the quadrupole $n = 2$ describes a symmetric saddle, $n = 3, 4, \ldots$ describe monkey saddles and their many-tailed counterparts.

The monopole is generated by $f(Z) = \ln Z = \ln r + i\varphi$.

(1) The height function obtained from the real part is axially symmetric (the beginnings of a neck). The minimal surface spanning a large outer circular ring of radius R, and a small coaxial inner one raised above it by a height $h \ll R$ forms part of a catenoid. Indeed, the asymptotics of a catenoid reproduce the small gradient approximation.

(2) The imaginary part represents half of a helicoid, $h = p\,\varphi$, a spiral staircase or ramp (depending on the range of r).[8] Notice that while $\nabla_0^2 \varphi = 0$, $h = \varphi$ does describe

---

[7] For simplicity we will suppose that not only the surface height but also its normal vector are fixed on the boundary.

[8] The other half of helicoid is given by $h = c(\pi + \varphi)$.

an exact minimal surface. Curiously, this is the only (nonplanar) minimal surface described exactly by its linearization.[9] These two geometries, despite appearances, are symmetric saddle everywhere! Indeed, so also are the monkey saddles except at the origin. Notice that the Gaussian curvature on a monkey saddle vanishes at the origin. One can easily confirm that $\mathcal{K}_G$ is also axially symmetric in any one of these elementary geometries, wiggle as it may.

Minimal ramp and dipole: Let us examine, more closely, the ramp $h = p\,\varphi$ with inner radius r_0 extending out to some cutoff R_{cutoff}. Its pitch p provides a length scale. This is not an academic exercise because, recently, it was discovered that the sheets within the laminar stacks of the rough endoplasmic reticulum (ER) are connected by spiral ramps, Terasaki et al. (2013). However, a simple minimal ramp is costly energetically, with $E_{\mathrm{ramp}} \sim p^2\sigma \ln R_{\mathrm{Cutoff}}/r_0$; they also require a significant vertical force to hoist. Because its pitch is independent of r, its footprint does not decay as one moves away from the axis. Individual minimal ramps are thus inconsistent with planar stacking even if they are metrically flat ($\mathcal{K}_G \to 0$) far away.

One can, however, construct a ramp dipole by pairing parallel ramps with opposite chiralities: if the axes are separated by a distance $R > 2r_0$, the height function

$$\begin{aligned} h/p &= \mathrm{Im}\ln[(Z - R/2)(\bar{Z} + R/2)] \\ &= \arctan[Rr\sin\varphi/(r^2 - R^2/4)] \end{aligned} \tag{15}$$

represents the dipole illustrated in Fig. 1a centered on the origin and aligned along the x-axis.

Here one observes the beauty of small gradients: superposition holds; if h_1 and h_2 are minimal, so is their sum.

The corresponding energy $E_{\mathrm{dipole}} \sim p^2\sigma \ln R/r_0$ depends on the distance between axes; the pitch is now screened and the Gaussian curvature decays as $\mathcal{K}_G \sim -p^2R^2/r^6$ or faster, so the geometry rapidly becomes planar outside a core of size R.

Dipoles also stack much like a parking garage with the two ramps connecting parallel floors. It has been conjectured, Guven et al. (2014) that the basic element within the stacks of the rough ER is one of these dipoles. One still needs to explain what sets the distance between ramps, their pitch, their tilt, as well as the inner radius. Indeed, the sheets are not simple bilayers, but tetralayers consisting of a parallel pair of bilayers separated by a lumen; the inner boundaries are highly curved bilayers.

[9] Under a deformation $h(\mathbf{r}) \to h(\mathbf{r}) + \delta h(\mathbf{r})$, fixed on the boundary, the change in area A (6), is given by

$$\delta A = \int d\mathbf{r}\, \nabla_0 \cdot \mathbf{J}\, \delta h \tag{13}$$

where

$$\mathbf{J} = -\frac{\nabla_0 h}{(1 + |\nabla_0 h|^2)^{1/2}}\,, \tag{14}$$

so that $\nabla_0 \cdot \mathbf{J} = 0$ in equilibrium. One can evaluate $|\nabla_0 h|^2 = p^2/r^2$, so that $\mathbf{J} = p(-\sin\theta, \cos\theta)/r(1 + p^2/r^2)^{1/2}$ and $\nabla_0 \cdot \mathbf{J} = 0$.

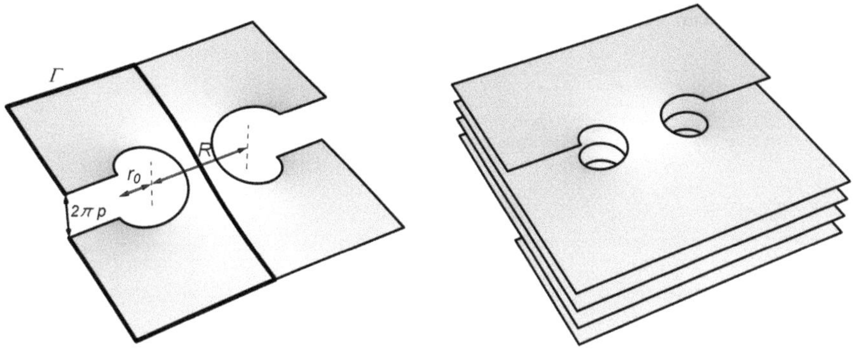

(a) Dipole (b) Ramp

Fig. 1 Ramp Dipole: **a** floor to floor and **b** stacked. The "square" contour Γ (heavy line) is used to determine the force between unpinned ramps, discussed in Sect. 5

In this chapter, the framework to address and answer some of these questions will be provided.

Biharmonic functions: Let us now examine solutions of the Biharmonic Equation $(\nabla^2)^2 h = 0$, describing the linearization of a membrane with $\Sigma = 0$. This is generally not true but we will see that there exist notable situations when it is. All solutions of $\nabla^2 h = 0$ are again solutions, as well as solutions with harmonic sources: $\nabla^2 h = J$, where $\nabla^2 J = 0$. Using the same method of complex variables used before, we look at harmonic sources in turn.

(1) For the constant function $J = 1$; $4\partial_Z \partial_{\bar{Z}} h = 1$, we find $h \sim |Z|^2$. This is the beginning of a sphere. Note that it is not analytic. Just as we did not possess the freedom to construct asymmetric saddles, this parabolic height function is axially symmetric.

(2) Next look at $J = \ln Z$. Now $4\partial_Z \partial_{\bar{Z}} h = \ln Z$ implies $h = |Z|^2(\ln Z - 1)$. This includes the singular axially symmetric surface $r^2 \ln r/r_0$ as well as the quadratically growing spiral $r^2\varphi$.[10] The surface $r^2 \ln r$ possesses a tangent plane at the origin. However, $\mathcal{K}_G$ diverges there. The parabolic appearance at the origin is deceiving. This is a danger of relying too heavily on visual cues. This curvature singularity signals the breakdown of the source free equation indicating that a necessity of a distributional external force at the origin.

(3) One also find solutions $h \sim |Z|^2 Z^{\pm n}$. In particular, if $n = -2$, we have the hairpins $h = e^{-2i\varphi}$, illustrated in Fig. 2, wiggling with finite amplitude independent of r but with asymptotically decaying curvature. The curvature is again singular at the origin.

Helmholtz equation: If $\Sigma \neq 0$, then instead of the Biharmonic equation, we need to solve the Helmholtz equation. Its elementary solutions are given by $J = K_0(\lambda r)$, $K_n(\lambda r)\cos n\varphi$, $I_0(\lambda r)$, $I_n(\lambda r)\cos n\varphi$ (as well as the sines), where (i) K_n are modified

[10]Just as $\ln|\mathbf{r} - \mathbf{r}'|$ is proportional to the Green's function for the Laplacian, $-|\mathbf{r} - \mathbf{r}'|^2 \ln|\mathbf{r} - \mathbf{r}'|$ is its counterpart for the bilaplacian.

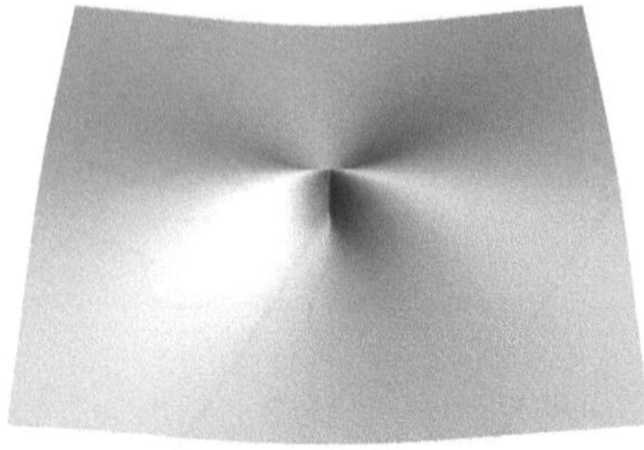

Fig. 2 Hairpin with $n = -2 : \cos 2\varphi$

Bessel functions diverging at the origin and monotonically decreasing as a function of r[11]; (ii) I_n are their monotonically increasing counterparts, diverging at infinity.

Now let $J = \nabla_0^2 h$. Then $h - J/\lambda^2$ satisfies Laplace's equation. Therefore there are no new solutions of Eq. (12) that are not already accounted for by forming linear combinations of harmonic functions and solutions of the Helmholtz equation.

Note that the axially symmetric function $K_0(\lambda r)$, the analogue of $-r^2 \ln r$, diverges at the origin. In this sense, it behaves like the harmonic $\ln r$.

One can check that there is an elementary Helmholtz spiral ramp, given by $K_0(\lambda r)\varphi$, forming a helix on its inner boundary, falling away remote from its axis. **Most general isotropic bending energy**: The bending energy, (9), is not the only isotropic energy of interest. Energies involving curvature quartics have been introduced to explain periodic *egg carton* structures, Goetz and Helfrich (1996); Dommersnes and Fournier (2002); Manyuhina et al. (2010); accounting for the relative softness of gel phases may also involve a square root dependence on the curvature, not captured by a simple higher order symmetric polynomial, Diggins IV et al. (2015). All such extensions can be accommodated, without prejudice, by considering an energy of the general form,

$$H[\mathbf{X}] = \int dA\, \mathcal{H}\,(g_{ab}, K_{ab})\ , \tag{16}$$

where the energy density $\mathcal{H}$ (a scalar), is some given function (or functional, if derivatives are entertained) of the two fundamental tensor fields, g_{ab} and K_{ab}, defined earlier.

4 Beyond Height Functions: The Nonlinear Shape Equation

Let us now examine how the energy (16) changes in response to a deformation of the surface

$$\mathbf{X}(u^1, u^2) \to \mathbf{X}(u^1, u^2) + \delta\,\mathbf{X}(u^1, u^2)\,. \tag{17}$$

[11]The Green's function of the Helmholtz operator is proportional to $K_0(|\mathbf{r} - \mathbf{r}'|)$.

One way to do this is to first track the response of g_{ab} and K_{ab} to this deformation. To this end, it is instructive to decompose the deformation vector $\delta\mathbf{X}$ into tangential and normal parts,

$$\delta\mathbf{X} = \Phi^a\,\mathbf{e}_a + \Phi\,\mathbf{n}\,. \tag{18}$$

Clearly the two parts play very different roles.

Normal Deformations: The induced normal deformations are given by, Capovilla et al. (2003):

$$\delta_\perp g_{ab} = 2K_{ab}\Phi\,; \tag{19a}$$

$$\delta_\perp K_{ab} = -(\nabla_a\nabla_b - K_{ac}K^c{}_b)\Phi\,. \tag{19b}$$

The former provides a reinterpretation of curvature as the response of the metric to normal deformations; the latter reproduces the identification of the curvature with the Hessian of the height function ($h \approx \Phi$), when gradients are small and the reference geometry is planar so that the quadratic in K_{ab} vanishes.

The normal deformation also provides a generalization of the height function when the reference geometry is not a plane, valid whenever $\Phi <<$ radius of curvature of this geometry. The curvature (19b) added to the surface standing at a height Φ is expressed in terms of the Hessian of Φ on the reference geometry screened (or anti-screened) by any pre-existing curvature.

Tangential deformations and reparametrizations: The tangential deformations of g_{ab} and K_{ab} are given by

$$\delta_\| g_{ab} = \nabla_a\Phi_b + \nabla_b\Phi_a\,; \tag{20a}$$

$$\delta_\| K_{ab} = (\nabla_c K_{ab} + K_{ac}\nabla_b + K_{bc}\nabla_a)\,\Phi^c\,. \tag{20b}$$

A mathematician may be aghast at the notation but should instantly recognize these two expressions as the Lie derivatives of the two tensors along the vector field defined by Φ^a. But this is not an accident: for the tangential deformations of a surface can be identified with the action of reparametrizations on the geometrical fields $\mathbf{X}$. Using the definition of the tangent vectors, the tangential deformation can be cast in the form $\delta_\|\mathbf{X} = \Phi^a\,\partial_a\mathbf{X}$, which is exactly how the embedding functions transform under an infinitesimal reparametrization of the surface, $u^a \to u^a - \Phi^a(u^1, u^2)$. The expressions for $\delta_\| g_{ab}$ and $\delta_\| K_{ab}$ thus should describe how the metric and any symmetric covariant tensor transform under this reparametrization: and this is by a Lie derivative. They are completely determined by the tensorial character of these variables; the details of their construction in terms of $\mathbf{X}$ are irrelevant. In this context, it should be noted that $\delta_\| K_{ab}$ or, for that matter, $\delta_\|$ of any surface tensor constructed from $\mathbf{X}$ does not depend on the metric: one can confirm that (and ditto for g_{ab} with the obvious replacement)

$$\delta_\| K_{ab} = (\partial_c K_{ab} + K_{ac}\partial_b + K_{bc}\partial_a)\,\Phi^c\,. \tag{21}$$

Using these results, it only takes a short step to see that the tangential deformation of any energy of the form (16) is given by the integral of a divergence. We first recall that, given any symmetric matrix B_{ab} (and, in particular, g_{ab}), $\partial\sqrt{\det B}/\partial B_{ab} = \sqrt{\det B}\, B^{-1ab}/2$. It then follows that, when the metric is varied, the area measure changes by

$$\delta\, dA = \frac{1}{2} dA\, g^{ab} \delta g_{ab} \,. \tag{22}$$

Using Eq. (20a), we have $g^{ab}\delta_\| g_{ab} = \nabla_a \Phi^a$[12], so that $\delta_\| dA = dA\, \nabla_a \Phi^a$; in addition, because $\mathcal{H}$ is a scalar depending only on $\mathbf{X}$, $\delta_\| \mathcal{H} = \nabla_a \mathcal{H}\, \Phi^a$. Summing terms, we find that any energy of the form (16) changes by a divergence

$$\delta_\| H = \int dA\, \nabla_a (\mathcal{H}\, \Phi^a) \,. \tag{23}$$

Using Stoke's theorem, the rhs can be cast as a boundary term $\oint ds\, \mathcal{H}\, l_a \Phi^a$ or equivalently $\oint ds\, \mathcal{H}\, \mathbf{l} \cdot \delta\mathbf{X}$, so that it makes no contribution to the response in the bulk and vanishes if the surface is closed. Here $l_a = g_{ab} l^b$ are the covariant components of $\mathbf{l} = l^a \mathbf{e}_a$, the surface tangent normal to its boundary (its conormal) pointing out of the surface. The upshot is that tangential deformations of the surface can be discounted everywhere except on boundaries where new surface is generated. Boundaries do get pushed about and, as Eq. (23) indicates, tangential deformations will play a role in understanding the behavior on them. There is a physically important corollary of the reparametrization invariance of the energy Eq. (16) and the identification of tangential deformations with reparametrizations: Because the tangential deformation is given by Eq. (23), derivatives of tangential surface deformations never occur as boundary terms. Nor can they arise when boundary energies are accommodated. Caveat: if there is a local constraint on the surface deformation, $\delta\mathbf{X}$, such as isometry, one needs to be more circumspect vis a vis the independence of Φ and Φ^a; for tangential deformations will tag along with their normal counterparts. Specifically, the constraint $\delta g_{ab} = 0$ is equivalent to

$$\nabla_a \Phi_b + \nabla_b \Phi_a + 2K_{ab}\Phi = 0 \,. \tag{24}$$

The Gauss–Weingarten equations: The derivations of Eqs. (19) and (20a) are simplified significantly using the structure equations that capture the connection between intrinsic and extrinsic geometry:

$$\nabla_a \mathbf{e}_b = -K_{ab}\, \mathbf{n} \,; \tag{25a}$$

$$\partial_a \mathbf{n} = K_a{}^b\, \mathbf{e}_b \,. \tag{25b}$$

[12] $\nabla_a \Phi^a = \partial_a(\sqrt{g}\Phi^a)/\sqrt{g}$.

The Weingarten equation (25b) captures directly the definition of K_{ab} (7). If Eq. (25a) appears mysterious, note that $\partial_a \mathbf{e}_b$, while not itself a surface tensor (or, more correctly, a triplet of tensors) may be expanded with respect to the basis vectors:

$$\partial_a \mathbf{e}_b = \Gamma^c_{ab} \mathbf{e}_c - K_{ab}\, \mathbf{n}\,. \tag{26}$$

One finds that $\Gamma^c_{ab} = \mathbf{e}^c \cdot \partial_a \mathbf{e}_b$ is the Christoffel connection (3) constructed using g_{ab}. While neither $\partial_a \mathbf{e}_b$ nor $\Gamma^c_{ab}\mathbf{e}_c$ is a tensor, their difference $\nabla_a \mathbf{e}_b$ is. And ∇_a is the covariant derivative introduced in Eq. (2).

To illustrate the utility of these equations, let us go back and derive Eqs. (19). Using the definition of g_{ab} (1) we have

$$\begin{aligned} \delta_\perp g_{ab} &= \mathbf{e}_a \cdot \partial_b(\Phi \mathbf{n}) + (a \leftrightarrow b) \\ &= \mathbf{e}_b \cdot \partial_b \mathbf{n}\; \Phi + (a \leftrightarrow b) \\ &= 2K_{ab}\, \Phi\,, \end{aligned} \tag{27}$$

where we use Eq. (25b) on the last line. On the other hand, to determine $\delta_\perp K_{ab}$, we require an intermediate result which follows from the orthogonality of $\mathbf{n}$ to the surface: $\mathbf{e}_a \cdot \mathbf{n} = 0$ and $\mathbf{n}^2 = 1$ together imply $\delta_\perp \mathbf{n} = -\partial_b \Phi\, \mathbf{e}^b$. Now using the definition of K_{ab}, given by Eq. (7), we have

$$\begin{aligned} \delta_\perp K_{ab} &= \mathbf{e}_a \cdot \partial_b(\delta_\perp \mathbf{n}) + \partial_a(\mathbf{n}\Phi) \cdot K_b{}^c \mathbf{e}_c \\ &= -\nabla_a \nabla_b\, \Phi + K_{ac} K^c{}_b \Phi\,. \end{aligned} \tag{28}$$

The Gauss–Codazzi and Codazzi–Mainardi equations: It is clear that the tensors g_{ab} and K_{ab} are not independent. Even naively the counting is wrong. Their interdependence is quantified by the Gauss–Codazzi (GM) and Codazzi-Mainardi (CM) equations, given by

$$\mathcal{R} = K^2 - K_{ab} K^{ab}\,; \tag{29a}$$

$$\nabla^b \left(K g_{ab} - K_{ab}\right) = 0\,, \tag{29b}$$

where $\mathcal{R}$ is the scalar curvature (defined in a moment). These three equations arise as integrability conditions on Eq. (25a). This is easy to show. Rewrite Eq. (25a) as $\mathbf{G}_{ab} = \nabla_a \mathbf{e}_b + K_{ab}\, \mathbf{n} = 0$. Now $\nabla_a \mathbf{G}_{bc} - \nabla_b \mathbf{G}_{ac} = 0$, or

$$[\nabla_a, \nabla_b]\, \mathbf{e}_c = \nabla_a(K_{bc}\mathbf{n}) - \nabla_b(K_{ac}\mathbf{n})\,. \tag{30}$$

Whereas

$$[\nabla_a, \nabla_b]\, \mathbf{e}_c = R_{abc}{}^d\, \mathbf{e}_d \tag{31}$$

as a consequence of the Ricci identities (4),

$$\nabla_a(K_{bc}\mathbf{n}) - \nabla_b(K_{ac}\mathbf{n}) = (\nabla_a K_{bc} - \nabla_b K_{ac})\,\mathbf{n} + (K_{ac}K_{bd} - K_{ad}K_{bc})\,\mathbf{e}^d \quad (32)$$

on account of Eq. (25b). Equating tangential and normal terms implies

$$\mathcal{R}_{abcd} - K_{ac}K_{bd} + K_{ad}K_{bc} = 0\,; \quad (33a)$$

$$\nabla_a K_{bc} - \nabla_b K_{ac} = 0\,. \quad (33b)$$

The first set of equations tells us that the (intrinsic) Riemann tensor induced on a surface embedded in Euclidean space is determined completely by the extrinsic curvature tensor. The second set of equations is a covariant statement of the fact that the extrinsic curvature is a Hessian.

Note that there are counterparts of the integrability conditions on the Weingarten equation (25b) but they provide no additional constraint, simply reproducing Eq. (33b).

For a two-dimensional geometry these equations simplify. It is easy to see that the Riemann tensor has a single independent component, say R_{1212}.[13] As a result, it is completely captured by the scalar $\mathcal{R}$:

$$\mathcal{R}_{abcd} = (g_{ac}g_{bd} - g_{ad}g_{bc})\mathcal{R}/2\,. \quad (35)$$

Equation (29a) follow on contraction. In addition, it is clear that there are only two independent Codazzi-Mainardi equations (33b) on a two-dimensional surface: $\nabla_1 K_{21} - \nabla_2 K_{11} = 0$ and $\nabla_1 K_{22} - \nabla_2 K_{12} = 0$. These two equations are equivalent to the contracted Codazzi- Mainardi equations. For two-dimensional surfaces, there are three integrability conditions.

The GC equation (29a) indicates that the Gaussian curvature is also an isometry invariant, depending as it does only on the metric: $2\mathcal{K}_G = \mathcal{R}$.[14] This is the content of Gauss's Theorema egregium. The CM equations (29b) is the statement that the tensor $K_{ab} - K\,g_{ab}$ is covariantly conserved, an identity that is very useful to recall when taking covariant derivatives of the extrinsic curvature.

Normal deformations and energy: We have looked at the response of the energy to tangential deformations. Let us now track its response to a normal deformation—one that pushes the surface outwards. Consider, to get started, the energy proportional to the area, $H = \sigma A$, describing any interface. Using Eqs. (19) and (22) we have

[13]Note that the Ricci identify (4) implies $R_{abcd} = -R_{bacd}$; whereas its application to the metric tensor implies $R_{abcd} = -R_{abdc}$:

$$0 = [\nabla_a, \nabla_b]g_{cd} = R_{abcd} + R_{abdc}\,. \quad (34)$$

These account for all the independent constraints on R_{abcd} on a two-dimensional surface.

[14]The identity $\det K^a{}_b = (K^2 - K_{ab}K^{ab})/2$ is true for the determinant of any two-dimensional symmetric matrix.

$$\delta_\perp A = \int dA\, K \Phi\,. \tag{36}$$

As a consequence, H is stationary for fixed boundaries when $K = 0$, representing a minimal surface or $\mathbf{n} \cdot \nabla^2 \mathbf{X} = 0$ where the Laplacian appearing here is the Laplacian on the surface constructed using g_{ab}. This is the nonlinear counterpart of Laplace's equation on the plane presented earlier. Note that the three Cartesian coordinates satisfy the Laplace equation, or $\nabla^2 \mathbf{X} = 0$. This is because the contracted Gauss equation (25a) implies that the tangential projections vanish identically, or $\mathbf{e_a} \cdot \nabla^2 \mathbf{X} = 0$.

Now look at $H_B = \frac{1}{2}\kappa \int dA\, K^2$. Using the definition $K = g^{ab} K_{ab}$, as well as the general identity for matrix inverses, $\delta g^{ab} = -g^{ac} g^{bd} \delta g_{ab}$, Eq. (19) now imply:

$$\delta_\perp K = -(\nabla^2 + K_{ab} K^{ab}) \Phi\,. \tag{37}$$

As a consequence:

$$\delta_\perp H_B = \kappa \int dA [-K(\nabla^2 + K_{ab} K^{ab}) + K^3/2] \Phi\,. \tag{38}$$

One may now use Stoke's theorem to perform two integrations by parts to peel derivatives appearing in the Laplacian off Φ and transfer them to K. Applying identical reasoning to the term linear in K appearing in the CH energy (9), the shape equation describing equilibrium membrane states in the absence of external forces is given by Capovilla and Guven (2002b)

$$\mathcal{E} = -\kappa \nabla^2 K + 2\kappa K \left(\mathcal{K}_G - K^2/4\right) - 2\kappa\, C_0\, \mathcal{K}_G + \Sigma\, K = 0\,. \tag{39}$$

In the presence of a pressure difference P across the surface, the surface geometry satisfies the equation: $\mathcal{E} = P$. This is a consequence of the intuitively simple identity

$$\delta_\perp V = \int dA\, \Phi\,. \tag{40}$$

The change in volume is the base area by the height! Thus, an energy contribution of the form $-PV$ has normal EL derivative P.

Early derivations of Eq. (39), notably in Refs. Ou-Yang and Helfrich (1987, 1989), tended not to exploit the fundamental tensors explicitly. Notice the cubic nonlinearity in the curvature appearing in Eq. (39); it has its source in the term quadratic in curvature appearing in $\delta_\perp K_{ab}$. It is simple to confirm that the linearization of Eq. (39) with respect to a planar reference plane reproduces Eq. (12).

A comment on height functions: Let us reexamine the variational principle in the Monge representation. We saw that, under a deformation $h(\mathbf{r}) \to h(\mathbf{r}) + \delta h(\mathbf{r})$, the change in area A (6), is given by Eq. (9). h is a scalar field, both on the base and on the surface, so it should be possible to express the rhs of Eq. (9) in a manifestly covariant form. Recall that, while $|\nabla_0 h|^2$ is a scalar on the reference plane, it is not a surface scalar. In fact, the density $\sqrt{g}$ is constructed in this parametrization in terms of $|\nabla_0 h|^2$.

The EL derivative of A appearing in Eq. (9) is clearly a divergence on the base plane. While not obvious, it should also be a divergence on the surface. If we use the identity $g^{ij}\partial_j h = \partial_i h/(1+|\nabla_0 h|^2) = \partial_i h/g$, we see that

$$\partial_i \left(\frac{\partial_i h}{(1+|\nabla_0 h|^2)^{1/2}} \right) = \partial_i (\sqrt{g} g^{ij} \partial_j h) = \sqrt{g} \nabla^2 h \,, \tag{41}$$

so that we can express δA in the manifestly reparametrization invariant form:

$$\delta A = - \int dA \, \nabla^2 h \, \delta h \,. \tag{42}$$

The Laplacian ∇^2 ($= g^{ab}\nabla_a \nabla_b$) is the Laplacian on the surface, not the base plane. One does not need to abandon surface reparametrization invariance when height functions are used. However note that $|\nabla h|^2 = |\nabla_0 h|^2/g$ or $g = 1/(1-|\nabla h|^2)$. Scalars do get conflated with densities.

In the Monge representation, $\mathbf{n} = (-\partial_i h, 1)/\sqrt{g}$, so that projecting (25a) onto the vertical $\mathbf{k}$, we identify

$$K_{ij}/\sqrt{g} = -\nabla_i \nabla_j h \,, \tag{43}$$

where ∇_i is the surface covariant derivative. Thus $K = -\sqrt{g}\nabla^2 h$.

Projecting Eq. (26) onto the Cartesian directions on the plane, one also finds that $\Gamma^k_{ij} = K_{ij}\nabla_k h/\sqrt{g}$; the connection is proportional to the extrinsic curvature. In the Monge representation the concepts of intrinsic and extrinsic geometry also get conflated.

Anisotropic energies: Other contributors discuss anisotropies. Consider the replacement of the bending energy Eq. (9) by an expression of the more general form

$$H = \int dA \, \mathcal{H}(C_1, C_2) \,, \tag{44}$$

say $\mathcal{H} = \kappa(C_1^2 + \alpha C_2^2)/2$, where $\alpha \neq 1$. It is convenient to know how C_1 and C_2 transform. One has $C_I = V_I^a K_{ab} V_I^b$, so that $\delta_\perp C_I = V_I^a V_I^b \delta_\perp K_{ab}$—an equation familiar in quantum mechanics as the first order perturbation in the eigenvalue due to a perturbation in the Hamiltonian. Now using Eq. (19b), we discover

$$\delta_\perp C_I = -V_I^a V_I^b \nabla_a \nabla_b \Phi + C_I^2 \Phi \,. \tag{45}$$

The peeling process can be performed exactly as before. Now, however, derivatives of the principal vector fields will appear in the first variation. On a surface only two scalars can be constructed using derivatives of a unit vector field, V^a: its divergence $I_1 = \nabla_a V^a$ and its curl, $I_2 = \epsilon^{ab}\nabla_a V_b$ (ϵ^{ab} is the antisymmetric

Levi–Cita tensor). Using the identity $\epsilon^{ab} = V^a V_\perp^b - V_\perp^a V^b$,[15] we find $I_2 = -V_\perp^a V^b \nabla_b V_a$, which is identified as the geodesic curvature along the integral curves of the vector field V^a.

The derivation of the shape equation presented in this section highlights the geometry. It is also perfectly adequate if one is interested in identifying membrane shapes. But it still has its limitations: for we are not only interested in shapes; indeed two very different models may predict qualitatively identical shapes. How then do we discriminate between them? What is missing is the distribution of stress underpinning the geometry, without which access to the forces acting on the membrane or transmitted by it is limited. This is simple enough, using a minimum of geometry, if gradients are small or the geometry is axially symmetric. We will have more to say about this approach. But we will now show that the little geometry we have introduced suggests a better way. We will also see there is a remarkably simple connection between the bending stress (as well as the torques) and the surface geometry, a consequence of the fact that the energy depends only on the geometry.

5 Stress and Geometry

The surface energy is invariant with respect to spatial translations. Using Noether's theorem, we know that this invariance implies the existence of a conserved current, identified as the stress tensor. This can be constructed by reassembling the normal and tangential boundary contributions to the energy associated with a translation. This was first done in Capovilla and Guven (2002b) (and even earlier in a relativistic context, in Arreaga et al. 2000).[16] The approach we describe here involves a refinement of the derivation in Capovilla and Guven (2002b), introduced a few years later by one of the authors, Guven (2004). There is no need to decompose deformations into normal and tangential parts; at the end of the calculation we will, however, interpret the conservation law by examining its projections.

We have seen that the energy (16) depends implicitly on the shape $\mathbf{X}$ through the two fundamental tensors. Thus far we have not fully exploited this dependence. While the metric and the extrinsic curvature are not independent, it is possible to treat them as though they were by making use of the method of Lagrange multipliers to record the steps in their construction in terms of $\mathbf{X}$ as local constraints and reformulating an unconstrained problem in the calculus of variations as a locally constrained one. We do this by replacing the energy H by the functional $H_C = H_C[\mathbf{X}, \mathbf{e}_a, \mathbf{n}, g_{ab}, K_{ab}, \mathbf{f}^a, f^a, f^n, T^{ab}, H^{ab}]$, defined by

[15] $V_\perp^a = \epsilon^{ab} V_b$ is orthogonal to V^a.

[16] A later derivation accommodating the finite thickness of the membrane is presented in Lomholt and Miao (2006).

$$\begin{aligned} H_C = H[g_{ab}, K_{ab}] &+ \int dA\, \mathbf{f}^a \cdot (\mathbf{e}_a - \partial_a \mathbf{X}) \\ &- \int dA\, f^a\, \mathbf{e}_a \cdot \mathbf{n} + \frac{1}{2}\int dA\, f^n (\mathbf{n}\cdot\mathbf{n} - 1) \\ &+ \frac{1}{2}\int dA\, T^{ab}(g_{ab} - \mathbf{e}_a \cdot \mathbf{e}_b) - \int dA\, H^{ab}(K_{ab} - \mathbf{e}_a \cdot \partial_b \mathbf{n}) \,. \end{aligned} \tag{46}$$

While it does appear that we have just taken a step in the wrong direction, this is not the case. Bear with us! We are now freed to treat H itself as a functional of two independent tensor fields, rather than of the embedding functions, i.e., $H[g_{ab}, K_{ab}]$. The construction of g_{ab} and K_{ab}, consigned to the constraints, is clearly independent of the specific choice of H.[17]

There is also an element of flexibility in this construction: we choose to introduce the tangent and normal vectors ($\mathbf{e}_a$ and $\mathbf{n}$), mediating the construction of g_{ab} and K_{ab} in terms of $\mathbf{X}$, as independent fields. As will be evident in a moment there is a good reason for doing this.

The tensorial character of the Lagrange multipliers reflects the constraint they enforce: $\mathbf{f}^a$, appearing in Eq. (46), is associated with the identification of $\mathbf{e}_a$ as the two tangent vectors adapted to the parametrization; f^a is associated with the implicit identification of the normal vector and f^n enforces its normalization. The fields T^{ab} and H^{ab}, completing the identification of g_{ab} and K_{ab} as the two fundamental tensors, are symmetric tensors.

It is now legitimate to vary independently each of the geometric fields $\mathbf{X}$, $\mathbf{e}_a$, $\mathbf{n}$, g_{ab} and K_{ab}. We will perform the variations in this same order. The first three of these fields appear only in the constraints, so their variations can be performed without reference to H. Significantly, the embedding functions appear only in the tangency constraint, imposed by the multiplier fields, $\mathbf{f}^a$. The translational invariance of the energy is captured by the fact that $\mathbf{X}$ also appears only through its derivative. One determines, almost trivially, the response of H_C to a deformation $\delta\mathbf{X}$:

$$\delta_{\mathbf{X}} H_C = - \int dA\, \mathbf{f}^a \cdot \partial_a \delta\mathbf{X} \,. \tag{47}$$

An integration by parts is now used to peel the derivative off the variation; the EL derivative with respect to $\mathbf{X}$ is then identified as a divergence:

$$\frac{\delta H_C}{\delta \mathbf{X}} = \frac{1}{\sqrt{g}}\, \partial_a \left(\sqrt{g}\, \mathbf{f}^a\right) = \nabla_a \mathbf{f}^a \,. \tag{48}$$

Thus, in equilibrium,

$$\nabla_a \mathbf{f}^a = 0 \,, \tag{49}$$

[17] In this approach, the deformation vector $\delta\mathbf{X}$ is never disassembled into normal and tangential parts, so that its reassembly is never necessary.

or $\mathbf{f}^a$ is conserved. Below, this tensor will be identified as the stress. We will be interested in the force per unit length transmitted across curves on the surface, given by the projection $\mathbf{f}_\perp := l_a \mathbf{f}^a$, where $\mathbf{l} = l^a \mathbf{e}_a$ is the conormal, introduced below Eq. (23).[18]

The EL equations for $\mathbf{e}_a$ and $\mathbf{n}$ identify the tangential and normal projections of $\mathbf{f}^a$ completely in terms of the two tensor-valued multipliers, T^{ab} and H^{ab}:

$$\mathbf{f}^a = f^{ab}\mathbf{e}_b + f^a \mathbf{n}\,, \tag{50}$$

where

$$f^{ab} = T^{ab} - H^{ac} K_c{}^b\,, \quad f^a = -\nabla_b H^{ab}\,. \tag{51}$$

The EL equations for g_{ab} and K_{ab} determine T^{ab} and H^{ab} in terms of the Euler–Lagrange derivatives of the energy density $\mathcal{H}$ with respect to g_{ab} and K_{ab}:

$$T^{ab} = -\frac{2}{\sqrt{g}}\frac{\delta(\sqrt{g}\mathcal{H})}{\delta g_{ab}}\,; \quad H^{ab} = \frac{\delta \mathcal{H}}{\delta K_{ab}}\,. \tag{52}$$

This completes the construction of $\mathbf{f}^a$. The normalization and sign of T^{ab} are chosen so that this tensor coincides with the metric stress tensor (see, for example, Wald 2010). However, unless $\mathcal{H}$ is independent of K_{ab} (so that $H^{ab} = 0$), T^{ab} is not the complete stress. Nor is it conserved.

The structure captured in Eqs. (51) and (52) is independent of the specific form of H. As promised, the stress is completely determined by the geometry. This is quite unlike the familiar situation in continuum mechanics where in-plane static shear—which is not supported by a two-dimensional incompressible fluid—generates stress.

Now let H be the CH energy given by Eq. (9). One then identifies, Capovilla and Guven (2002b); Guven (2004)

$$\mathbf{f}^a = \left[\kappa\,(K - C_0)\left(K^{ab} - \frac{1}{2}(K - C_0)g^{ab}\right) - \sigma g^{ab}\right]\mathbf{e}_b - \kappa\,\nabla^a K\,\mathbf{n}\,. \tag{53}$$

To confirm this, note that $\mathcal{H}$ involves a sum of terms proportional to $\mathcal{H}_n = K^n/n$, $n = 0, 1, 2$. For each n, straightforward calculus gives for the corresponding tensors defined by Eq. (52), $T_n^{ab} = K^{n-1}\left(2K^{ab} - K g^{ab}/n\right)$ and $H_n^{ab} = K^{n-1} g^{ab}$, so that the contribution to the stress (51) is

$$\mathbf{f}_n^a = K^{n-1}(K^{ab} - K g^{ab}/n)\,\mathbf{e}_b - \nabla^a K^{n-1}\,\mathbf{n}\,. \tag{54}$$

The Gaussian energy is not of this form. However, one can use the identity $\mathcal{K}_G = (K^2 - K_{ab}K^{ab})/2$ to show that $T_G^{ab} = K K^{ab} - K^{ac} K_c{}^b$, and $H_G^{ab} = K g^{ab} - K^{ab}$. As a result, the tangential stress, $T^{ab} - H^{ac} K_c{}^b$, appearing in Eq. (51) vanishes identically; its normal counterpart vanishes on account of the CM equations, (29b).

---

[18]If $\mathbf{t} = t^a \mathbf{e}_a$ is the unit tangent vector to the curve, $\mathbf{l} \cdot \mathbf{t} = 0$ or $g_{ab} l^a t^b = 0$ or $l_a t^a = 0$.

The upshot is that there is no Gaussian stress, and the Gaussian modulus $\bar{\kappa}$ does not feature in $\mathbf{f}^a$, a reflection of the topological nature of this energy.

In the case of an interface or a soap film, described by an energy proportional to area, $H = \sigma A$, the stress is tangential with $f^{ab} = -\sigma\, g^{ab}$ and $f^a = 0$. The proportionality to g^{ab} indicates that the tangential stress is isotropic; it is also homogeneous because σ is constant.

The stress in a fluid membrane is quadratic in curvature; as a consequence the stress is generally neither homogeneous nor isotropic. We note, however, that the tangential stress f^{ab} is a polynomial in K^{ab} and g^{ab}; this implies that the orthogonal eigenvectors of K^{ab}, $\mathbf{V}_1$ and $\mathbf{V}_2$, are also eigenvectors of f^{ab}. For pure bending (with $C_0 = 0$, $\sigma = 0$), the eigenvalues of f^{ab} are now easily identified as $f_1 = \kappa\,(C_1^2 - C_2^2)/2 = -f_2$. The tangential bending stress therefore is bounded by f_1 and f_2 which it assumes along these directions. If $C_1 < C_2$, then the membrane is under tension along $\mathbf{V}_1$, and under an equal compression along the orthogonal direction, $\mathbf{V}_2$. Because $f_1 + f_2 = g_{ab} f^{ab} = 0$, this is equivalent to the statement that the tangential bending stress is traceless. This is also not an accident. It can be understood to be a consequence of the scale invariance of the two-dimensional bending energy. Indeed, invariance of the bending energy under a rescaling $\delta\mathbf{X} = \lambda\mathbf{X}$ implies $f^a{}_a = 0$, Capovilla and Guven (2002b); Guven (2005). This is very different from the behavior we observe in an interface where the stress is tensile everywhere. More generally, let H be scale invariant. Equation (47) then implies that the contribution to δH due to $\mathbf{X}$ is proportional to

$$\int dA\, \mathbf{f}^a \cdot \nabla_a \mathbf{X} = \int dA\, f^a{}_a = 0\,, \tag{55}$$

and, for energy densities of the form $\mathcal{H}(g_{ab}, K_{ab})$, there are no boundary additions (these will be discussed below). As a consequence $f^a{}_a = 0$ pointwise. If, however, the energy involves higher derivatives of either g_{ab} or K_{ab}, boundary additions may show up, implying that the trace does not necessarily vanish but is a divergence: $f^a{}_a = \nabla_a g^a$, where g^a is some vector field. A scale invariant energy with this property is easy to construct, given by $\int dA\, \sqrt{\nabla_a K \nabla^a K}$, but without any physical application that we are aware of. For the curious, we remark that one needs to look at energies in higher dimensions involving derivatives to encounter polynomial examples displaying a $g^a \neq 0$.

Notice also that the bending stress vanishes on minimal surfaces, with $K = C_1 + C_2 = 0$, and on spheres, with $C_1 = C_2$. Normal stress is not supported in either case, for this requires nonvanishing gradients in K. On any other surface, the bending stress may change from tension to compression along a given direction. In fact, the integrated force may vanish but the torque need not. We will encounter this behavior in toroidal vesicles.

The force per unit length $\mathbf{f}_\perp$ transmitted across any curve can be expanded with respect to the orthonormal basis, $\{\mathbf{t}, \mathbf{l}, \mathbf{n}\}$, adapted to the curve, $\mathbf{f}_\perp = f_{\perp\perp}\mathbf{l} + f_{\perp\parallel}\mathbf{t} + f_\perp \mathbf{n}$, where we introduce the notation, $A_{\perp\perp} := l^a l^b A_{ab}$, $A_{\perp\parallel} := l^a t^b A_{ab}$, and

$A_{\|\|} := t^a t^b A_{ab}$, for any symmetric tensor, A_{ab}. For the CH energy (9), with stress given by Eq. (53), we identify the forces transmitted along the three directions as

$$f_{\perp\perp} = \kappa/2\left(K_{\perp\perp}^2 - (K_{\|\|} - C_0)^2\right) - \sigma\,, \tag{56a}$$

$$f_{\perp\|} = \kappa\ (K - C_0)\ K_{\perp\|}\,, \tag{56b}$$

$$f_{\perp} = -\kappa\,\nabla_\perp K\,, \quad \text{where} \quad \nabla_\perp = l^a\nabla_a\,. \tag{56c}$$

Note that $f_{\|\|}$ is given by $f_{\perp\perp}$ with $\|$ and $\perp$ interchanged; if the tangential bending stress is tensile along one direction (not necessarily a principal direction), it will be compressive along the orthogonal direction. There will generally be a geometrical in-plane shear $f_{\perp\|}$ if $K_{\perp\|} \neq 0$. There is, of course, no inconsistency with the fluid character of the membrane.

Whereas a fluid sphere may be stress free, a cylinder will generally be under tension along the axial direction (the curvature along the axis $K_{\perp\perp} = 0$ in Eq. (56a)). A cylinder thus needs to be supported by an external axial force to prevent its collapse along the axis. By scale invariance, this will necessarily be accompanied by contraction along the radial direction.

Like surface tension, spontaneous curvature breaks the scale invariance of the energy. Its presence introduces an additional contribution to the tangential stress, $f_S^{ab} = -\frac{1}{2}\kappa\, C_0^2\, g^{ab} - \kappa C_0\left(K^{ab} - K\, g^{ab}\right)$, but no normal stress. On a flat membrane, only the first term remains, so spontaneous curvature contributes isotropically to the tension, Lipowsky (2013); if the membrane is not flat, however, it makes an additional curvature dependent and generally non-isotropic contribution to the principal tangential stresses.

As Eq. (56) indicates, spontaneous curvature biases only the contribution of the transverse curvature to the forces transmitted across curves. A positive spontaneous curvature will thus reduce the axial tension required to support a cylinder. An appropriate spontaneous curvature will even allow tethers to form in the absence of external forces, Lipowsky (2013); Deserno (2015).

An alternative natural appearance of a term linear in K is in the bilayer couple model, which accommodates a fixed area difference between the two layers within the bilayer, Svetina and Žekž (1989); Svetina and Žekš (2014). Indeed, if the bilayer has a constant thickness t, the difference in area between its two sides is given by Eq. (22) to be $t\int dA\, K$. This model has been repurposed recently in the context of tetralayers consisting of pairs of parallel bilayers with different areas. As such, it could play a role in explaining the morphology of the rough endoplasmic reticulum or the nuclear envelope, Guven et al. (2014).

Historical note: the concept of a stress tensor for fluid membranes was first explored, in the small gradient approximation, some time ago by physicists, Evans and Skalak (1980); it was also examined by applied mathematicians and engineers, Jenkins (1977); Steigmann (1999) from a continuum mechanical point of view. It would be fair to say, however that its origins in geometry were overlooked. The conservation laws implied by Euclidean invariance appears to have been first understood by Kusner in the context of minimal and constant mean curvature surfaces,

Kusner (1991). Inexplicably, as recounted in Bernard (2015), pure mathematicians working on the Willmore functional (our symmetric bending energy), Willmore (1982) (proven in 2014, Marques and Neves (2014a, b)) were late to appreciate the implications of Euclidean invariance in this context.

Shape equation from the conservation law: We have discussed stress but we have yet to unpack the contents of the conservation law for $\mathbf{f}^a$. Naively, there appears to be a discrepancy: for whereas there is a single shape equation, we possess three conservation laws (indeed there are more to come). We first show that the shape equation is implied by the conservation of the stress tensor. Using the notation introduced in Eq. (50), we note that the projection onto the normal vector of the conservation law Eq. (49) gives

$$\mathcal{E} := \mathbf{n} \cdot \nabla_a \mathbf{f}^a = \nabla_a f^{\,a} - K_{ab} f^{ab} = 0 \,. \tag{57}$$

Using the expression (53), this reproduces the shape equation for a fluid membrane, (39). The divergence is no longer evident. Even if we were to stop here, we now possess a better understanding of the shape equation: in equilibrium, the coupling of the tangential stress to curvature is the source of the normal stress.

External forces or normal constraints on the geometry will introduce sources on the right hand side of the conservation law. For instance, the source associated with an osmotic pressure is normal, given by $P\,\mathbf{n}$, so Eq. (39) is replaced by $\mathcal{E} = P$.

The projections of Eq. (49) along tangent directions implies

$$\mathcal{E}_b := \mathbf{e}_b \cdot \nabla_a \mathbf{f}^a = \nabla_a f^a{}_b + K_{ab} f^a = 0 \,. \tag{58}$$

This is another constraint between the tangential and normal stresses: and it appear to suggest a symmetry between Eqs. (58) and (57) with tangential and normal stresses interchanged. The character of Eq. (58), however, is very different. In general, if the only degrees of freedom are geometric, this equation amounts always to a geometric identity; it hold for each term in the energy independently of the shape equation. To understand why this is so, recall that infinitesimal tangential deformations are identified with reparametrizations; using the fact that reparametrization acts by Lie derivation along the tangent vector field, this identity is reproduced for any geometrical invariant of the form (16) whether or not the geometry is equilibrated, Guven and Vázquez-Montejo (2013a).

More explicitly, using Eqs. (20a), we can write

$$\begin{aligned} \delta_\parallel H &= \int dA \left(-\frac{1}{2} T^{ab}\, \delta_\parallel g_{ab} + H^{ab} \delta_\parallel K_{ab} \right) \\ &= \int dA \left(\nabla_a f^{ab} + K^{ab} f_a \right) \Phi_b + \int dA\, \nabla_a (\mathcal{H} \Phi^a) \,, \end{aligned} \tag{59}$$

where we have used the definitions of the tangential and normal stresses given by Eq. (51). These are analogues of the contracted Bianchi identities in general relativity which follow from the general covariance of the Hilbert Einstein action.

Matters are less straightforward if local constraints are imposed on the geometry or material degrees of freedom interact with it. It also bears remarking that if the manifest reparametrization invariance is broken, choosing a parametrization adapted to the geometry such as the harmonic parametrization used in the Weierstrass–Enneper representation of a surface, which constrains the metric, Guven and Vázquez-Montejo (2010), the counterparts of Eq. (58) are no longer trivially satisfied. Instead they determine the additional Lagrange multipliers associated with this choice of gauge. It is only modulo this input that the counterpart of Eq. (57) reproduces the shape equation.

Isometric bending: A surprising application of the framework presented in Sect. 5 has been to the description of thin unstretchable sheets (think paper if $\mathcal{K}_G = 0$), an idealization that may be relevant in cell biology when proteins condense on a membrane. Whereas static fluid membranes shear freely; shear is impossible without stretching. The two limits, nonetheless, are described by the geometrical degrees of freedom of the surface. Unstretchability translates geometrically into the constraint that the metric be isometric to some fixed metric. This local congruence is accommodated in the variational principle by introducing a tensor-valued Lagrange multiplier $\mathcal{T}^{ab}$, and replacing H_C in Eq. (46) by[19]

$$H_C - \frac{1}{2}\int dA\,\mathcal{T}^{ab}(u^1, u^2)(g_{ab} - g^{(0)}_{ab})\,, \tag{60}$$

where $g^{(0)}_{ab}$ is this fixed metric. As a consequence, the tangential stress f^{ab} is replaced by $f^{ab} + \mathcal{T}^{ab}$, whereas the normal stress is unchanged. The significant point is that, even though no extra fields are introduced, the stress is no longer completely determined by the local geometry even though the degrees of freedom remain geometrical. The EL equation (57) is replaced by $\mathcal{E} - K_{ab}\mathcal{T}^{ab} = 0$. The presence of the isometry constraint also breaks the identification of tangential deformations with reparametrizations: The tangential projection of the conservation law (58) implies that the multiplier field $\mathcal{T}^{ab}$ is conserved: $\nabla_a \mathcal{T}^{ab} = 0$, Guven and Müller (2008); Guven et al. (2012).

In this context, it is worth looking at the weaker constraint, local incompressibility. The constraint (60) is replaced by

$$H_C - \frac{1}{2}\int d^2u\,\mathcal{A}(u^1, u^2)(\sqrt{g} - \sqrt{g^{(0)}})\,, \tag{61}$$

Now instead of $f^{ab} + \mathcal{T}^{ab}$, we have $f^{ab} + \mathcal{A}g^{ab}$, with an inhomogeneous stress. However, in equilibrium, the tangential conservation law (58) implies that $\mathcal{A}$ is constant; which is the same stress as that associated with a globally constrained area.

Tangential EL with material fields: Let there be material fields: this could be a scalar $S(u^1, u^2)$ or a vector field $V^a(u^1, u^2)$, so that the total energy density is replaced by $\mathcal{H}[g_{ab}, K_{ab}, V^a, S]$. The EL equations for these fields can be determined

[19] The local parametrization is fixed.

conventionally: $\mathcal{V}_a = 0$, $\mathcal{S} = 0$, where $\mathcal{V}_a = \delta H/(\delta V^a \sqrt{g})$ and $\mathcal{S} = \delta H/(\delta S \sqrt{g})$. Neither V^a nor S depends on $\mathbf{X}$, so that the identification of tangential deformations with reparametrizations breaks down for this $\mathcal{H}$. Now Eq. (58) is not reproduced by the argument leading to Eq. (59).

Suppose that V^a and S interact only with the intrinsic geometry and that $\mathcal{H}$ decomposes as $\mathcal{H} = \mathcal{H}(g_{ab}, K_{ab}) + \mathcal{H}_{\text{int}}(g_{ab}, V^a, S)$, with correspond metric stress, $T^{ab} + T^{ab}_{\text{int}}$. Now Eq. (58) implies $\nabla_a T^{ab}_{\text{int}} = 0$, a nontrivial conclusion, Capovilla and Guven (2004a, b). The total stress associated with the fields on the Riemannian manifold described by the metric tensor g_{ab} is conserved. The surface itself does not even need to be in equilibrium.

Note that the conservation law does not necessarily imply the corresponding EL equations for the individual fields. If, however, the two fields do not couple directly, their tangential stresses decouple and are separately conserved. For example, suppose for simplicity that the scalar field is minimally coupled, described by the gradient energy plus a potential:

$$\mathcal{H}_I(g_{ab}, S) = \frac{1}{2} c\, g^{ab} \nabla_a S \nabla_b S + V(S)\,. \tag{62}$$

Now

$$T^{ab}_{\text{int}}[g_{ab}, S] = c\left(\nabla^a S \nabla^b S - \frac{1}{2} g^{ab} g^{cd} \nabla_c S \nabla_d S\right) - g^{ab} V(S) \tag{63}$$

is conserved. This implies the EL equation for S,

$$-c\nabla^2 S + \partial V/\partial S = 0\,. \tag{64}$$

This equation involves the surface geometry only through its metric. The corresponding EL equation for the surface is modified by the addition of a source: $\mathcal{E} = 0$ in (57) is replaced by $\mathcal{E} - T^{ab}_{\text{int}} K_{ab} = 0$. If instead of one scalar field we had two noninteracting fields, say S_1 and S_2, they will be separately conserved and their EL equations of the form (64) are uncoupled. Nonetheless the surface geometry mediates an interaction between these two fields through the shape equation: $\mathcal{E} - T^{ab}_{\text{int}}[g_{ab}, S_1, S_2] K_{ab}$, where $T^{ab}_{\text{int}}[g_{ab}, S_1, S_2] = T^{ab}_{\text{int}}[g_{ab}, S_1] + T^{ab}_{\text{int}}[g_{ab}, S_2]$. In particular in a three-component membrane (described by two relative concentration fields), one can anticipate nontrivial behavior associated with this indirect interaction. There is clearly a lot of nice physics yet to be unearthed, nevermind explored on this topic. A simple non-minimal coupling to curvature, such as a term of the form, $f(S)K$, will introduce a source in the EL equation for S.

Laplace pressure as effective surface stress: If there is a pressure difference P across the membrane, or the volume is fixed, there is an additional term in H given (modulo a possible constant) by $-PV$. Using Stokes theorem for the volume integral of the spatial identity, $\mathbf{div} \cdot \mathbf{x} = 3$, the volume can be expressed as a surface integral,[20]

[20]On a surface with boundary, this identity yields the volume of the cone standing on the surface patch, with its apex located at the origin.

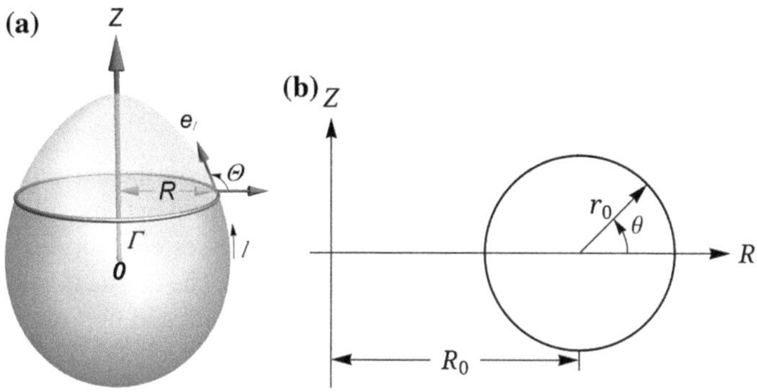

Fig. 3 **a** Parametrization of an axially symmetric surface. **b** Generating curve for a torus

$V = \frac{1}{3}\int dA\, \mathbf{X}\cdot\mathbf{n}$. However, the translational invariance of V implies that $\int dA\, \mathbf{n} = 0$, which in turn implies that $\mathbf{n}$ itself is a surface divergence. It is simple to verify that

$$\mathbf{n} = \nabla_a \mathbf{f}_L^a, \quad \text{where} \quad \mathbf{f}_L^a = \frac{1}{2}\mathbf{X}\times\left(\mathbf{n}\times\mathbf{e}^a\right). \tag{65}$$

It follows that the Laplace force across the surface can, itself, always be treated as an effective surface stress tensor $-P\,\mathbf{f}_L^a$, Guven (2006).

Axially symmetric vesicles: In general, the EL equation cannot be integrated. Axial symmetry, however, implies the existence of a first integral of the shape equation which facilitates the identification of equilibrium states. The traditional way to identify this first integral is to adapt the variational principle to this symmetry, Zheng and Liu (1993); Jülicher and Seifert (1994); Podgornik et al. (1995). However, the conservation law provides an instructive alternative derivation in terms of the stresses shaping an axially symmetric geometry.

First, let us review a few essential properties of axially symmetric surfaces. The parallels and meridians on the surface form the principle directions. The curvatures along these directions are given respectively by $C_\| = \sin\Theta/R$ and $C_\perp = \dot{\Theta}$, where R is the polar radius, Θ is the angle that the tangent along the meridian makes with the polar direction, and a dot represents a derivative with respect to arc length along the meridian, $\dot{} = d/dl = \nabla_\perp$, as indicated in Fig. 3a. Substituting into Eq. (56), we identify the nonvanishing tangential and normal forces per unit length transmitted across a parallel from the membrane below it, $f_{\perp\perp}$ and $f_\perp$ respectively.

We now use the conservation law to identify the first integral: first project $\nabla_a \mathbf{f}^a = P\mathbf{n}$ onto the symmetry axis; now integrate over the source-free region bounded below by a given parallel circle. Using Stokes theorem on the left-hand side, we find that the linear combination of stresses

$$\mathcal{L} := \sin\Theta f_{\perp\perp} - \cos\Theta f_\perp\,, \tag{66}$$

satisfies

$$\mathcal{L} = PR/2 + C/(2\pi R)\,, \tag{67}$$

on this circle, where C is a constant of integration. Note how the derivative is peeled off $f_\perp$ in this construction. This equation, first written down in Capovilla and Guven (2002b) (but without the important constant C), expresses equilibrium in terms of an algebraic balance of tangential and normal stresses. The constant is identified as the total external axial force acting from above; an equal and opposite force must counteract it somewhere else. In the absence of such a force, $C = 0$. We will examine one situation where it vanishes (even though radial external forces act on the vesicle) and two where it does not: one for topological reasons; the other due to external axial forces bearing down on the poles of a vesicle.

Stress and Torque conservation from Euclidean invariance: It is straightforward using the auxiliary framework to show that the change in energy under a deformation of the surface $\delta\mathbf{X}$ is given by

$$\delta H[\mathbf{X}] = \int dA\,\mathcal{E}\,\mathbf{n}\cdot\delta\mathbf{X} + \int dA\,\nabla_a\left[-\mathbf{f}^a\cdot\delta\mathbf{X} + H^{ab}\mathbf{e}_b\cdot\delta\mathbf{n}\right]. \tag{68}$$

This identity follows from the first variation of Eq. (46): the first term on the right in Eq. (68) involves the EL derivative with respect to $\mathbf{X}$ and it vanishes in equilibrium if the region is source-free; the second term collects in a divergence the two terms linear in derivatives, $\partial_a\delta\mathbf{X}$ and $\partial_a\delta\mathbf{n}$, appearing in δH_C when $\mathbf{X}$ and $\mathbf{n}$ are varied. On any patch of free surface, it depends only on the boundary behavior of the stress, $\mathbf{f}^a$, and the response to changes in the curvature, H^{ab}; this is not an accident.

Using Eq. (68), it is simple to reproduce the conservation law for the stress tensor, Eq. (49). For under a constant translation $\delta\boldsymbol{a}$, the energy is unchanged within any surface patch so that $\delta H = 0$; as a result, $\delta\boldsymbol{a}\cdot\int dA\,\nabla_a\mathbf{f}^a = 0$. Because the patch is arbitrary, the integrand must vanish pointwise, reproducing the conservation law. This is not surprising: translational invariance was already manifest in the variational principle.

The rotational invariance, on the other hand, was not. Under a constant rotation $\delta\boldsymbol{\omega}$, one has $\delta\mathbf{X} = \delta\boldsymbol{\omega}\times\mathbf{X}$ and $\delta\mathbf{n} = \delta\boldsymbol{\omega}\times\mathbf{n}$ so that, in equilibrium, $\delta\boldsymbol{\omega}\cdot\int dA\,\nabla_a\mathbf{m}^a = 0$, where

$$\mathbf{m}^a = \mathbf{X}\times\mathbf{f}^a + H^{ab}\,\mathbf{e}_b\times\mathbf{n}\,. \tag{69}$$

Thus, $\mathbf{m}^a$, which is identified as the surface torque tensor, Capovilla and Guven (2002b); Müller et al. (2007), is also conserved: $\nabla_a\mathbf{m}^a = 0$. The first term appearing in $\mathbf{m}^a$ represents the moment of the local stress, whereas the second term—the bending moment—is position independent, a contribution originating in the curvature dependence of the energy.

Consider the effect of a scaling $\delta\mathbf{X} = \lambda\mathbf{X}$ on any functional of the form (16): now Eq. (68) implies

$$\delta H[\mathbf{X}] = \lambda \int dA\, \mathcal{E}\, \mathbf{n} \cdot \delta \mathbf{X} - \lambda \int ds\, \mathbf{f}_\perp \cdot \mathbf{X} . \tag{70}$$

Suppose that H has a consistent scaling dimension. If $H = A$, then $H[\Lambda \mathbf{X}] = \Lambda^2 H[\mathbf{X}]$. furthermore $\mathcal{E} = K$. and $\mathbf{f}^a = -g^{ab}\mathbf{e}_b$. One identifies

$$2A = \int dA\, K\, \mathbf{n} \cdot \mathbf{X} + \int ds\, \mathbf{l} \cdot \mathbf{X} . \tag{71}$$

This is the Jellett–Minkowski identity, identified by Jellett mid-nineteenth century but usually attributed only to the latter mathematician, Jellett (1853). A corollary is that there do not exist closed minimal surfaces.

Stresses in the Monge representation: It is instructive to compare the manifestly covariant framework described here with its counterpart using the height function representation of the surface, treating the energy as a functional of this scalar field on the reference plane. We now examine the forces and torques along this plane and orthogonal to it.

One can exploit the Euclidean invariance of the energy with respect to translations on the base plane, exactly as one does for a classical scalar field (electrostatics say), to identify the conserved horizontal stress. But we do not need to: because we can also project $\mathbf{f}^a$ onto the base plane. In the quadratic approximation, this stress is given by, Fournier (2007)

$$T_{ij} = \kappa T^B_{ij} + \Sigma T^0_{ij} , \tag{72}$$

where

$$T^B_{ij} \approx \nabla^2 h \left(\partial_i \partial_j h - \frac{1}{2} \nabla_0^2 h \delta_{ij} \right) - \partial_i (\nabla_0^2 h) \partial_j h , \tag{73}$$

and

$$T^0_{ij} \approx -\left(1 + (\nabla_0 h)^2/2\right) \delta_{ij} + \partial_i h\, \partial_j h . \tag{74}$$

Not surprisingly, modulo the constant term associated with the area of the base plane, T^0_{ij} assumes the form of the stress tensor of a free scalar field h on the plane. Compare Eq. (74) with (63). We observe also that T^B_{ij} is not symmetric so that, whereas $\partial_i T_{ij} = 0$, if the indices are switched the divergence does not vanish: $\partial_i T^B_{ji} \neq \partial_i T^B_{ij}$. There is no such ambiguity in the reparametrization invariant approach. Notice also that in T^0_{ij} neither homogeneity nor isotropy are manifest. This apparent spatial variation is an artifact of the planar projection.[21] Already in this simple setting, one can see the advantage of possessing the covariant description. Even if one decides to perform calculations in the height function representation, the covariant approach provides

[21] Intriguingly, the quadratic contribution to T^B_{ij} is trace-free in this approximation, a property we would associate with scale invariance. Yet the area itself is clearly not scale invariant. The source of this peculiarity is that, in the quadratic approximation in gradients of h, the area is represented by a massless two-dimensional scalar field on the plane which is scale invariant if the plane is scaled, but not if h is. On the other hand, T^B_{ji} is not trace-free but should not have been expected to be.

an unambiguous statement about the nature of the underlying stress. The apparent discrepancies using height functions arise because T_{ij} is not the physically significant tangential stress but its projection onto a plane and the fact that heights are treated differently from locations on this plane.

The conserved normal force density is $N_i = \kappa \partial_i (\nabla_0^2 h) - \sigma \partial_i h$. The conservation laws for T_{ij} and N_i together encode the information content of Eq. (49), correct to quadratic order. As we saw, they are not independent.

The invariance of the energy with respect to rotations about an axis orthogonal to the base plane implies the conservation of the vertical torque,

$$M_i = \kappa \left(\nabla^2 h - C_0\right) \varepsilon_{ij} \partial_j h + \bar{k} \left(\nabla^2 h \varepsilon_{ij} - \partial_i \partial_k h \varepsilon_{kj}\right) \partial_j h \,. \tag{75}$$

To account for the full rotational invariance of the surface energy, one needs to consider rotations about two orthogonal axes lying in the plane. This involves the rotation of the base plane itself, conflating the scalar field and the reference geometry, a symmetry without any analogue in the theory of a scalar field on the plane. This is probably why this was not considered in the height function until recently, Fournier (2007), several years after the covariant description. Taking the appropriate projections, we identify the horizontal torques

$$M_{ij} = \kappa \left(\nabla^2 h - C_0\right) \varepsilon_{ij} + \bar{\kappa} \left(\varepsilon_{ij} \nabla^2 h - \varepsilon_{ik} \partial_k \partial_j h\right) \,. \tag{76}$$

Forces and torques without gauges: In biology, membranes are invariably shaped by external forces or constraints; often these act locally: for example, the final stage of endocytosis may involve the constriction of membrane necks by dynamin spirals, Kozlov (2001); Morlot and Roux (2013); McDargh et al. (2016). There has also been a considerable amount of work on the interactions between membrane bound particles (read proteins) that are mediated by the deformed membrane geometry, Goulian et al. (1993); Kralj-Iglič et al. (1996, 1999); Weikl et al. (1998); Kim et al. (1998); Yolcu et al. (2011, 2012); Yolcu and Deserno (2012); Fournier (2014); Haussman and Deserno (2014); Božič et al. (2015); Fournier and Galatola (2015); Schweitzer and Kozlov (2015). The covariant stress tensor has also been shown to provide insight into these processes, permitting one to understand non-perturbative behavior, without the need to resort to triangulations or simulations, Müller et al. (2005a, b); it also provides a rigorous framework guiding the design of the computational setup and the interpretation of results, Reynwar et al. (2007). The subject of membrane fluctuations or Casimir forces has received considerable attention and has been the subject of reviews, Deserno (2009); Yolcu et al. (2014). It would also appear that the covariant language is the natural one to use in order to progress beyond the quadratic or Gaussian approximation in height functions in this context. The geometrical nature of the problem indicates that these corrections will involve geometrical invariants.

In this section, we show how the forces and torques acting on the membrane (or transmitted by it) are identified in our framework. Consider a number of localized sources acting on the membrane. These could also be particles interacting with the

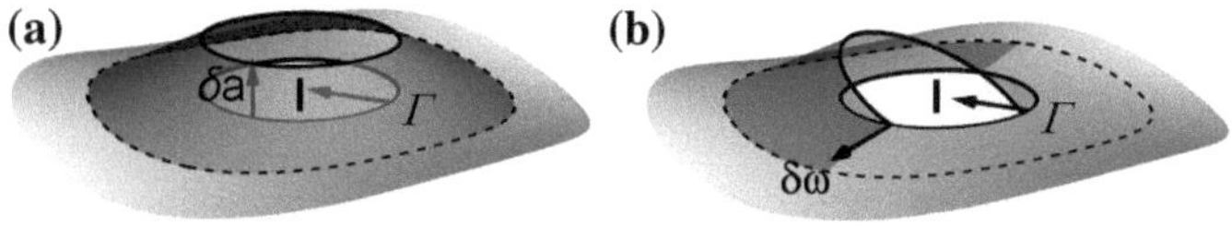

Fig. 4 Local **a** translation and **b** rotation of the contours enclosing localized sources on the membrane. The vector **l** is the normal to the contour Γ pointing into the source. Their magnitude has been exaggerated for illustration purposes

membrane. To identify the force on the membrane associated with any particular source, one needs to determine the change in the energy when this source alone is displaced. This sounds like a complicated operation because its displacement will drag the membrane along with it. Fortunately, we are only interested in deformations around equilibrium, so to first order the membrane deformation turns out to be irrelevant. We thus let $\delta\mathbf{X}$ be *any* deformation reducing to a constant vector $\delta\mathbf{a}$ on the curve Γ bounding one of these sources while vanishing on all other boundaries, as illustrated in Fig. 4a. Now we use Stokes theorem in Eq. (68) to recast the divergence as an integral along Γ. The change in the energy of the membrane is then given by the intuitively simple expression:

$$\delta H = -\delta\mathbf{a}\cdot\mathbf{F}\,; \quad \mathbf{F} = \int_\Gamma ds\; \mathbf{f}_\perp\,. \tag{77}$$

where $\mathbf{f}_\perp$ was defined below Eq. (49). This is the work done on the membrane by the source inside Γ when it is displaced a distance $\delta\mathbf{a}$; as such, the vector $\mathbf{F}$ is identified as the force on the source (or particle), or equivalently minus the force on the membrane, Müller et al. (2005a, b); importantly, it is determined completely by the surface geometry in the neighborhood of this boundary. What exactly is occurring inside Γ is irrelevant, it may be treated as a black box. If there is membrane inside, and we decide to look inside, the distribution of the normal force acting on it is given by $\mathcal{E}\,\mathbf{n}$, and the total force (not necessarily normal) is determined by integrating over the interior of Γ covered by membrane, Phillips et al. (2009). Of course, any geometry subjected to appropriate sources represents an equilibrium: simply evaluate $\mathcal{E}$ to determine them. There is, however, no guarantee that such sources will be physical.

Note that the integrated conservation law over the free surface implies Newton's third law: if there are N boundaries with associated forces $\mathbf{F}_I$, $I = 1, \ldots, N$, then

$$0 = \int dA\, \nabla_a \mathbf{f}^a = \sum_{I=1}^{N} \mathbf{F}_I\,. \tag{78}$$

The torque associated with an external source can be identified by examining the response of the energy to a rigid rotation of the source. Under a rigid infinitesimal rotation of Γ through an angle $\delta\boldsymbol{\omega}$, the corresponding change in the energy is

$$\delta H = -\delta\boldsymbol{\omega} \cdot \mathbf{M}\,, \quad \mathbf{M} = \int_\Gamma ds\, \mathbf{m}_\perp\,, \tag{79}$$

where $\mathbf{m}_\perp = l_a \mathbf{m}^a$ and $\mathbf{m}^a$ was defined in Eq. (69). The Gaussian term in the CH energy contributes through H_G^{ab} to $\mathbf{m}^a$, Fournier (2007). It also plays an important role in the local boundary conditions. However, it does not contribute to the total torque: this is because the corresponding contribution to $\mathbf{m}_\perp$ is proportional to the arc length derivative of the surface normal vector along the boundary, $\mathbf{m}_\perp = \bar{\kappa}\mathbf{n}'$, and thus integrates to zero.

In a manner analogous to that for $\mathbf{f}_\perp$, with an obvious notation, we can express $\mathbf{m}_\perp = m_{\perp\parallel}\mathbf{T} + m_{\perp\perp}\mathbf{l} + m_{\perp}\mathbf{n}$. Along a parallel circle on an axially symmetric geometry, the only surviving component is $m_{\perp\parallel}$, given by

$$m_{\perp\parallel} = f_{\perp\perp}\mathbf{X}\cdot\mathbf{n} - f_\perp \mathbf{X}\cdot\mathbf{l} - H_{\perp\perp}\,. \tag{80}$$

The conservation laws for stress and torque imply that the integrals in (77) and (79) will be identical on any contour homotopically equivalent to Γ outside of sources. In particular, if the geometry possesses symmetries, the contour can also be deformed to exploit these symmetries. This stratagem was used to determine the forces and torques mediated by the membrane between identical particles on a membrane, Müller et al. (2005a, b). One has, perhaps unsurprisingly, nonlinear analogues of Gauss' law in electrostatics.

Horizontal force on the ramp dipole: The line integral (77) permits one to determine the horizontal force between two elements of the minimal dipole discussed earlier. One can show that it is always attractive, and given in the small gradient approximation by (T^0_{ij} is defined in Eq. (74))

$$F = \sigma \int dy\, T^0_{xx} = \frac{1}{2}\Sigma \int dy\, (\partial_y h)^2 = 2\pi p^2 \Sigma / R\,. \tag{81}$$

Geometrically, this is the length added to the midline, Guven et al. (2014). Minimal ramps of opposite chiralities attract; if they were the same, they would repel, Müller et al. (2005a). To prove this note that the contour can be deformed so as to coincide with the "square" Γ illustrated in Fig. 1a. As shown in Guven et al. (2014), the minimal dipole is stabilized by nonharmonic corrections. The simplest example involves the local addition to each helicoid of the solution to the Helmholtz equation, $h \approx K_0(\lambda r)\,\varphi$, described earlier. The integrated stress associated with such a correction is always negative which implies repulsion. The dipole size is set by the competition between this short range repulsion (associated with bending) and the long range attraction associated with tension.

Note that the behavior of the minimal ramp persists in the nonlinear theory, Müller et al. (2005a). For a minimal dipole, the force on either ramp is given by $\mathbf{F} = \Sigma \int_\Gamma ds\, \mathbf{l}$. By symmetry, the force is given by $\Sigma\Delta L$, where ΔL is—as before—the length added to the midline.

It bears emphasizing that, even in the linearized theory, this approach always outperforms the approach still overwhelmingly used in this field to determine forces which requires integrating the energy density over the entire surface and then differentiating with respect to the placement of the sources. Here, it is sufficient to know the geometry in the neighborhood of a single bounding curve: avoiding the unnecessary integration and subsequent differentiation.

Boundary conditions: If the free surface terminates on a free boundary, with an associated line tension or its own bending energy, the second term in Eq. (68) provides the appropriate boundary conditions, circumventing the necessity to reevaluate them anew every time we have a boundary to contend with. The explicit use of the stress tensor in this context was first made in Capovilla et al. (2002). An elegant derivation of these boundary conditions, including not only boundary tension but also boundary bending energy, using the methods of exterior differential calculus was provided in Tu and Ou-Yang (2003, 2004).

If an interface separates two phases with distinct physical parameters, the difference in the contributions from the two permits one to identify the appropriate matching conditions, Müller (2007). If the membrane adheres to a substrate, with a contact potential, (68) facilitates the identification of the discontinuity in the normal curvature at the boundary of the region of contact, Capovilla and Guven (2002a). A rather more comprehensive treatment of the adhesion process using the framework presented here is given in Deserno et al. (2007).

Surfaces as emergent: An alternative to the auxiliary route to the shape equation is again to focus on g_{ab} and K_{ab} as independent tensor fields but, instead of the structure equations, to impose the Gauss–Codazzi and Codazzi–Mainardi equations as constraints, Guven and Vázquez-Montejo (2013a). This approach, as we will see, has some surprising consequences.

Consider a Riemannian manifold with a metric g_{ab} (there is a whiff of gravity here), coupling to a symmetric tensor K_{ab}. If these two tensor fields satisfy the Gauss–Codazzi and Codazzi–Mainardi equations (29), they describe the induced metric and extrinsic curvature on a surface embedded in three-dimensional Euclidean space, Spivak (1999). We thus see that these equations are both necessary and sufficient conditions for forming a surface.

It is now possible, in principle, to address geometric questions about surfaces without any explicit reference to its environment; the surface itself is an emergent equilibrium entity. In this approach one does not have a surface to speak of away from equilibrium.

In contrast with the auxiliary approach, there is even no need to introduce the embedding, $\mathbf{X}$, explicitly in the variational principle. Let us replace $H[\mathbf{X}] = \int dA\, \mathcal{H}[g_{ab}, K_{ab}]$ by

$$H_c[g_{ab}, K_{ab}, \Lambda, \lambda^a] = H[g_{ab}, K_{ab}] + I[g_{ab}, K_{ab}, \Lambda, \lambda^a] , \tag{82}$$

where

$$I = \frac{1}{4}\int dA\, \Lambda\, \mathcal{C}_\perp - \frac{1}{2}\int dA\, \lambda^a \mathcal{C}_a , \tag{83}$$

with $\mathcal{C}_\perp := \mathcal{R} - K^2 + K_{ab}K^{ab}$ and $\mathcal{C}^a := \nabla_b(K^{ab} - g^{ab}K)$. The multiplier fields Λ and λ^a enforce Eq. (29) permitting g_{ab} and K_{ab} to be treated as independent variables. Notice that K_{ab} will tag along even if H depends only on g_{ab}, such as it does in an interfacial energy. For the reader familiar with general relativity, I has all the appearance of the Arnowitt–Deser–Misner (ADM) action in the Hamiltonian formulation of the theory, Arnowitt et al. (1959). This is a two-dimensional accident! For if we were genuinely working in four dimensions, we would need to replace the single Gauss–Codazzi equation by the twenty equations (33), and Λ by a tensor-valued Λ^{abcd}, with the symmetries of the Riemann tensor. Curiously, the Codazzi–Mainardi equations in four and higher dimensions are completely determined by their Gauss–Codazzi counterparts (Thomas) so that the corresponding multipliers are redundant: $\lambda^a \to \lambda^{abc} = 0$.

Remarkably, one never needs to identify these multiplier fields explicitly in the derivation of the shape equation.

Instead of $\mathbf{X} \to \mathbf{X} + \delta\mathbf{X}$, we have

$$\delta(H + I) = \int dA \left[-\frac{1}{2}(T^{ab} + \mathcal{T}^{ab})\, \delta g_{ab} + (H^{ab} + \mathcal{H}^{ab})\, \delta K_{ab} \right] + BT \quad (84)$$

where T^{ab} and H^{ab} are the functional derivatives of H wrt g_{ab} and K_{ab} defined earlier; $\mathcal{T}^{ab}$ and $\mathcal{H}^{ab}$ are the counterparts for the constraint term I. BT represents terms collected in a divergence after integration by parts. The equilibrium states of the surface are described by the coupled pdes on the two-dimensional Riemannian manifold:

$$T^{ab} + \mathcal{T}^{ab} = 0\,; \quad (85a)$$
$$H^{ab} + \mathcal{H}^{ab} = 0\,, \quad (85b)$$

supplemented with $\mathcal{C}_\perp = 0$ and $\mathcal{C}_a = 0$. Equation (85a) are the analogues of the Einstein equations. Equation (85b) are their counterparts for K_{ab}.

The EL derivatives, $\mathcal{T}^{ab}$ and $\mathcal{H}^{ab}$, originating in the constraints are model independent. Both are linear in $\mathcal{L}_\Lambda g_{ab}$ and $\mathcal{L}_\Lambda K_{ab}$:

$$\begin{aligned}\mathcal{T}^{ab} &= \frac{1}{4}\left(g^{ab}K^{cd} - g^{cd}K^{ab}\right) \mathcal{L}_\Lambda g_{cd} \\ &\quad + \frac{1}{2}\left(g^{ac}g^{bd} - g^{ab}g^{cd}\right) \mathcal{L}_\Lambda K_{cd}\,; \quad (86a)\\ \mathcal{H}^{ab} &= \frac{1}{4}\left(g^{ac}g^{bd} - g^{ab}g^{cd}\right) \mathcal{L}_\Lambda g_{cd}\,, \quad (86b)\end{aligned}$$

where

$$\mathcal{L}_\Lambda g_{ab} = 2K_{ab}\Lambda + \mathcal{L}_\lambda g_{ab}\,; \quad (87a)$$
$$\mathcal{L}_\Lambda K_{ab} = \left(-\nabla_a\nabla_b + K_{ac}K^c{}_b\right)\Lambda + \mathcal{L}_\lambda K_{ab}\,. \quad (87b)$$

If we consult Eqs. (19) and (20a), and restore our Euclidean background, then $\mathbf{\Lambda} = \Lambda\, \mathbf{n} + \lambda^a\, \mathbf{e}_a$ is identified as the generator of a surface displacement with the identification of Λ with Φ and λ^a with Φ^a in (18). But remember that here λ^a and Λ are the generalized forces coupling the two fields to form a surface. Their role is not to displace.

To identify the shape equation, the contraction of Eq. (86) provides a very useful identity:

$$K_{ab}\mathcal{T}^{ab} = \frac{1}{4}\left(g^{ab}\nabla^2 - \nabla^a\nabla^b + K^{ac}K_c{}^b - g^{ab}K_{cd}K^{cd}\right)\mathcal{L}_{\mathbf{\Lambda}}g_{ab}\,, \tag{88}$$

equating the contraction $K_{ab}\mathcal{T}^{ab}$ to a differential expression linear in $\mathcal{L}_{\mathbf{\Lambda}}g_{ab}$. Significantly, $\mathcal{L}_{\mathbf{\Lambda}}K_{ab}$ does not appear.

Now let us apply this framework to *gravitational impostors*, described by a Hamiltonian depending only on the metric, $H = H[g_{ab}]$. Now $H^{ab} = \delta H/\delta K_{ab} = 0$ so that Eq. (85b) implies that $\mathcal{H}^{ab} = 0$ as well. But the identity (86b) then implies that $\mathcal{L}_{\mathbf{\Lambda}}g_{cd} = 0$ or that $\mathbf{\Lambda}$ generates surface isometries. The identity (88) now implies that $K_{ab}\mathcal{T}^{ab} = 0$. The Einstein equations Eq. (85a) finally imply that $-K_{ab}\,T^{ab} = 0$: an unusually short story.

For the trivial example of an interface, $H = \sigma A$, with $T^{ab} = -\sigma g^{ab}$, we reproduce the equation, $K = 0$. The stationary states are minimal surfaces.

This is somewhat mysterious. The Lagrange multipliers appear to have been very obliging: in the derivation of the surface EL equations, they do their job but we never even need to identify $\mathbf{\Lambda}$ explicitly. But let's look at them. For the area, $H = \sigma A$, the trace of T^{ab}, $T^a{}_a = -2\sigma$. Equation (86) implies that $\mathcal{T}^a_{\ a} = -\frac{1}{2}\,\mathcal{L}_{\mathbf{\Lambda}}K$. But $\mathcal{L}_{\mathbf{\Lambda}}K = \left(-\nabla^2 + \mathcal{R}\right)\Lambda$, so that

$$\left(-\nabla^2 + \mathcal{R}\right)\Lambda = -4\sigma \tag{89}$$

The determination of Λ decouples from that of λ^a. One can show that the appropriate boundary conditions are $\Lambda = 0$. Because Eq. (89) is inhomogeneous, the isometry will be nontrivial! It is also uniquely determined by the equilibrium geometry.

The differential operator $\mathcal{D} = -\nabla^2 + \mathcal{R}$ also appears in the second variation of the area:

$$\delta^2 A = \int dA\, \Phi\, \mathcal{D}\, \Phi\,, \tag{90}$$

where Φ is the normal deformation of surface. To see this use Eq. (36) to obtain for the second variation of the area about an equilibrium, $\delta^2 A = \int dA\, \Phi\, \delta_\perp K$. Using Eq. (37) for $\delta_\perp K$ we recover Eq. (90). Negative eigenvalues signal instability.

Now let us look at solutions of Eq. (89). In particular, consider a catenoid of neck radius R_0, parametrized $R(l)/R_0 = \sqrt{1 + (l/R_0)^2}$, $Z(l) = \operatorname{arcsinh} l/R_0$, bounded between 2 rings separated a distance $2L$ along the meridian. There exists an exact solution for Λ, negative everywhere, vanishing on the boundaries, with a minimum on the neck, as illustrated in Fig. 5b, Guven and Vázquez-Montejo (2013a). The solution diverges as $L \to 1.5088 R_0$.

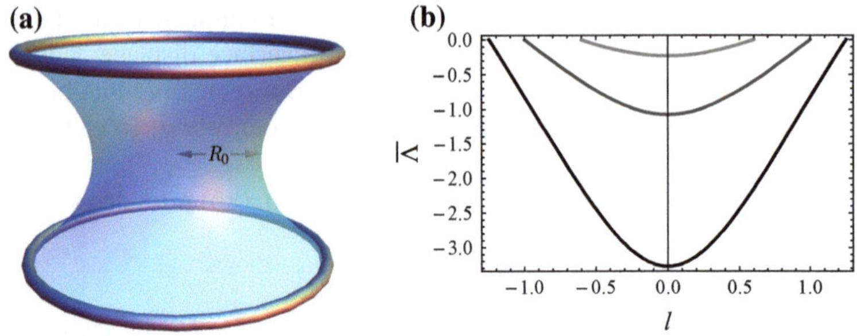

Fig. 5 **a** Soap Bridge **b** Λ versus l

Now look at the spectrum of the operator $\mathcal{D}$: $(-\nabla^2 + \mathcal{R})\Phi_n = E_n\Phi_n$, $E_0 < E_1 < E_2 < \cdots$. If L is small, $E_0 > 0$. The spectral expansion of Λ, $\Lambda = \sum \Lambda_n \Phi_n$, indicates that $\Lambda \to \infty$ correlates with $E_0 \to 0$. Singularities in Λ correlate with the onset of instability. A new criterion is identified for the onset of surface instability. This is an intuitive result: instabilities are reflected in our inability to find an equilibrium pair, g_{ab} and K_{ab}, satisfying Eq. (29).

6 External Forces and Nontrivial Topology

In this section we present examples of surface states minimizing the CH energy when the geometry is subjected to localized external forces or topological constraints, analyzing the connection between the stress and the geometry using the framework presented here. The first example considers the response of a spherical vesicle, of fixed area, to the radial constriction of its equator. We next present new insight into an old problem: the equilibrium of toroidal vesicles; we show how the topology provides sources for both forces and torques in the vesicle, and describe the distribution of stress associated with these sources. Finally, we demonstrate how the conformal invariance of the bending energy can be exploited to examine the morphologies of a vesicle (not necessarily axially symmetric) subjected to localized external forces bringing two points (or small patches) into contact.

6.1 Constriction of a Spherical Vesicle

If the membrane possesses spherical topology, and is free of axial forces, then $C = 0$ on the right side of Eq. (67). Let the vesicle have a fixed area $A_0 = 4\pi R_0^2$, and be subjected to an equatorial constriction provided by an external rigid ring of radius r_0. To keep matters simple, we do not admit spontaneous curvature or fix the volume. There are no axial forces, so $\mathcal{L} = 0$ (with $C_0 = 0$) in Eq. (67) everywhere except

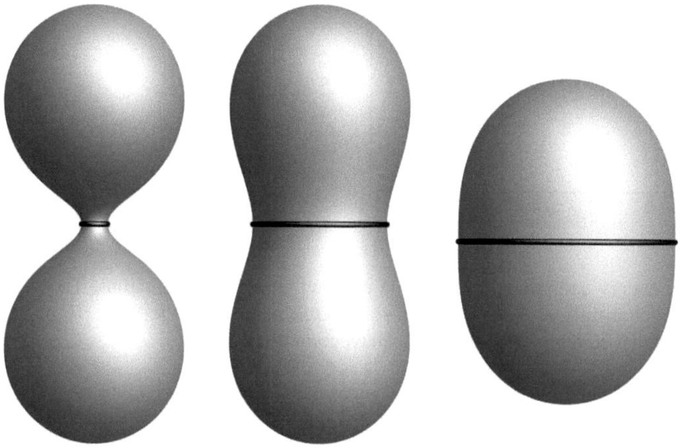

Fig. 6 Equatorial constriction of a spherical vesicle

along the equator where there is a radial source. The equation $\mathcal{L} = 0$ is solved by the shooting method for values of $0 \leq r_0 \leq R_0$. The one free parameter is σ, which is tuned to fix the area to A_0. As r_0 is decreased one observes the following morphological sequence: if the constriction is moderate ($0.84 \leq r_0/R_0 < 1$) the deformation is prolate, represented in Fig. 6c; if $0.2 \leq r_0/R_0 \leq 0.84$ the vesicle develops a waist as illustrated in Fig. 6b[22]; if r_0 is reduced further, so that $r_0/R_0 \leq 0.2$, the geometry morphs into two spherical lobes connected by an increasingly narrow neck (Fig. 6a); in the limit $r_0 \to 0$, it resembles two touching spheres of radius $R_0/\sqrt{2}$. Various questions suggest themselves. What is the geometry in the small neck connecting these two spheres and what is the force squeezing this neck? This force is nonvanishing, i.e., there are sources; thus the neck cannot be a catenoid. How then does it differ?

If the traction along the edge instead were outward, the vesicle would tend to become increasingly oblate as r_0 is increased, tending to a limiting geometry formed by two flat disks of radius $\sqrt{2}\, R_0$, glued together along their common perimeters with diverging energy and force. The force constricting the membrane is completely encoded in the membrane geometry in the neighborhood of the equator. To see this, consider the change in energy under a radial deformation of the equator, $\delta \mathbf{X} = \delta c\, \hat{\mathbf{r}}$. Using Eq. (68), one finds that $\delta H = -\delta c\, F$, where the equatorial constriction F is given by the jump across the equator at $l = 0$:

$$\frac{F}{2\pi r_0 \kappa} = \Big[-\sin\Theta\, f_\perp + \cos\Theta\, f_{\perp\perp} \Big]_{-\epsilon}^{\epsilon} . \tag{91}$$

[22]Unlike the prolate, this geometry is stable with respect to membrane slippage under the ring.

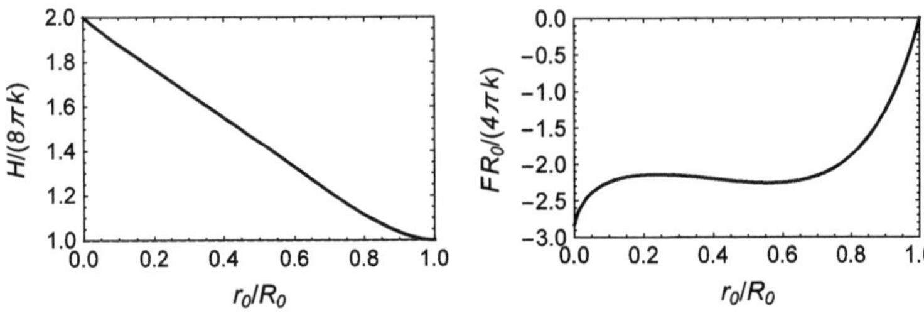

Fig. 7 Total energy and radial compression as functions of the ring radius. The bending energy increases monotonically as the equator is constricted, tending to the value $16\pi\kappa$ as $r_0 \to 0$; in contrast, the radial compression, does not behave monotonically

The tangent angle $\Theta = \pi/2$ so that the two contributions are equal and opposite:

$$\frac{F}{2\pi r_0 \kappa} = C'_{\perp 0}\Big|_{-\epsilon}^{\epsilon} = 2\Theta''_0 \,. \tag{92}$$

The presence of the force is signaled by a discontinuity in the derivative of the normal (meridian) curvature. along the equator. An alternative derivation is provided in Božič et al. (2014). Contrast this with line tension where the discontinuity is in the curvature $C_\perp$ itself.

Note that total equatorial force on the vesicle vanishes, in much the same way as the total Laplace force vanishes on a closed vesicle does, $\int dA\, \mathbf{n} = 0$. There is also an intriguing duality between the functional form of the expression within parenthesis in Eq. (91) and the first integral, $\mathcal{L}$ given by Eq. (66).

Notice that, whereas the energy increases almost linearly with r_0, the dependence of F on r_0 illustrated in Fig. 7b is non-monotonic; its behavior displays a striking correlation with the morphological changes in the membrane, increasing in magnitude as the membrane becomes prolate; relaxing slowly as the belt tightens but increasing again as the two lobes develop. In particular, not only is it nonvanishing in the limit $r_0 \to 0$, its magnitude is a global maximum. Using the equation, $\mathcal{L} = 0$, it is simple to show that the polar radius can be expanded $R(l)/r_0 \approx 1 + (\ell/r_0)^2 + F_0\, \pi r_0 |\ell/r_0|^3/12$ at the neck, when $r_0/R_0 \ll 1$, indicating explicitly the curvature derivative singularity (the third derivative) proportional to the limit force F_0. Significantly, it is not approximated by a catenoid of neck radius r_0, which is given exactly by $R(l)/r_0 = 1 + (\ell/r_0)^2$.

We have already seen that the scale invariance of the bending energy has physical implications. The constricted vesicle involves two scales, R_0 defined by the area, and the equatorial radius, r_0. The corresponding constraints can be introduced explicitly into the variational principal so that the unconstrained functional to be minimized is

$$H[\mathbf{X}] = H_{CH}[\mathbf{X}] + \sigma\,(A - 4\pi R_0^2) + \int ds\, \mathcal{F}(s)\,(|\mathbf{X}| - r_0)\,. \tag{93}$$

The function $\mathcal{F}(s)$ is a new local Lagrange multiplier enforcing the constraint on the equatorial radius. If axial symmetry is relaxed it will not be constant. If this constraint is removed, then $\sigma = 0$, a consequence of the scale invariance of bending energy. In equilibrium, one identifies $F := \int ds\, \mathcal{F}(s) = \partial H_B / \partial r_0$. This is the familiar expression involving H_B, and thus requiring knowledge of the complete vesicle geometry.

On casual inspection, Fig. 7a would suggest that H_B depends linearly on r_0. The wiggles may be small but they are real, as comparison with Fig. 7b indicates.

Consider now the effect of a membrane rescaling $\mathbf{X} \to \Lambda \mathbf{X}$ on H. One has

$$H[\Lambda \mathbf{X}] = H_{CH}[\mathbf{X}] + \sigma\,(\Lambda^2 A - 4\pi R_0^2) + \int ds\, \mathcal{F}(s)\, \Lambda(\Lambda|\mathbf{X}| - r_0)\,. \qquad (94)$$

In equilibrium $dH/d\Lambda = 0$ when $\Lambda = 1$. This implies that $r_0\, F = -2\sigma A$: σ thus also determines the force; its sign correlating with it. Note that σ vanishes in the limit $r_0 \to 0$. Elementary calculus then implies that the limiting traction is given by $F_0 = \lim_{r_0 \to 0} \partial\sigma/\partial r_0$, reflecting the scale-free nontrivial neck geometry lurking in this limit.[23]

The model presented here is a very simplified description of the physics: the ring is assumed rigid. If the ring is elastic, it need not remain circular, and the contraction process will involve non-axially symmetric deformations of the vesicle. Modeling the constriction of a membrane neck by a dynamin spiral will necessarily involve both the breaking of axial symmetry as well as the deformation of the spiral, Nam et al. (2012); McDargh et al. (2016).

6.2 *Topology as a Source of Stress*

Nontrivial topology can also provide a source of stress. On a torus, closed curves along both the wheel and tube are homotopically nontrivial. Such geometries are not only of academic interest: toroidal vesicles were first observed experimentally some time ago, not only the common or garden single-holed variety, Mutz and Bensimon (1991), but also genus two geometries, Michalet and Bensimon (1995). Shape transitions in toroidal vesicles have also recently been examined both experimentally and numerically in Noguchi and Imai (2015). Indeed, topology plays a role in almost all intracellular membranes—the Golgi, the Endoplasmic reticulum (rough and smooth alike), as well as the inner membrane of the mitochondrion—exhibit highly nontrivial topologies.

Here we will limit our discussion to an axially symmetric toroidal membrane in order to demonstrate how the topology of a membrane can, itself, provide a source

[23] This is well known in the context of global constraints. In a symmetric closed fluid membrane subject to area and volume constraints, the identity $2\sigma A - 3PV = 0$ is a consequence of the scale invariance of the bending energy, Svetina and Žekž (1989).

of stress in the membrane. Of course, axially symmetric torii have been studied extensively; yet extraordinarily, to our knowledge, never from this more physical point of view. Let the torus have a wheel radius R_0 and tube radius r_0 (see Fig. 3b). Let $\theta = \ell/r_0$ be the angle made along the tube with the outer radial direction, so that $R = R_0 + r_0 \cos\theta$ is the polar radius (the tangent angle is $\Theta = \theta + \pi/2$). The curvature across the tube is constant, $C_\perp = 1/r_0$; whereas that along the wheel is θ dependent: $C_\| = \cos\theta/R$. Notice that $C_\perp - C_\| = R_0/(r_0 R)$. The Gaussian curvature is positive (negative) on the outer (inner) tube, vanishing on the upper and lower parallels. The mean curvature, K, on the other hand, is positive everywhere unless $2r_0 > R_0$, where it is negative on the interior band of angular width given by $2\theta_0$, where $\cos\theta_0 = -R_0/2r_0$. On a Clifford torus, with $R_0 = \sqrt{2}r_0$, $2\theta_0 = \pi/2$. The sign of K will be reflected in the distribution of stress. Substituting into Eq. (67), one finds that H_{CH} is minimized for a Clifford torus, independent of the physical parameters; the latter do need to be tuned appropriately, Willmore (1965); Ou-Yang (1990); Ou-Yang et al. (1999):

$$P = 2\kappa\, C_0/r_0^2\,, \sigma = \kappa\, C_0(2/r_0 - C_0/2)\,, \tag{95}$$

see also Seifert (1991, 1997). Moreover, the magnitude of the total axial force on a parallel appearing in Eq. (67) is

$$C = -\kappa(1/r_0 + 2C_0)\,. \tag{96}$$

Its origin will be traced to the topology.

In the absence of spontaneous curvature, P and σ vanish and the vertical force is determined by the bending modulus $C = -\kappa/r_0$. Now only the stresses and torques due to bending are relevant. Their nonvanishing projections are given by Eqs. (56) and (80):

$$f_{\perp\perp} = -f_{\|\|} = \frac{\kappa\, K}{\sqrt{2}\, R}\,, \quad f_\perp = -\frac{\sqrt{2}\kappa}{R^2}\sin\theta\,, \tag{97a}$$

$$m_{\perp\|} = -\frac{\kappa r_0}{R^2}\left(2\cos^2\theta + \sqrt{2}\cos\theta + 1\right)\,. \tag{97b}$$

$f_{\|\|}$ correlates directly with mean curvature. It is plotted in Fig. 8a.

Thus the torus is under tension everywhere along the wheel except within the band of angular width $\pi/2$ on the inner tube where K is negative, and tension is replaced by compression. The parallels at $\theta = \pm 3\pi/4$ are free of tangential stress marking the boundary along which tension turns to compression. Because $f_{\perp\perp} = -f_{\|\|}$, across the tube, tension and compression are interchanged. This behavior, implied by scale invariance is not so intuitively clear.

Topological Torque $\mathbf{M}_\|$ *closing the wheel*: The total force closing the tubular cylinder (evaluated, say, on any meridional circle) vanishes, $\mathbf{F}_\| = r_0 \int d\theta\, f_{\|\|}\mathbf{l} = 0$. The moments of the local forces, however, do not vanish. The corresponding torque, closing the torus, is given by $\mathbf{M}_\| = 2\pi\kappa\hat{\mathbf{z}}$. It is topological in origin. This behavior contrasts with a cylinder, where the axial force is nonvanishing and there is no torque.

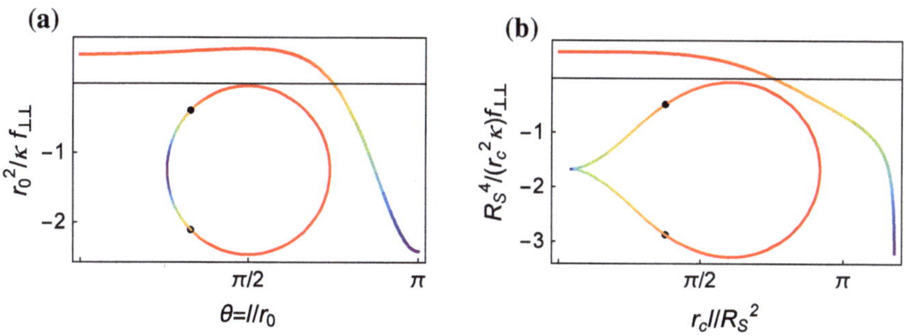

Fig. 8 Components $f_{\perp\perp}$ as function of the arc length l measured from the outer parallel for the **a** Clifford torus **b** discocyte (constructed in Sect. 7). The black dots represent the parallels along which $f_{\perp\perp}$ vanish. Outside of these parallels, the two geometries are essentially indistinguishable; inside the stress diverges in the discocyte reflecting its source in external forces, whereas in the torus it is finite everywhere reflecting its origin in the topology

Topological Force $\mathbf{F}_\perp$ *closing the tube*: The topological force across the tube $\mathbf{F}_\perp$ may be evaluated on any parallel circle (such as the wheel outer circle $\theta = 0$), giving $\mathbf{F}_\perp = \int d\varphi\, R\,(f_{\perp\perp}\mathbf{l} + f_\perp \mathbf{n}) = 2\pi C\,\hat{\mathbf{z}}$. It is also completely determined by the normal forces on appropriate parallels: evaluate $\mathbf{F}_\perp$ along $\theta = \pi/2$.

The corresponding topological torque vanishes, $\mathbf{M}_\perp = 0$. These results quantify the connection between the topology and the internal stresses established in the toroidal membrane. Analogous results will describe higher genus surfaces. But it is not obvious how one would even access this information without knowledge of the local stress and torque in the membrane.

In the next section we will examine some of the consequences of the conformal invariance of bending energy. As we will see, the stress itself is not invariant; and nor are the conserved quantities. And just as well. Under a conformal transformation, an axially symmetric torus will map to a Dupin cyclide, Pinkall (1986). Unlike the energy, which is invariant, the magnitude of the principal stresses increase as the geometry deviates from axial symmetry. As we will describe, the conformally deformed membrane of equal energy may rupture.

7 Conformal Invariance as Probe of Highly Deformed States

An extraordinary feature of the two-dimensional symmetric isotropic bending energy is its invariance under conformal transformations, Willmore (1982, 1996). These are the transformations preserving angles: in addition to its invariance under Euclidean motions it is also scale invariant; less obvious is the fact that it is invariant under inversion in spheres. If this sphere has a radius R_S and is centered at the origin, inversion is represented by the mapping of points in space: $\mathcal{I} : \mathbf{x} \to \bar{\mathbf{x}} = R_S^2\, \mathbf{x}/|\mathbf{x}|)^2$, where $|\mathbf{x}|^2 := \mathbf{x} \cdot \mathbf{x}$. This induces an inversion of the surface by the replacement of $\mathbf{x}$ by $\mathbf{X}$.

The symmetric bending energy alone rarely provides an accurate description of the physics; typically material fields, constraints, or even a spontaneous curvature will be inconsistent with conformal symmetry. It would be a curiosity were it not for the fact that it can be consistent with nontrivial physically relevant constraints.

Usually one looks at conformal transformations at linear order; let $\mathbf{x} \to \bar{\mathbf{x}} = \mathbf{x} + \delta\mathbf{x}$. Angles are preserved if $d\bar{\mathbf{x}} \cdot d\bar{\mathbf{x}} = \Omega^2(\mathbf{x}) d\mathbf{x} \cdot d\mathbf{x}$. This implies that, correct to linear order, $\delta\mathbf{x}$ satisfies $\partial_i \delta x_j + \partial_j \delta x_i = 2\mathbf{div} \cdot \delta\mathbf{x}\, \delta_{ij}/3$. The most general solution is the sum of a Euclidean motion, a scaling, and a special conformal transformation: $\delta_\mathbf{c}\mathbf{x} = |\mathbf{x}|^2\, \mathrm{R}_\mathbf{x}\, \delta\mathbf{c}$, where $\mathrm{R}_\mathbf{x}$ is the linear operator on three-dimensional space defined by $\mathrm{R}_\mathbf{x} = \mathbb{1} - 2\hat{\mathbf{x}} \otimes \hat{\mathbf{x}}$, where $\mathbb{1}$ is the identity transformation and $\hat{\mathbf{x}} = \mathbf{x}/|\mathbf{x}|$; $\mathrm{R}_\mathbf{x}$ represents a reflection in the plane perpendicular to $\mathbf{x}$ passing through the origin, so that $\mathrm{R}_\mathbf{x}^2 = \mathbb{1}$. The constant space vector $\delta\mathbf{c}$ has dimensions of inverse length squared.

This transformation exponentiates to give (for finite $\mathbf{c}$)

$$\mathbf{x} \to \bar{\mathbf{x}} = \frac{\mathbf{x} + |\mathbf{x}|^2\, \mathbf{c}}{|\mathbf{c}|^2\, |\mathbf{x}|^2 + 2\, \mathbf{c} \cdot \mathbf{x} + 1}\,. \tag{98}$$

This can be recast (we set the radius of inversion equal to one)

$$\bar{\mathbf{x}} = \left(\frac{\mathbf{x}}{|\mathbf{x}|^2} + \mathbf{c}\right) \Big/ \left|\frac{\mathbf{x}}{|\mathbf{x}|^2} + \mathbf{c}\right|^2. \tag{99}$$

Thus a finite special conformal transformation can be represented as the composition of an inversion, a translation $\mathbf{c}$, and another inversion: $\mathcal{I} \circ \mathbf{c} \circ \mathcal{I}$. Any conformal transformation is a composition of inversions in spheres with Euclidean motions and scalings, Kreyszig (1991). If we understand conformal inversion, we are done.

As was known to the mathematicians of ancient Greece, spheres map to spheres or planes. But, because distances to the center of inversion get inverted, $|\bar{\mathbf{X}}| = R_S^2/|\mathbf{X}|$, points on a sphere will get moved around unless it coincide with the sphere used for inversion. And its center will not remain the center. Other geometries, as we will see, suffer less recognizable distortions.

To understand the conformal symmetry of the bending energy, one needs to know how the two fundamental tensors on a surface transform under inversion. The adapted tangent vectors and the normal vector transform as follows: $\bar{\mathbf{e}}_a \to R_S^2\, \mathrm{R}_\mathbf{X}\, \mathbf{e}_a/|\mathbf{X}|^2$, $\mathbf{n} \to -\mathrm{R}_\mathbf{X}\, \mathbf{n}$. As a consequence of the former, the induced metric (1) transforms by $g_{ab} \to (R_S/|\mathbf{X}|)^4\, g_{ab}$. Thus in particular, the area measure on the inverted surface is $(R_S/|\mathbf{X}|)^4\, dA$. As for the extrinsic curvature, for our purposes it will suffice to know how the two principal curvatures, C_1, C_2, transform. It follows from the fact that circles map to circles that[24]

---

[24] $K_{ab} \to K_{ab} = -|\mathbf{X}|^2 \left(K_{ab} - 2\,(\mathbf{X} \cdot \mathbf{n}) g_{ab}/|\mathbf{X}|^{-2}\right)$.

$$C_I \to -(|\mathbf{X}|/R_S)^2 \left(C_I - 2\,\mathbf{X}\cdot\mathbf{n}/|\mathbf{X}|^2\right)\,, I = 1, 2; \tag{100}$$

their difference thus transforms multiplicatively, or

$$C_1 - C_2 \to -(|\mathbf{X}|/R_S)^2\,(C_1 - C_2)\,. \tag{101}$$

This result, together with the transformation of area, implies that the energy

$$H_W = \frac{1}{2}\int dA\,(C_1 - C_2)^2 \tag{102}$$

is manifestly invariant. It follows that the unadorned quadratic bending energy, $H_B = \kappa/2\,\int dA\,(C_1 + C_2)^2 + \bar\kappa\int dA\,C_1C_2$ is also invariant—unless the topology changes.

Two physically significant consequences are immediate:
(1) any two geometries related by a conformal transformation possess the same bending energy if the topology is unchanged.
(2) if one of these geometries is an equilibrium state of this energy, then the other is also, modulo possible pointlike singularities. We will show that these singularities can be interpreted as external forces acting on the membrane.

In the 90s Seifert and coworkers, Seifert (1991); Jülicher et al. (1993); Jülicher (1996), observed a degeneracy in the ground states of higher genus vesicles, which they dubbed conformal diffusion, associated with the existence of conformal transformations preserving constraints. This involved examining the behavior of the constrained energy under small special conformal transformations. Such transformations induce continuous deformations of the surface. We will focus on conformal inversion, a non-perturbative feature of conformal symmetry that cannot be probed by exponentiation. It will provide a window, albeit a narrow one of its choosing, into the non-perturbative response of membranes to external forces.

Suppose that the geometry we begin with is not compact. Its image under inversion generally will be compact. To illustrate this point, let us examine the inversion of a catenoid, a minimal surface. The catenoid is described in polar coordinates by $R(Z) = r_0\,\cosh(Z/r_0)$, where r_0 here is its neck radius. Under an inversion centered at the origin, this catenoid maps into the axisymmetric geometry with radial and height coordinates,

$$\bar R(Z) = R_S^2 R(Z)/(R(Z)^2 + Z^2)\,; \quad \bar Z(Z) = R_S^2\, Z/(R(Z)^2 + Z^2)\,, -\infty < Z < \infty\,. \tag{103}$$

Its image resembles a discocyte (see Fig. 9), but this coincidence should not be taken too literally: there are many physical mechanisms producing superficially identical morphologies.[25] While the catenoid and its inverted image are linked mathematically, they are topologically different and clearly describe very different physics, Castro-Villarreal and Guven (2007a, b). Whereas the catenoid is well known, the second

[25] Fixing the discocyte area at $4\pi r_0^2$ determines $R_S = 1.089\,r_0$.

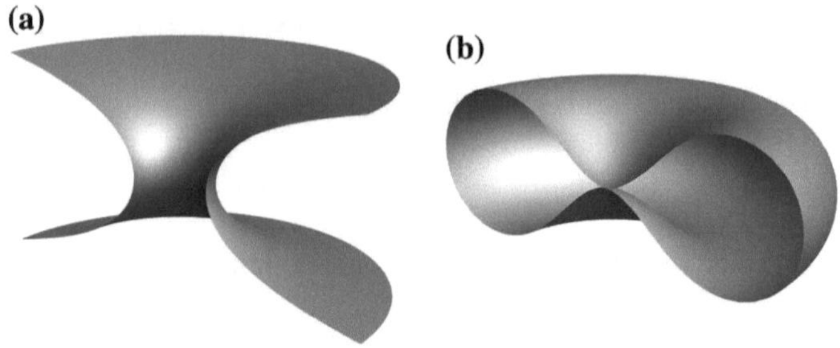

Fig. 9 **a** Catenoid and **b** discocyte obtained by its inversion in a sphere centered at the origin

geometry is not and, had it not been for the stratagem provided by inversion, it is unlikely that one would have guessed that this simple discocyte describes an equilibrium, nevermind lending itself to an exact analytical treatment.

Let us examine the discocyte geometry more closely. Under inversion, the two ends of the catenoid map to the origin, forming the north and south poles of the discocyte which touch with a common tangent plane. The existence of a tangent plane, however, belies the fact that curvature singularities are present at these points: the source-free EL equation breaks down implying the presence of localized distributional external forces. The mathematical origin of these singularities is the compactification of the exponential ends into a bounded region, Castro-Villarreal and Guven (2007a, b); Guven and Vázquez-Montejo (2013b). Note that, in contrast with a catenoid, the inversion of a hyperboloid of revolution—which is asymptotically conical—gives a pair of conical singularity at the origin, without tangent planes to hide behind. We will have more to say about singularities in a moment. Notice also that whereas $\int dA\, C_1 C_2 = -4\pi$ for a catenoid, it is given by 4π for a discocyte, which is topologically a sphere. That the poles touch is irrelevant.

Whereas the distribution of stress associated with bending vanishes in a catenoid, it does not in the discocyte. The tangential stress along the meridian $f_{\perp\perp} = -f_{\|\|}$ is plotted in Fig. 8b as a functions of archlength, l, measured from the outer parallel. It is observed that it is strongly localized within the neighborhoods of the two touching poles, where it diverges.

If we now translate the center of inversion along the axis of the catenoid, a one-parameter family of equilibrium states is generated: as illustrated for a northward movement in Fig. 10, the symmetric discocyte morphs smoothly into a stomatocyte. The limiting shape is a sphere within a sphere, touching at the bottom, connected by a microscopic catenoidal neck at the top. Unlike the constricted sphere discussed previously, this time the neck is a catenoid because there are no sources acting within it.

Now let us place the center of inversion off axis, Guven and Vázquez-Montejo (2013b). Consider, for example, inversion in a sphere centered on a point located

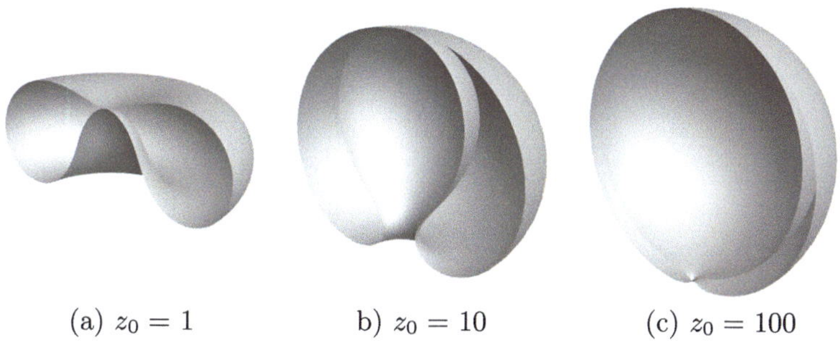

Fig. 10 Surfaces generated by inversion of a catenoid in spheres centered at height z_0 along its axis of symmetry

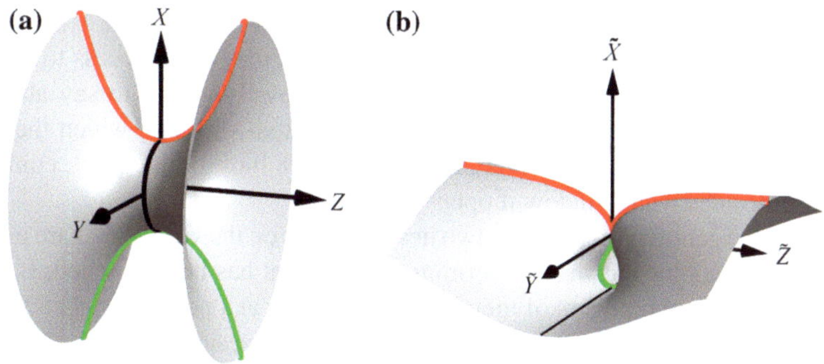

Fig. 11 The *red* and *green* catenaries on the mirror plane $Y = 0$ map to a single curve. The circular neck maps to a straight line

along the X axis. This preserves the mirror symmetries in the XZ and XY planes.[26] In Fig. 12, we illustrate four geometries in this sequence. Of special interest are those generated when this point lies close to the neck of the catenoidal geometry, say $\mathbf{x}_0 = (1 + \epsilon)r_0\mathbf{i}$, with ϵ small. Now, the neighborhood of the point $r_0\mathbf{i}$ on the neck gets inflated into a large spherical region: the geometry is spherical almost everywhere, with a *defect*, formed by the two points held together, localized upon it as indicated in Fig. 12b, c.

To facilitate the visualization of this construction, it is useful to follow the fate of the catenary meridians and the neck of the catenoid as illustrated in Fig. 11.

Physically, one can interpret these geometries as the end point of a procedure bringing two nearby points on an almost spherical vesicle, of fixed area and fixed enclosed volume, into contact by applying normal forces (as opposed to tangential ones which would not disturb the initial geometry, fluid in its tangent plane), Guven

[26] The bilateral symmetry (not necessarily in the original XZ plane) is preserved if the point strays off this axis; however, the up-down symmetry is broken, just as it was in the axially symmetric family.

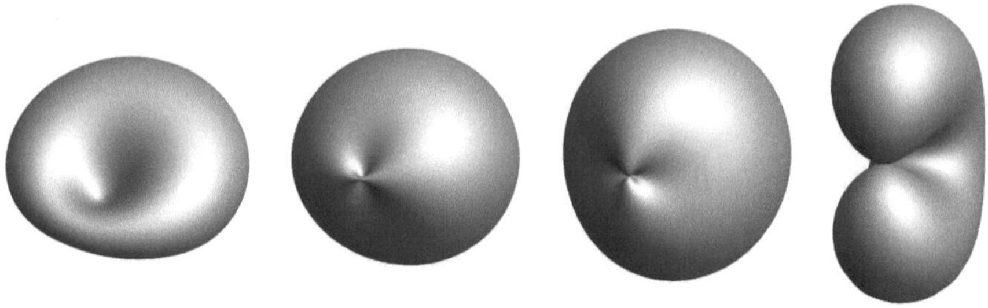

Fig. 12 Surfaces generated by inversion of a catenoid in a sphere centered at increasing distances along the radial direction

and Vázquez-Montejo (2013b). Conformal invariance may grant access to the non-pertubative equilibrium end state, but it does not have anything to say about the intermediate states—with the two points still separated—through which the membrane has to pass on its way to this final state. As we have pointed out, conformal invariance decides what window it opens.

If $\epsilon < 0$, the geometry represents two nearby points on the vesicle that are pinched together; it resembles the wrinkle forming on skin that has been pinched. If $\epsilon > 0$, on the other hand, two fingers of membrane—touching at a point—project out from the vesicle. In the latter the force is directed out of the vesicle, not into it as in the former. Despite the apparent differences in the two descriptions, however, it takes only a moment's thought to appreciate that the two limiting geometries are mirror images in the neighborhood of the defect.

These geometries are also stable. The constraints on the area and on the volume reduce the conformal symmetry to a single nontrivial degree of freedom. This is reflected in a zero mode of the operator controlling fluctuations corresponding to deformations which break the up-down mirror symmetry in the XY plane but leave fixed the geodesic distance s along the surface separating the two touching points.[27] As a consequence, once formed, the membrane cannot slip out of this defect. If the volume constraint were to be relaxed, however, there would be nothing to prevent the two points from approaching each other along the membrane.

More exotic increasingly deflated geometries are generated by increasing the magnitude of $\mathbf{x}_0$. The surface first morphs continuously into a folded sausage, as illustrated in Fig. 12d; beyond a critical value of x_0, the waist on the sausage begins to constrict. As x_0 increases further, the geometry evolves, just like its axially symmetry counterpart, into a geometry that is spherical almost everywhere: this time it resembles two spherical lobes, touching at one point and connected by a vanishingly small catenoidal neck adjacent to it.

[27] This is the shortest distance between the two points on the surface. They are, of course, in contact in space.

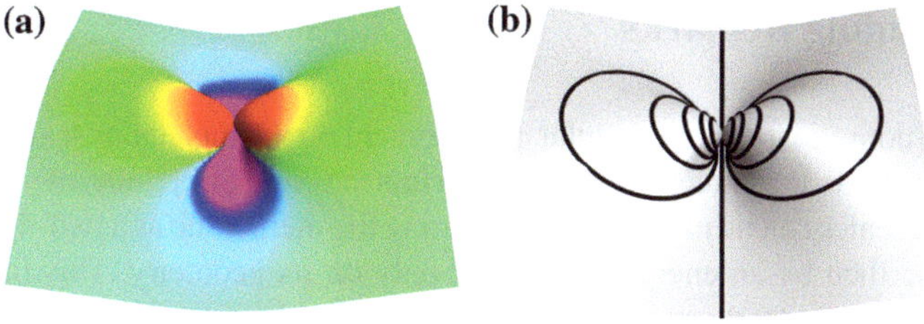

Fig. 13 **a** Stress across the images of parallel circles. **b** Images of parallel circles represented by closed curves

On a catenoid, $\mathbf{f}^a = 0$. In the inverted geometry, $K \neq 0$ and equilibrium involves a nontrivial balance of normal and tangential stresses. These stresses possess a pair of external sources ($\pm\mathbf{F}$ say) normal to the surface localized at the points of contact; this is what holds them together. It is easy to show that in the height function description of the geometry with respect to the local tangent plane at these points, one has $h \approx \mp r^2 \ln r$, where r is the polar radius.[28,29] The presence of a source is signaled geometrically by the logarithmic curvature singularity at these points.

To identify the contact force $\mathbf{F}$, we can deform the contour surrounding the pole so that it coincide with the line of symmetry running along the valley (or ridge) between the two poles. Using the expression for the force given in Eq. (77), one determines $\mathbf{F} \approx 9.65\pi\kappa/s\, \mathbf{k}$. Note that the detailed distribution of stress is not required to determine the contact force, Guven and Vázquez-Montejo (2013b). As the vesicle is inflated, s decreases and the contact force increases. One can show, however, that it remains below the rupture tension of the membrane until $s \approx 20\,\text{nm}$, a separation so small that the mesoscopic modeling in terms of a surface is no longer reliable (Fig. 13).

There is also an important cautionary point to be emphasized here: had one used the Monge representation for the horizontal stress with respect to the local tangent plane, an incorrect answer would have been obtained.

[28]An unexpected duality between the weak field behavior in one geometry and the strong field behavior in the other is evident: asymptotically, the catenoid is accurately described by the height function $h \sim \ln r$, $r \gg r_0$; this asymptotic region is mapped into the neighborhood of the poles described by $h \sim -r^2 \ln r$, $r \ll s$. Inversion provides a connection between the harmonic behavior in the former and the biharmonic behavior in the latter, Guven and Vázquez-Montejo (2013b). To understand this duality between harmonic and biharmonic function, look at the inversion in the origin $\mathbf{x} \to \mathbf{x}/|\mathbf{x}|^2$ (for transparency set the scale to one), described in the height function representation by $(r, h) \to (r, h)/((r^2 + h^2)$, so that $h \approx \ln r \to \frac{h}{r^2+h^2} = \ln[r/(r^2 + h^2)]$. Now, if $h \ll r$, then $\frac{h}{r^2} \approx -\ln r$, and as claimed the Green function of the Laplacian is mapped to its biharmonic counterpart.

[29]In this context, note also that the symmetric saddle with $h \sim r^2 \cos 2\theta$ maps to the biharmonic dipole $h \sim \cos 2\theta$.

8 Concluding Remarks

Despite the complexity of the cellular environment on the molecular scale, on zooming out, one finds that much of the equilibrium physics of cellular membranes on mesoscopic scales can be understood by treating the membrane as a two-dimensional surface described by an energy depending only on its geometry. This implies that this physics is completely encoded in the membrane geometry. We have presented a number of simple examples to illustrate the nature of this connection. Even if additional structure is relevant, it can be treated in terms of fields interacting with this geometry.

There are other examples falling within the scope of this chapter which we have not treated for want of space, or because they will be treated thoroughly by other contributors. Some of these are treated, as commented previously, in a nice recent review, Deserno (2015). Here this framework is used to determine the Gaussian bending modulus of a buckled membrane. The interested reader can find a statistical mechanical treatment of the stress tensor, important but absent in our presentation, in Shiba et al. (2016). Membranes in a viscous fluid are examined by Powers (2010), membrane viscosity itself is accommodated in Arroyo and DeSimone (2009).

To explore how confinement can shape a membrane, Müller and his coworkers have looked at the confinement of a topologically spherical membrane within another membrane of smaller area, Kahraman et al. (2012a, b). This can be thought of as a model of the inner membrane of the mitochondrion, capturing the geometrical aspect of the physics. As the area is increased, in addition to the external forces associated with confinement, self-contacts will occur breaking the symmetry of the ground state. If membrane fusion occurs, the topology will change accompanied by a large scale reorganization of the ground state geometry, Bouzar et al. (2015).

Deflated high-genus geometries are exhibited by the nuclear envelope and the Golgi apparatus. In the limit where the enclosed volume becomes very small, in the absence of additional agents, the ground state of a closed fluid membrane with a fixed genus g approximates two concentric spherical bilayers connected by $g + 1$ catenoidal necks (Kusner and Yu, private communication). Recent simulations, involving additional players with their own degrees of freedom, controlling the pore radius provide a more accurate description of the nuclear envelope, Noguchi (2016a). In this context, the stability of the inner ramp boundaries in the rough ER described in Sect. 4. appears to involve the condensation along their length of membrane-shaping proteins, Guven et al. (2014); Schweitzer and Kozlov (2015); Noguchi (2016b). While getting the biological details right always involves elaborate numerical analysis; understanding how the membrane geometry gets shaped by external forces and constraints will inform how we go about it.

Acknowledgements JG would like to thank David Steigmann for the invitation to lecture at the CISM Summer School held in Udine, Italy during July of 2016. This chapter is based on these lectures. We would like also to thank Markus Deserno, Martin Müller and Saša Svetina for their valuable input. This work was partially supported by CONACyT grant 180901.

References

L. Amoasii, K. Hnia, G. Chicanne, A. Brech, B.S. Cowling, M.M. Müller, Y. Schwab, P. Koebel, A. Ferry, B. Payrastre, J. Laporte, Myotubularin and ptdins3p remodel the sarcoplasmic reticulum in muscle in vivo. J. Cell Sci. **126**(8), 1806–1819 (2013). doi:10.1242/jcs.118505

R. Arnowitt, S. Deser, C.W. Misner, Dynamical structure and definition of energy in general relativity. Phys. Rev. **116**(5), 1322–1330 (1959). doi:10.1103/PhysRev.116.1322

G. Arreaga, R. Capovilla, J. Guven, Noether currents for bosonic branes. Ann. Phys. **279**(1), 126–158 (2000). doi:10.1006/aphy.1999.5979

M. Arroyo, A. DeSimone, Relaxation dynamics of fluid membranes. Phys. Rev. E **79**(3), 031915 (2009). doi:10.1103/PhysRevE.79.031915

P. Bassereau, B. Sorre, A. Lévy, Bending lipid membranes: experiments after w. helfrich's model. Adv. Colloid Interface Sci. **208**, 47–57 (2014). doi:10.1016/j.cis.2014.02.002. Special issue in honour of Wolfgang Helfrich

Y. Bernard, Noether's theorem and the willmore functional. Adv. Calc. Var. (2015). doi:10.1515/acv-2014-0033

L. Bouzar, F. Menas, M.M. Müller, Toroidal membrane vesicles in spherical confinement. Phys. Rev. E **92**, 032721 (2015). doi:10.1103/PhysRevE.92.032721

B. Božič, J. Guven, P. Vázquez-Montejo, S. Svetina, Direct and remote constriction of membrane necks. Phys. Rev. E **89**, 052701 (2014). doi:10.1103/PhysRevE.89.052701

B. Božič, S.L. Das, S. Svetina, Sorting of integral membrane proteins mediated by curvature-dependent protein-lipid bilayer interaction. Soft Matter **11**, 2479–2487 (2015). doi:10.1039/C4SM02289K

P.B. Canham, The minimum energy of bending as a possible explanation of the biconcave shape of the human red blood cell. J. Theor. Biol. **26**(1), 61–81 (1970). doi:10.1016/S0022-5193(70)80032-7

R. Capovilla, J. Guven, Geometry of lipid vesicle adhesion. Phys. Rev. E **66**, 041604 (2002a). doi:10.1103/PhysRevE.66.041604

R. Capovilla, J. Guven, Stresses in lipid membranes. J. Phys. A Math. Gen. **35**(30), 6233 (2002b). doi:10.1088/0305-4470/35/30/302

R. Capovilla, J. Guven, Stress and geometry of lipid vesicles. J. Phys.-Condens. Mat. **16**, S2187–S2191 (2004a). doi:10.1088/0953-8984/16/22/018

R. Capovilla, J. Guven, Second variation of the Helfrich-Canham Hamiltonian and reparametrization invariance. J. Phys. A Math. Gen. **37**(23), 5983 (2004b). doi:10.1088/0305-4470/37/23/003

R. Capovilla, J. Guven, J.A. Santiago, Lipid membranes with an edge. Phys. Rev. E **66**, 021607 (2002). doi:10.1103/PhysRevE.66.021607

R. Capovilla, J. Guven, J.A. Santiago, Deformations of the geometry of lipid vesicles. J. Phys. A Math. Gen. **36**(23), 6281 (2003). doi:10.1088/0305-4470/36/23/301

P. Castro-Villarreal, J. Guven, Axially symmetric membranes with polar tethers. J. Phys. A Math. Theor. **40**(16), 4273 (2007a). doi:10.1088/1751-8113/40/16/002

P. Castro-Villarreal, J. Guven, Inverted catenoid as a fluid membrane with two points pulled together. Phys. Rev. E **76**, 011922 (2007b). doi:10.1103/PhysRevE.76.011922

M. Deserno, Membrane elasticity and mediated interactions in continuum theory: a differential geometric approach, in *Biomembrane Frontiers*, ed. by R. Faller, M.L. Longo, S.H. Risbud, T. Jue. Handbook of Modern Biophysics (Humana Press, New York, 2009), pp. 41–74. doi:10.1007/978-1-60761-314-5_2

M. Deserno, Fluid lipid membranes: from differential geometry to curvature stresses. Chem. Phys. Lipids **185**, 11–45 (2015). doi:10.1016/j.chemphyslip.2014.05.001. Membrane mechanochemistry: From the molecular to the cellular scale

M. Deserno, M.M. Müller, J. Guven, Contact lines for fluid surface adhesion. Phys. Rev. E **76**, 011605 (2007). doi:10.1103/PhysRevE.76.011605

P. Diggins IV, Z.A. McDargh, M. Deserno, Curvature softening and negative compressibility of gel-phase lipid membranes. J. Am. Chem. Soc. **137**(40), 12752–12755 (2015). doi:10.1021/jacs.5b06800

M. Do Carmo, *Differential Geometry of Curves and Surface* (Prentice Hall, Upper Saddle River, 1976)

M. Do Carmo. *Riemannian Geometry*. (Birkhauser, Basel, 1992)

P.G. Dommersnes, J.-B. Fournier, The many-body problem for anisotropic membrane inclusions and the self-assembly of saddle defects into an egg carton. Biophys. J. **83**, 2898–2905 (2002). doi:10.1016/S0006-3495(02)75299-5

E.A. Evans, Bending resistance and chemically induced moments in membrane bilayers. Biophys. J. **14**, 923–931 (1974). doi:10.1016/S0006-3495(74)85959-X

E.A. Evans, R. Skalak, *Mechanics and Thermodynamics of Biomembranes* (CRC Press, Boca Raton, 1980)

J.-B. Fournier, On the stress and torque tensors in fluid membranes. Soft Matter **3**, 883–888 (2007). doi:10.1039/B701952A

J.-B. Fournier, Dynamics of the force exchanged between membrane inclusions. Phys. Rev. Lett. **112**, 128101 (2014). doi:10.1103/PhysRevLett.112.128101

J.-B. Fournier, P. Galatola, High-order power series expansion of the elastic interaction between conical membrane inclusions. Eur. Phys. J. E **38**(8) (2015). doi:10.1140/epje/i2015-15086-3

R. Goetz, W. Helfrich, The egg carton: theory of a periodic superstructure of some lipid membranes. J. Phys. II Fr. **6**(2), 215–223 (1996). doi:10.1051/jp2:1996178

M. Goulian, R. Bruinsma, P. Pincus, Long-range forces in heterogeneous fluid membranes. EPL (Europhysics Letters) **22**(2), 145 (1993). doi:10.1209/0295-5075/22/2/012

J. Guven, Membrane geometry with auxiliary variables and quadratic constraints. J. Phys. A Math. Gen. **37**(28), L313 (2004). doi:10.1088/0305-4470/37/28/L02

J. Guven, Conformally invariant bending energy for hypersurfaces. J. Phys. A Math. Gen. **38**(37), 7943 (2005). doi:10.1088/0305-4470/38/37/002

J. Guven, Laplace pressure as a surface stress in fluid vesicles. J. Phys. A Math. Gen. **39**(14), 3771 (2006). doi:10.1088/0305-4470/39/14/019

J. Guven, M.M. Müller, How paper folds: bending with local constraints. J. Phys. A Math. Theo. **41**(5), 055203 (2008). doi:10.1088/1751-8113/41/5/055203

J. Guven, M.M. Müller, P. Vázquez-Montejo, Conical instabilities on paper. J. Phys. A Math. Theo. **45**(1), 015203 (2012). doi:10.1088/1751-8113

J. Guven, P. Vázquez-Montejo, Spinor representation of surfaces and complex stresses on membranes and interfaces. Phys. Rev. E **82**, 041604 (2010). doi:10.1103/PhysRevE.82.041604

J. Guven, P. Vázquez-Montejo, Constrained metric variations and emergent equilibrium surfaces. Phys. Lett. A **377**(23–24), 1507–1511 (2013a). doi:10.1016/j.physleta.2013.04.031

J. Guven, P. Vázquez-Montejo, Force dipoles and stable local defects on fluid vesicles. Phys. Rev. E **87**, 042710 (2013b). doi:10.1103/PhysRevE.87.042710

J. Guven, G. Huber, D.M. Valencia, Terasaki spiral ramps in the rough endoplasmic reticulum. Phys. Rev. Lett. **113**, 188101 (2014). doi:10.1103/PhysRevLett.113.188101

R.C. Haussman, M. Deserno, Effective field theory of thermal casimir interactions between anisotropic particles. Phys. Rev. E **89**, 062102 (2014). doi:10.1103/PhysRevE.89.062102

W. Helfrich, Elastic properties of lipid bilayers, theory and possible experiments. Z. Naturforsch. C **28**, 693–703 (1973). http://zfn.mpdl.mpg.de/data/Reihe_C/28/ZNC-1973-28c-0693.pdf

J.H. Jellett. Sur la surface dont la courbure moyenne est constante. Journal de Mathematiques Pures et Appliquees, 163–167 (1853)

J.T. Jenkins, The equations of mechanical equilibrium of a model membrane. SIAM J. Appl. Math. **32**(4), 755–764 (1977). doi:10.1137/0132063

F. Jülicher, The morphology of vesicles of higher topological genus: conformal degeneracy and conformal modes. J. Phys. II Fr. **6**(12), 1797–1824 (1996). doi:10.1051/jp2:1996161

F. Jülicher, U. Seifert, Shape equations for axisymmetric vesicles: a clarification. Phys. Rev. E **49**, 4728–4731 (1994). doi:10.1103/PhysRevE.49.4728

F. Jülicher, U. Seifert, R. Lipowsky, Conformal degeneracy and conformal diffusion of vesicles. Phys. Rev. Lett. **71**, 452–455 (1993). doi:10.1103/PhysRevLett.71.452

O. Kahraman, N. Stoop, M.M. Müller, Morphogenesis of membrane invaginations in spherical confinement. EPL (Europhysics Letters) **97**(6), 68008 (2012a). doi:10.1209/0295-5075/97/68008

O. Kahraman, N. Stoop, M.M. Müller, Fluid membrane vesicles in confinement. New J. Phys. **14**(9), 095021 (2012b). doi:10.1088/1367-2630/14/9/095021

K.S. Kim, J. Neu, G. Oster, Curvature-mediated interactions between membrane proteins. Biophys. J. **75**(5), 2274–2291 (1998). doi:10.1016/S0006-3495(98)77672-6

M.M. Kozlov, Fission of biological membranes: interplay between dynamin and lipids. Traffic **2**(1), 51–65 (2001). doi:10.1034/j.1600-0854.2001.020107.x

V. Kralj-Iglič, S. Svetina, B. Žekž, Shapes of bilayer vesicles with membrane embedded molecules. Eur. Biophys. J. **24**(5), 311–321 (1996). doi:10.1007/BF00180372

V. Kralj-Iglič, V. Heinrich, S. Svetina, B. Žekž, Free energy of closed membrane with anisotropic inclusions. Eur. Phys. J. B - Condens. Matter Complex Syst. **10**(1), 5–8 (1999). doi:10.1007/s100510050822

E. Kreyszig, *Differential Geometry* (Dover Publications, New York, 1991)

R. Kusner, Geometric analysis and computer graphics, in *Mathematical Sciences Research Institute Publications*, vol. 17, ed. by P. Concus, R. Finn, D.A. Hoffman (Springer, New York, 1991), pp. 103–108. doi:10.1007/978-1-4613-9711-3_11

R. Lipowsky, Spontaneous tubulation of membranes and vesicles reveals membrane tension generated by spontaneous curvature. Faraday Discuss. **161**, 305–331 (2013). doi:10.1039/C2FD20105D

M.A. Lomholt, L. Miao, Descriptions of membrane mechanics from microscopic and effective two-dimensional perspectives. J. Phys. A Math. Gen. **39**(33), 10323 (2006). doi:10.1088/0305-4470/39/33/005

O.V. Manyuhina, J.J. Hetzel, M.I. Katsnelson, A. Fasolino, Non-spherical shapes of capsules within a fourth-order curvature model. Eur. Phys. J. E **32**(3), 223–228 (2010). doi:10.1140/epje/i2010-10631-2

F.C. Marques, A. Neves, Min-Max theory and the Willmore conjecture. Ann. Math. Second Series **179**(2), 683–782 (2014a). doi:10.4007/annals.2014.179.2.6

F.C. Marques, A. Neves, The Willmore conjecture. Jahresbericht der Deutschen Mathematiker-Vereinigung **116**(4), 201–222 (2014b). doi:10.1365/s13291-014-0104-8

Z. McDargh, P. Vázquez-Montejo, J. Guven, M. Deserno. Constriction by dynamin: Elasticity vs. adhesion. Biophy. J. **111**(11), 2470–2480 (2016). doi:10.1016/j.bpj.2016.10.019

X. Michalet, D. Bensimon, Observation of stable shapes and conformal diffusion in genus 2 vesicles. Science **269**(5224), 666–668 (1995). doi:10.1126/science.269.5224.666

S. Morlot, A. Roux, Mechanics of dynamin-mediated membrane fission. Ann. Rev. Biophys. **42**(1), 629–649 (2013). doi:10.1146/annurev-biophys-050511-102247

M.M. Müller, Theoretical studies of fluid membrane mechanics, Ph.D. thesis, University of Mainz (Germany), 2007

M.M. Müller, M. Deserno, J. Guven, Geometry of surface-mediated interactions. Europhys. Lett. **69**(3), 482 (2005a). doi:10.1209/epl/i2004-10368-1

M.M. Müller, M. Deserno, J. Guven, Interface-mediated interactions between particles: a geometrical approach. Phys. Rev. E **72**, 061407 (2005b). doi:10.1103/PhysRevE.72.061407

M.M. Müller, M. Deserno, J. Guven, Balancing torques in membrane-mediated interactions: exact results and numerical illustrations. Phys. Rev. E **76**, 011921 (2007). doi:10.1103/PhysRevE.76.011921

M. Mutz, D. Bensimon, Observation of toroidal vesicles. Phys. Rev. A **43**, 4525–4527 (1991). doi:10.1103/PhysRevA.43.4525

G.-M. Nam, N.-K. Lee, H. Mohrbach, A. Johner, I.M. Kulić, Helices at interfaces. EPL (Europhysics Letters) **100**(2), 28001 (2012). doi:10.1209/0295-5075/100/28001

H. Noguchi, Construction of nuclear envelope shape by a high-genus vesicle with pore-size constraint. Biophy. J. **111**(4), 824–831 (2016a). doi:10.1016/j.bpj.2016.07.010

H. Noguchi, Membrane tubule formation by banana-shaped proteins with or without transient network structur. Sci. Rep. **6**, 20935 (2016b). doi:10.1038/srep20935

A.S.H. Noguchi, M. Imai, Shape transformations of toroidal vesicles. Soft Matter **11**, 193–201 (2015)

Z.-C. Ou-Yang, Anchor ring-vesicle membranes. Phys. Rev. A **41**, 4517–4520 (1990). doi:10.1103/PhysRevA.41.4517

Z.-C. Ou-Yang, W. Helfrich, Instability and deformation of a spherical vesicle by pressure. Phys. Rev. Lett. **59**, 2486–2488 (1987). doi:10.1103/PhysRevLett.59.2486

Z.-C. Ou-Yang, W. Helfrich, Bending energy of vesicle membranes: General expressions for the first, second, and third variation of the shape energy and applications to spheres and cylinders. Phys. Rev. A **39**, 5280–5288 (1989). doi:10.1103/PhysRevA.39.5280

Z.C. Ou-Yang, J.X. Liu, Y.Z. Xie, X. Yu-Zhang, *Geometric Methods in the Elastic Theory of Membranes in Liquid Crystal Phases, Advanced series on theoretical physical science* (World Scientific, Singapore, 1999)

R. Phillips, T. Ursell, P. Wiggins, P. Sens, Emerging roles for lipids in shaping membrane-protein function. Nature **459**, 379–385 (2009). doi:10.1038/nature08147

U. Pinkall, Cyclides of Dupin, in *Mathematical Models from the Collections of Universities and Museums*, ed. by E.G. Fischer. Advanced Lectures in Mathematics Series (Friedrick Vieweg & Son, Braunschweig, 1986), pp. 28–30. Chap. 3.3

R. Podgornik, S. Svetina, B. Žekš, Parametrization invariance and shape equations of elastic axisymmetric vesicles. Phys. Rev. E **51**, 544–547 (1995). doi:10.1103/PhysRevE.51.544

T.R. Powers, Dynamics of filaments and membranes in a viscous fluid. Rev. Mod. Phy. **82**(2), 1607–1631 (2010). doi:10.1103/RevMod-Phys.82.1607

B.J. Reynwar, G. Illya, V.A. Harmandaris, M.M. Müller, K. Kremer, M. Deserno, Aggregation and vesiculation of membrane proteins by curvature-mediated interactions. Nature **447**, 461–464 (2007). doi:10.1038/nature05840

Y. Schweitzer, M. Kozlov, Membrane-mediated interaction between strongly anisotropic protein scaffolds. PLoS Comput. Biol. **11**, 1004054 (2015). doi:10.1371/journal.pcbi.1004054

U. Seifert, Conformal transformations of vesicle shapes. J. Phys. A Math. Gen. **24**(11), 573 (1991). doi:10.1088/0305-4470/24/11/001

U. Seifert, Vesicles of toroidal topology. Phys. Rev. Lett. **66**, 2404–2407 (1991). doi:10.1103/PhysRevLett.66.2404

U. Seifert, Configurations of fluid membranes and vesicles. Adv. Phys. **46**(1), 13–137 (1997). doi:10.1080/00018739700101488

U. Seifert, R. Lipowsky, Morphology of vesicles, in *Structure and Dynamics of Membranes From Cells to Vesicles*, ed. by R. Lipowsky, E. Sackmann. Handbook of Biological Physics, vol. 1 (North-Holland, Amsterdam, 1995), pp. 403–463. doi:10.1016/S1383-8121(06)80025-4

P. Sens, L. Johannes, P. Bassereau, Biophysical approaches to protein-induced membrane deformations in trafficking. Current Opinion Cell Biol. **20**(4), 476–482 (2008). doi:10.1016/j.ceb.2008.04.004

H. Shiba, H. Noguchi, J.-B. Fournier, Monte carlo study of the frame, fluctuation and internal tensions of fluctuating membranes with fixed area. Soft Matter **12**, 2373–2380 (2016). doi:10.1039/C5SM01900A

M. Spivak, *A Comprehensive Introduction to Differential Geometry*, vol. 1–5, 3rd edn. (Publish or Perish, Inc., Houston, 1999)

D.J. Steigmann, Fluid films with curvature elasticity. Arch. Rational Mech. Anal. **150**(2), 127–152 (1999). doi:10.1007/s002050050183

S. Svetina, B. Žekž, Membrane bending energy and shape determination of phospholipid vesicles and red blood cells. Eur. Biophys. J. **17**(2), 101–111 (1989). doi:10.1007/BF00257107

S. Svetina, B. Žekš, Nonlocal membrane bending: a reflection, the facts and its relevance. Adv. Colloid Interface Sci. **208**, 189–196 (2014). doi:10.1016/j.cis.2014.01.010. Special issue in honour of Wolfgang Helfrich

M. Terasaki, T. Shemesh, N. Kasthuri, R.W. Klemm, R. Schalek, K.J. Hayworth, A.R. Hand, M. Yankova, G. Huber, J.W. Lichtman, T.A. Rapoport, M.M. Kozlov, Stacked endoplasmic reticulum sheets are connected by helicoidal membrane motifs. Cell **154**, 285–296 (2013). doi:10.1016/j.cell.2013.06.031

Z.C. Tu, Z.C. Ou-Yang, Lipid membranes with free edges. Phys. Rev. E **68**, 061915 (2003). doi:10.1103/PhysRevE.68.061915

Z.C. Tu, Z.C. Ou-Yang, A geometric theory on the elasticity of bio-membranes. J. Phys. A Math. Gen. **37**(47), 11407 (2004). doi:10.1088/0305-4470/37/47/010

Z.C. Tu, Z.C. Ou-Yang, Recent theoretical advances in elasticity of membranes following helfrich's spontaneous curvature model. Adv. Colloid Interface Sci. **208**, 66–75 (2014). doi:10.1016/j.cis.2014.01.008. Special issue in honour of Wolfgang Helfrich

R.M. Wald, *General Relativity* (University of Chicago Press, Chicago, 2010)

T.R. Weikl, M.M. Kozlov, W. Helfrich, Interaction of conical membrane inclusions: effect of lateral tension. Phys. Rev. E **57**, 6988–6995 (1998). doi:10.1103/PhysRevE.57.6988

T.J. Willmore, Note on embedded surfaces. An. St. Univ. Iasi, sIa Mat. B **12**, 493–496 (1965)

T.J. Willmore, *Total Curvature in Riemannian Geometry* (Ellis Horwood, Chichester, 1982)

T.J. Willmore, *Riemannian Geometry* (Oxford University Press, Oxford, 1996)

C. Yolcu, M. Deserno, Membrane-mediated interactions between rigid inclusions: an effective field theory. Phys. Rev. E **86**, 031906 (2012). doi:10.1103/PhysRevE.86.031906

C. Yolcu, I.Z. Rothstein, M. Deserno, Effective field theory approach to casimir interactions on soft matter surfaces. EPL (Europhysics Letters) **96**(2), 20003 (2011). doi:10.1209/0295-5075/96/20003

C. Yolcu, I.Z. Rothstein, M. Deserno, Effective field theory approach to fluctuation-induced forces between colloids at an interface. Phys. Rev. E **85**, 011140 (2012). doi:10.1103/PhysRevE.85.011140

C. Yolcu, R.C. Haussman, M. Deserno, The effective field theory approach towards membrane-mediated interactions between particles. Adv. Colloid Interface Sci. **208**, 89–109 (2014). doi:10.1016/j.cis.2014.02.017. Special issue in honour of Wolfgang Helfrich

W.-M. Zheng, J. Liu, Helfrich shape equation for axisymmetric vesicles as a first integral. Phys. Rev. E **48**, 2856–2860 (1993). doi:10.1103/PhysRevE.48.2856

On the Computational Modeling of Lipid Bilayers Using Thin-Shell Theory

Roger A. Sauer

Abstract This chapter discusses the computational modeling of lipid bilayers based on the nonlinear theory of thin shells. Several computational challenges are identified and various theoretical and computational ingredients are proposed in order to counter them. In particular, C^1-continuous, NURBS-based, LBB-conforming surface finite element discretizations are discussed. The constitutive behavior of the bilayer is based on in-plane viscosity and (near) area-incompressibility combined with the Helfrich bending model. Various shear stabilization techniques are proposed for quasi-static computations. All ingredients are formulated in the curvilinear coordinate system characterizing general surface parameterizations. The consistent linearization of the formulation is presented, and several numerical examples are shown.

List of Important Symbols

$\mathbf{1}$	identity tensor in $\mathbb{R}^3$
a	determinant of matrix $[a_{\alpha\beta}]$
A	determinant of matrix $[A_{\alpha\beta}]$
$\boldsymbol{a}_\alpha$	covariant tangent vectors of surface $\mathcal{S}$ at point $\boldsymbol{x}$; $\alpha = 1, 2$
$\boldsymbol{A}_\alpha$	covariant tangent vectors of surface $\mathcal{S}_0$ at point $\boldsymbol{X}$; $\alpha = 1, 2$
$\boldsymbol{a}^\alpha$	contra-variant tangent vectors of surface $\mathcal{S}$ at point $\boldsymbol{x}$; $\alpha = 1, 2$
$\boldsymbol{A}^\alpha$	contra-variant tangent vectors of surface $\mathcal{S}_0$ at point $\boldsymbol{X}$; $\alpha = 1, 2$
$\boldsymbol{a}_{\alpha,\beta}$	parametric derivative of $\boldsymbol{a}_\alpha$ w.r.t. ξ^β
$\boldsymbol{a}_{\alpha;\beta}$	covariant derivative of $\boldsymbol{a}_\alpha$ w.r.t. ξ^β
$a_{\alpha\beta}$	covariant metric components of surface $\mathcal{S}$ at point $\boldsymbol{x}$
$A_{\alpha\beta}$	covariant metric components of surface $\mathcal{S}_0$ at point $\boldsymbol{X}$
$a^{\alpha\beta\gamma\delta}$	derivative of $a^{\alpha\beta}$ w.r.t. $a_{\gamma\delta}$
$\mathbf{a}$	class of stabilization methods based on artificial shear viscosity
$\mathbf{A}$	class of stabilization methods based on artificial shear stiffness
$b_{\alpha\beta}$	covariant curvature tensor components of surface $\mathcal{S}$ at point $\boldsymbol{x}$

R.A. Sauer (✉)
Aachen Institute for Advanced Study in Computational Engineering Science (AICES), RWTH Aachen University, Templergraben 55, 52056 Aachen, Germany
e-mail: sauer@aices.rwth-aachen.de

D.J. Steigmann (ed.), *The Role of Mechanics in the Study of Lipid Bilayers*,
CISM International Centre for Mechanical Sciences 577,
DOI 10.1007/978-3-319-56348-0_5

$B_{\alpha\beta}$	covariant curvature tensor components of surface $\mathcal{S}_0$ at point $\boldsymbol{X}$
$b^{\alpha\beta\gamma\delta}$	derivative of $b^{\alpha\beta}$ w.r.t. $a_{\gamma\delta}$
$\boldsymbol{b}$	curvature tensor of surface $\mathcal{S}$ at point $\boldsymbol{x}$
$\boldsymbol{B}$	left surface Cauchy–Green tensor
$\boldsymbol{C}$	right surface Cauchy–Green tensor
$c^{\alpha\beta\gamma\delta}$	derivative of $\tau^{\alpha\beta}$ w.r.t. $a_{\gamma\delta}$
γ	surface tension of $\mathcal{S}$
$\Gamma^{\gamma}_{\alpha\beta}$	Christoffel symbols of the second kind of surface $\mathcal{S}$
$d_{\alpha\beta}$	covariant components of the symmetric surface velocity gradient
$\mathcal{D}$	dissipation per current surface area
$\mathcal{D}_0$	dissipation per reference surface area
$\mathrm{d}a$	differential surface element on $\mathcal{S}$
$\mathrm{d}A$	differential surface element on $\mathcal{S}_0$
$d^{\alpha\beta\gamma\delta}$	derivative of $\tau^{\alpha\beta}$ w.r.t. $b_{\gamma\delta}$
$\delta\ldots$	variation of ...
$\Delta\ldots$	increment of ... that is required for linearization
Δ_{s}	Laplace operator on surface $\mathcal{S}$
$\mathrm{div}_{\mathrm{s}}$	divergence operator on surface $\mathcal{S}$
e	index numbering the finite elements; $e = 1, \ldots, n_{\mathrm{el}}$
$e^{\alpha\beta\gamma\delta}$	derivative of $M_0^{\alpha\beta}$ w.r.t. $a_{\gamma\delta}$
ϵ	penalty parameter
$\boldsymbol{E}$	surface Green–Lagrange strain tensor
$f^{\alpha\beta\gamma\delta}$	derivative of $M_0^{\alpha\beta}$ w.r.t. $b_{\gamma\delta}$
$\boldsymbol{f}$	'body' force acting on $\mathcal{S}$
$\mathbf{f}^e$	finite element force vector of element Ω^e
g	expression for the area-incompressibility constraint
G	expression for the weak form
G^e	contribution to G from finite element Ω^e
$\mathbf{g}^e$	finite element 'force vector' of element Ω^e due to constraint g
∇_{s}	gradient operator on surface $\mathcal{S}$
H	mean curvature of $\mathcal{S}$ at $\boldsymbol{x}$
H_0	spontaneous curvature prescribed at $\boldsymbol{x}$
η	in-plane surface viscosity
I	index numbering the finite element nodes
I_1, I_2	invariants of the surface Cauchy–Green tensors
$\boldsymbol{i}$	surface identity tensor on $\mathcal{S}$
$\boldsymbol{I}$	surface identity tensor on $\mathcal{S}_0$
J	area change between $\mathcal{S}_0$ and $\mathcal{S}$
J_a	area change between $\mathcal{P}$ and $\mathcal{S}$
J_A	area change between $\mathcal{P}$ and $\mathcal{S}_0$
k	bending modulus
k_{g}	Gaussian modulus
K	initial surface bulk modulus (=area compression modulus)
K_{eff}	effective surface bulk modulus

$\mathbf{k}^e$	finite element tangent matrix associated with $\mathbf{f}^e$ and $\mathbf{g}^e$
κ	Gaussian curvature of surface $\mathcal{S}$ at $\boldsymbol{x}$
κ_1, κ_2	principal curvatures of surface $\mathcal{S}$ at $\boldsymbol{x}$
L_I	pressure shape function of finite element node I
$\mathcal{L}$	interface between two NURBS patches
λ_1, λ_2	principal surface stretches of $\mathcal{S}$ at $\boldsymbol{x}$
m_e	number of pressure nodes of finite element Ω^e
m_ν, m_τ	bending moment components acting at $\boldsymbol{x} \in \partial\mathcal{S}$
$\bar{m}_\nu, \bar{m}_\tau$	prescribed bending moment components
$M^{\alpha\beta}$	contra-variant bending moment components
$M_0^{\alpha\beta}$	$= J M^{\alpha\beta}$
μ	initial in-plane membrane shear stiffness
μ_{eff}	effective in-plane membrane shear stiffness
n_{no}	total number of finite element nodes used to discretize $\mathcal{S}$
n_{el}	total number of finite elements used to discretize $\mathcal{S}$
n_{mo}	total number of finite element nodes used to discretize pressure q
n_e	number of displacement nodes of finite element Ω^e
$N^{\alpha\beta}$	total, contra-variant, in-plane membrane stress components
N_I	displacement shape function of finite element node I
$\boldsymbol{n}$	surface normal of $\mathcal{S}$ at $\boldsymbol{x}$
$\boldsymbol{N}$	surface normal of $\mathcal{S}_0$ at $\boldsymbol{X}$
$\mathbf{N}$	array of the shape functions for element Ω^e
$\boldsymbol{\nu}$	normal vector on $\partial\mathcal{S}$
ξ^α	convective surface coordinates; $\alpha = 1, 2$
$\mathcal{P}$	parametric domain spanned by ξ^1 and ξ^2
$\mathbf{P}$	class of stabilization methods based on normal projection; projcction matrix
ψ	Helmholtz free energy per unit mass
Ψ_0	Helmholtz free energy per reference area
q	Lagrange multiplier associated with area-incompressibility
$\mathbf{q}$	array of all Lagrange multipliers q_I in the system; $I = 1, \ldots, n_{\text{mo}}$
$\mathbf{q}_e$	array of all Lagrange multipliers q_I for finite element Ω^e; $I = 1, \ldots, m_e$
$\mathcal{R}$	arbitrary subregion of $\mathcal{S}$
ρ	surface density of $\mathcal{S}$ at $\boldsymbol{x}$
ρ_0	surface density of $\mathcal{S}_0$ at $\boldsymbol{x}$
S^α	contra-variant, out-of-plane shear stress components
$\mathcal{S}$	current configuration of the surface
$\mathcal{S}_0$	initial configuration of the surface
$\boldsymbol{\sigma}$	Cauchy stress tensor of the shell
$\sigma^{\alpha\beta}$	stretch-related, contra-variant, in-plane stress components
$\boldsymbol{t}$	effective traction acting on the boundary $\partial\mathcal{S}$ normal to $\boldsymbol{\nu}$
$\bar{\boldsymbol{t}}$	prescribed boundary tractions on Neumann boundary $\partial_t\mathcal{S}$
$\boldsymbol{T}$	traction acting on the boundary $\partial\mathcal{S}$ normal to $\boldsymbol{\nu}$
$\boldsymbol{T}^\alpha$	traction acting on the boundary $\partial\mathcal{S}$ normal to $\boldsymbol{a}^\alpha$

$\tau^{\alpha\beta}$	$= J\sigma^{\alpha\beta}$
$\mathcal{V}$, $\mathcal{Q}$	admissible function spaces
φ	deformation map of surface $\mathcal{S}$
$\bar{\varphi}$	prescribed boundary deformations on boundary $\partial_x\mathcal{S}$
w	hyperelastic stored surface energy density (per current surface area)
W	hyperelastic stored surface energy density (per reference surface area)
$\boldsymbol{x}$	current position of a surface point on $\mathcal{S}$
$\boldsymbol{X}$	initial position of $\boldsymbol{x}$ on the reference surface $\mathcal{S}_0$
$\boldsymbol{x}_I$	position vector of finite element node I lying on $\mathcal{S}$
$\boldsymbol{X}_I$	initial position of finite element node I on $\mathcal{S}_0$
$\mathbf{x}$	stacked array of all $\boldsymbol{x}_I$ of the discretized surface; $I = 1, ..., n_{\text{no}}$
$\mathbf{x}_e$	stacked array of all $\boldsymbol{x}_I$ for finite element Ω^e; $I = 1, ..., n_e$
$\mathbf{X}_e$	stacked array of all $\boldsymbol{X}_I$ for finite element Ω^e_0; $I = 1, ..., n_e$
Ω^e	current configuration of finite element e
Ω^e_0	reference configuration of finite element e

Part I: Introduction

The aim of this work is to present the computational treatment of lipid bilayers using the framework of isogeometric finite element analysis and nonlinear shell theory. The presentation follows earlier work on membranes (Sauer et al. 2014) and shells (Sauer and Duong 2017; Duong et al. 2017; Sauer et al. 2017). It thus presents a condensed and combined version of earlier work by focussing on the most important aspects that are required for the computational description of lipid bilayers. Additionally, several new parts have been incorporated into the presentation. Those are:

- a summary and discussion of the computational challenges
- an extension of the theory to include surface differential operators, surface contact and surface viscosity
- the discretization and linearization of the viscosity term
- an investigation of the LBB condition for mixed shell finite elements
- a computational example on lipid bilayer indentation.

The remainder of Part I gives an overview of the ingredients and challenges of the computational modeling of lipid bilayers (Sect. 1), and surveys related literature (Sect. 2). Part II (Sects. 4–9) and Part III (Sects. 10–13) then discuss the theoretical background and the computational modeling in detail. Readers familiar with shell theory may directly jump to Part III and revisit relevant sections of Part II as they are addressed. Section 14 concludes this work.

1 Computational Ingredients and Challenges

The modeling of lipid bilayer shells is a challenging task due to a variety of reasons. Lipid bilayers are liquid shells that are characterized by in-plane flow and out-of-plane bending elasticity (Fig. 1a). The mechanics of such shells can lead to very complex surface shapes (Fig. 1b). Table 1 gives an overview of the computational modeling challenges of lipid bilayers and lists corresponding ingredients to deal with them. The remainder of this section provides a short discussion on those challenges.

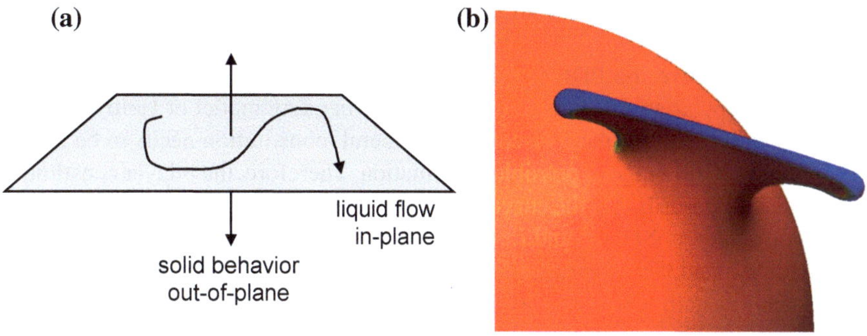

Fig. 1 Lipid bilayer deformations: **a** combined solid-like and liquid-like behavior; **b** complex bud shapes (Sauer et al. 2017)

Table 1 Lipid bilayer modeling: computational challenges and corresponding model ingredients (and the sections where they are addressed)

Challenge	Ingredient	Sects.
Surface description	Curvilinear coordinates	3
Liquid- and solid-like behavior	In-plane flow + out-of-plane bending	4
Geometric PDE's	Surface balance laws	5
Bilayer constitution	Helfrich model + in-plane viscosity	6 and 7
Nonlinearity	Consistent linearization	8 and 10
Smooth discretization	NURBS-based surface finite elements	10
Area-incompressibility	LBB-conforming mixed methods	11
Zero shear stiffness	In-plane shear stabilization	12
Complex surface flow	Surface ALE	–
Coupled problems	Coupled balance laws	–
Local refinement	LR-B-splines, LR-NURBS	–
Tilt, interlayer sliding	Additional degrees-of-freedom	–

In order to deal with the solid- and liquid-like behavior of lipid bilayers, a very general model formulation is required that is capable of describing the kinematics of large bending deformations and surface flows. This requires a very general surface description that can capture large deformations and rotations. Such a formulation is offered by curvilinear surface coordinates. It is presented in Sects. 3 and 4. Curvilinear coordinates offer the extra advantage that they can be used to define the finite element shape functions. In consequence this leads to a straightforward finite element description of the problem.

The bilayer deformation is governed by so-called geometric PDE's. These are partial differential equations that live on evolving surfaces. For mechanical systems, these PDE's follow from the balance laws of mass and momentum. This is presented in Sect. 5.

In order to solve the PDE's, the constitutive behavior of the bilayer has to be defined. A popular approach is to use the elastic bending model of Helfrich (1973) and combine it with in-plane viscosity. In general, constitution needs to be able to account for the full range of possible deformation. Therefore, the bilayer constitution should also be described in the curvilinear coordinate system of the evolving surface. This is presented in Sects. 6 and 7.

The PDE's and their corresponding weak form are strongly nonlinear. In order to solve such a system within implicit finite element methods, the consistent linearization of the formulation is required. This is presented in Sects. 8 and 10.

Lipid bilayers are very thin structures, and it is appropriate to describe them with thin-shell theory. Thin-shell theory leads to a high-order weak form that requires a surface description that is at least C^1-continuous. Such a formulation is provided by NURBS-based finite element spaces. They are presented in Sect. 10.

The surface flow of lipid bilayers can be considered to be area-incompressible. Area-incompressibility is a constraint that introduces new unknowns. The discretization of those needs to conform with the discretization of the surface and its velocity according to the LBB condition. This is discussed in Sect. 11.

Under quasi-static conditions, the bilayer offers no resistance to shear deformations. To solve such cases computationally, numerical shear stabilization is required. Several stabilization techniques can be used, as is presented in Sect. 12.

Under dynamic conditions, viscosity offers resistance to shear flow. However, surface flow can lead to very large surface deformations that cannot be tracked by a pure Lagrangian (i.e., material) mesh. Also pure Eulerian (i.e., fixed) meshes cannot be used, since the surface shape can change. Thus an arbitrary Lagrangian-Eulerian (ALE) surface formulation is required.

The mechanics of lipid bilayers may be coupled to other phenomena, such as diffusion, phase transitions, and protein binding reactions. To account for these, the surface balance laws have to be extended by the energy and mass balance of multiple species. A recent theory for this has been provided by Sahu et al. (2017).

The surface deformation can become very localized. For such cases local mesh refinement is desirable. Classical NURBS do not offer this, but there is recent work on locally refined NURBS (Zimmermann and Sauer 2017).

Classical thin-shell theory does not account for tilting of the lipids. Also, they do not account for sliding between the two lipid layers. In order to describe these aspects the kinematic description of the bilayer deformation has to be generalized. This effectively adds degrees-of-freedom to the formulation. Lipid tilt and interlayer sliding are addressed in other chapters of this book.

2 Literature Survey

This section gives an overview of existing literature that is related to the computational modeling of lipid bilayers based on nonlinear shell theory. The presentation focuses on finite element models and follows Sauer et al. (2017).

In the past, several computational models have been proposed for cell membranes. Depending on how the membrane is discretized, two categories can be distinguished: Models based on an explicit surface discretization, and models based on an implicit surface discretization. In the second category, the surface is captured by a phase field (Du and Wang 2007) or level set function (Salac and Miksis 2011) that is defined on the surrounding volume mesh. In the first category, the surface is captured directly by a surface mesh. The approach is particularly suitable if only surface effects are studied, such that no surrounding volume mesh is needed. This is the approach taken here. An example is to use Galerkin surface finite elements: The first corresponding 3D FE model for lipid bilayer membranes seems to be the formulation of Feng and Klug (2006) and Ma and Klug (2008). Their FE formulation is based on so-called subdivision surfaces (Cirak and Ortiz 2001), which provide C^1-continuous FE surface discretizations. Such discretizations are advantageous, since they do not require additional degrees of freedom as C^0-continuous FE formulations do. Still, C^0-continuous FEs have been considered to model red blood cell (RBC) membranes and their supporting protein skeleton (Dao et al. 2003; Peng et al. 2010), phase changes of lipid bilayers (Elliott and Stinner 2010), and viscous cell membranes (Tasso and Buscaglia 2013). Subdivision finite elements have been used to study confined cells (Kahraman et al. 2012). Lipid bilayers can also be modeled with so-called 'solid shell' (i.e., classical volume) elements instead of surface shell elements (Kloeppel and Wall 2011). Using solid elements, C^0-continuity is sufficient, but the formulation is generally less efficient. For two-dimensional and axisymmetric problems also C^1-continuous B-spline and Hermite finite elements have been used to study membrane surface flows (Arroyo and DeSimone 2009; Rahimi and Arroyo 2012), cell invaginations (Rim et al. 2014), and cell tethering and adhesion (Rangarajan and Gao 2015). The latter work also discusses the generalization to three-dimensional B-spline FE. For some problems it is also possible to use specific, Monge-patch FE discretizations (Rangamani et al. 2013, 2014).

The computational framework considered here is based on isogeometric finite elements (Hughes et al. 2005; Cottrell et al. 2009). Those provide C^1-continuity through the use of splines. Isogeometric FE formulations have been applied to solid shells (Kiendl et al. 2009, 2010, 2015; Benson et al. 2011; Nguyen-Thanh et al.

2011) based on rotation-free FE discretizations (Flores and Estrada 2007; Linhard et al. 2007; Dung and Wells 2008). In Duong et al. (2017) a new isogeometric FE formulation is proposed using curvilinear shell theory (Naghdi 1982; Pietraszkiewicz 1989; Libai and Simmonds 1998). The isogeometric shell model has been extended to liquid shells (Sauer et al. 2017) based on the shell formulation of Steigmann (1999) and the bilayer models of Canham (1970) and Helfrich (1973).

There are also several works that do not use finite element approaches. Examples are numerical ODE integration (Agrawal and Steigmann 2009), Monte Carlo methods (Ramakrishnan et al. 2010), molecular dynamics (Li and Lykotrafitis 2012), finite difference methods (Lau et al. 2012; Gu et al. 2014) and mesh-free methods (Rosolen et al. 2013). There are also non-Galerkin FE approaches that use triangulated surfaces, e.g., see Jarić et al. (1995), Jie et al. (1998).

Ideal liquids lack shear stiffness. Under quasi-static conditions, liquid membranes, and shells therefore do not provide any resistance to in-plane shear deformations and thus need to be stabilized. Various stabilization methods have been proposed in the past, considering artificial viscosity (Ma and Klug 2008; Sauer 2014), artificial stiffness (Kahraman et al. 2012), and normal offsets – either as a projection of the solution (with intermediate mesh update steps) (Sauer 2014), or as a restriction of the surface variation (Rangarajan and Gao 2015). The instability problem is absent, if shear stiffness is present, e.g., due to an underlying cytoskeleton, like in RBCs (Dao et al. 2003; Peng et al. 2010; Kloeppel and Wall 2011).

Part II: Theoretical Description

Part II discusses the theoretical description of lipid bilayers that is required for the computational formulation following in Part III.

3 Surface Description

This section discusses the description of curved surfaces based on the general framework of curvilinear coordinates. The description is based on a surface parameterization (Sect. 3.1), from which the surface decomposition (Sect. 3.2), surface differentiation (Sect. 3.3), surface curvature (Sect. 3.4) and the surface Cayley–Hamilton theorem (Sect. 3.5) follow.

3.1 Surface Parameterization

The bilayer surface, denoted by $\mathcal{S}$, can be described by the parametric description

$$\boldsymbol{x} = \boldsymbol{x}(\xi^\alpha)\,, \tag{1}$$

where ξ^α, $\alpha = 1, 2$ are coordinates associated with a parameter domain $\mathcal{P}$. Equation (1) corresponds to a mapping from point $(\xi^1, \xi^2) \in \mathcal{P}$ to the surface point $\boldsymbol{x} \in \mathcal{S}$, see Fig. 2. The mapping reflects the property that the surface is a 2D object embedded within 3D space. Mapping (1) fully characterizes the surface geometry. Coordinates ξ^α are known as *curvilinear coordinates*. The tangent vector to coordinate ξ^α is given by

$$\boldsymbol{a}_\alpha = \frac{\partial \boldsymbol{x}}{\partial \xi^\alpha}\,. \tag{2}$$

The two vectors $\boldsymbol{a}_1$ and $\boldsymbol{a}_2$ are generally not orthonormal, i.e., the four numbers

$$a_{\alpha\beta} = \boldsymbol{a}_\alpha \cdot \boldsymbol{a}_\beta\,, \tag{3}$$

generally give $[a_{\alpha\beta}] \neq [1\ 0;\ 0\ 1]$. The object $a_{\alpha\beta}$ is an important characteristic of the surface, known as the *surface metric*. To restore orthonormality, a second set of tangent vectors $\boldsymbol{a}^1$ and $\boldsymbol{a}^2$ is introduced such that

$$\boldsymbol{a}_\alpha \cdot \boldsymbol{a}^\beta = \delta^\beta_\alpha\,, \tag{4}$$

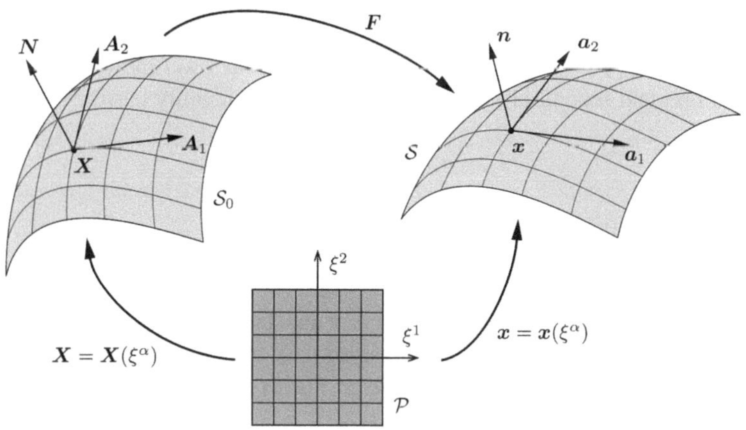

Fig. 2 Mapping between parameter domain $\mathcal{P}$, reference surface $\mathcal{S}_0$ and current surface $\mathcal{S}$ (Sauer et al. 2014)

where $[\delta^\beta_\alpha] = [1\ 0;\ 0\ 1]$. δ^β_α is known as the *Kronecker delta*. Multiplication by δ^β_α simply exchanges indices, e.g., $\boldsymbol{a}^\alpha\, \delta^\beta_\alpha = \boldsymbol{a}^\beta$. It follows that

$$\boldsymbol{a}^\alpha = a^{\alpha\beta}\boldsymbol{a}_\beta\,, \tag{5}$$

where $[a^{\alpha\beta}] = [a_{\alpha\beta}]^{-1}$. Tangent vectors $\boldsymbol{a}_1$ and $\boldsymbol{a}_2$ are also called the *covariant tangent vectors*, while $\boldsymbol{a}^1$ and $\boldsymbol{a}^2$ are also known as the *contra-variant tangent vectors*. Analogously, $a_{\alpha\beta}$ is called the *covariant surface metric* and $a^{\alpha\beta}$ the *contra-variant surface metric*. Equation (5) uses index notation, i.e., summation is implied on repeated indices. By construction, repeated indices always appear as covariant/contra-variant pairs.

The normal vector to surface $\mathcal{S}$ can be defined as

$$\boldsymbol{n} = \frac{\boldsymbol{a}_1 \times \boldsymbol{a}_2}{\|\boldsymbol{a}_1 \times \boldsymbol{a}_2\|}\,. \tag{6}$$

The quantity $J_a := \|\boldsymbol{a}_1 \times \boldsymbol{a}_2\|$ gives the area enclosed by vectors $\boldsymbol{a}_1$ and $\boldsymbol{a}_2$. It can be shown that $J_a = \sqrt{\det[a_{\alpha\beta}]}$.

3.2 Surface Decomposition

The triads $\{\boldsymbol{a}_1, \boldsymbol{a}_2, \boldsymbol{n}\}$ and $\{\boldsymbol{a}^1, \boldsymbol{a}^2, \boldsymbol{n}\}$ form bases that can be used to decompose vectors $\boldsymbol{v} \in \mathbb{R}^3$ into their in-plane and out-of-plane components, i.e.,

$$\boldsymbol{v} = \boldsymbol{v}_\mathrm{s} + \boldsymbol{v}_\mathrm{n}\,, \quad \boldsymbol{v}_\mathrm{s} = v^\alpha\,\boldsymbol{a}_\alpha = v_\alpha\,\boldsymbol{a}^\alpha\,, \quad \boldsymbol{v}_\mathrm{n} = v\,\boldsymbol{n}\,, \tag{7}$$

where $v = \boldsymbol{v}\cdot\boldsymbol{n}$ is the vector component along $\boldsymbol{n}$, and $v_\alpha = \boldsymbol{v}\cdot\boldsymbol{a}_\alpha$ and $v^\alpha = \boldsymbol{v}\cdot\boldsymbol{a}^\alpha$ are the vector components along $\boldsymbol{a}^\alpha$ and $\boldsymbol{a}_\alpha$, respectively. v_α is also called the covariant and v^α the contra-variant vector component. Applying (5)–(7) yields $v^\alpha = a^{\alpha\beta}\,v_\beta$. Likewise $v_\alpha = a_{\alpha\beta}\,v^\beta$. Generally, $a^{\alpha\beta}$ and $a_{\alpha\beta}$ raise and lower indices, respectively.

Two important second-order tensors are the surface identity tensor,

$$\boldsymbol{i} := \boldsymbol{a}_\alpha \otimes \boldsymbol{a}^\alpha = \boldsymbol{a}^\alpha \otimes \boldsymbol{a}_\alpha\,, \tag{8}$$

and the full identity in $\mathbb{R}^3$,

$$\boldsymbol{1} := \boldsymbol{i} + \boldsymbol{n}\otimes\boldsymbol{n}\,. \tag{9}$$

With those follow, $\boldsymbol{i}\boldsymbol{v} = \boldsymbol{v}_\mathrm{s}$, $\boldsymbol{i}\boldsymbol{v}_\mathrm{s} = \boldsymbol{v}_\mathrm{s}$, $\boldsymbol{1}\boldsymbol{v} = \boldsymbol{v}$ and $\boldsymbol{1}\boldsymbol{v}_\mathrm{s} = \boldsymbol{v}_\mathrm{s}$. Thus $\boldsymbol{i}$, can be viewed as a projection operator, that extracts $\boldsymbol{v}_\mathrm{s}$ from $\boldsymbol{v}$. In the same fashion $\boldsymbol{i}$ can be used to extract the in-plane contents of a tensor. For example, the surface part of the second-order tensor $\boldsymbol{c} \in \mathbb{R}^3 \times \mathbb{R}^3$ is

$$\boldsymbol{c}_\mathrm{s} := \boldsymbol{i}\cdot\boldsymbol{c}\,\boldsymbol{i}\,. \tag{10}$$

From (8) follows

$$\begin{aligned} \boldsymbol{c}_{\mathrm{s}} &= c^{\alpha\beta}\,\boldsymbol{a}_\alpha \otimes \boldsymbol{a}_\beta\,, & c^{\alpha\beta} &= \boldsymbol{a}^\alpha \cdot \boldsymbol{c}\,\boldsymbol{a}^\beta\,, \\ &= c^{\alpha}_{\ \beta}\,\boldsymbol{a}_\alpha \otimes \boldsymbol{a}^\beta\,, & c^{\alpha}_{\ \beta} &= \boldsymbol{a}^\alpha \cdot \boldsymbol{c}\,\boldsymbol{a}_\beta\,, \\ &= c_{\alpha\beta}\,\boldsymbol{a}^\alpha \otimes \boldsymbol{a}^\beta\,, & c_{\alpha\beta} &= \boldsymbol{a}_\alpha \cdot \boldsymbol{c}\,\boldsymbol{a}_\beta\,, \\ &= c_{\alpha}^{\ \beta}\,\boldsymbol{a}^\alpha \otimes \boldsymbol{a}_\beta\,, & c_{\alpha}^{\ \beta} &= \boldsymbol{a}_\alpha \cdot \boldsymbol{c}\,\boldsymbol{a}^\beta\,. \end{aligned} \tag{11}$$

If $\boldsymbol{c}$ is symmetric, then $c_{\beta}^{\ \alpha} = c^{\alpha}_{\ \beta} =: c^{\alpha}_{\beta}$. Apart from $\boldsymbol{c}_{\mathrm{s}}$, tensor $\boldsymbol{c}$ also has components along $\boldsymbol{n} \otimes \boldsymbol{n}$, $\boldsymbol{a}_\alpha \otimes \boldsymbol{n}$ and $\boldsymbol{n} \otimes \boldsymbol{a}_\alpha$.

Based on these definitions, three important tensor functions can be defined. The first is the surface trace, defined by

$$\mathrm{tr}_{\mathrm{s}}\,\boldsymbol{c} := \boldsymbol{i} : \boldsymbol{c}\,. \tag{12}$$

It is related to the regular trace operator $\mathrm{tr}\,\boldsymbol{c} := \boldsymbol{1} : \boldsymbol{c}$, by $\mathrm{tr}_{\mathrm{s}}\,\boldsymbol{c} := \mathrm{tr}\,\boldsymbol{c}_{\mathrm{s}}$. Further $\mathrm{tr}_{\mathrm{s}}\,\boldsymbol{c} = c^{\alpha}_{\alpha}$. The second important tensor function is the surface determinant, defined by

$$\mathrm{det}_{\mathrm{s}}\,\boldsymbol{c} := \det[c^{\alpha}_{\beta}]\,, \tag{13}$$

i.e., as the usual matrix-determinant[1] of the 2×2 matrix $[c^{\alpha}_{\beta}]$. Since $c^{\alpha}_{\beta} = a^{\alpha\gamma} c_{\gamma\beta}$, the surface determinant can also be written as

$$\mathrm{det}_{\mathrm{s}}\,\boldsymbol{c} := \det[a^{\alpha\beta}]\,\det[c_{\alpha\beta}] = \det[c_{\alpha\beta}]/\det[a_{\alpha\beta}]\,. \tag{14}$$

Note that this expression does not contain any summation on α or β, since det[...] is a scalar. The third tensor function is the surface inverse $\boldsymbol{c}_{\mathrm{s}}^{-1}$, defined from

$$\boldsymbol{c}_{\mathrm{s}}^{-1}\boldsymbol{c}_{\mathrm{s}} = \boldsymbol{i}\,. \tag{15}$$

$\boldsymbol{c}_{\mathrm{s}}^{-1}$ is a surface tensor with the contra-variant components

$$c^{\alpha\beta}_{\mathrm{inv}} :- \frac{1}{c}\,e^{\alpha\gamma}\,c_{\delta\gamma}\,e^{\beta\delta}\,, \quad c :- \det[c_{\alpha\beta}]\,, \tag{16}$$

since $c^{\alpha\beta}_{\mathrm{inv}}\,c_{\beta\gamma} = \delta^{\alpha}_{\gamma}$. Here

$$[e^{\alpha\beta}] = \begin{bmatrix} 0 & 1 \\ -1 & 0 \end{bmatrix} \tag{17}$$

is the so-called unit alternator. In particular, (16) yields

$$a^{\alpha\beta} = \frac{1}{a}\,e^{\alpha\gamma}\,a_{\gamma\delta}\,e^{\beta\delta}\,, \quad a := \det[a_{\alpha\beta}]\,. \tag{18}$$

[1]Note that $\det[c^{\alpha}_{\beta}] = \det[c^{\alpha}_{\ \beta}] = \det[c_{\beta}^{\ \alpha}]$ even if $c^{\alpha}_{\ \beta} \neq c_{\beta}^{\ \alpha}$.

Note that in general $\mathrm{tr_s}\,\boldsymbol{c} \neq \mathrm{tr}\,\boldsymbol{c}$, $\det_s \boldsymbol{c} \neq \det \boldsymbol{c}$ and $\boldsymbol{c}_s^{-1} \neq \boldsymbol{c}^{-1}$. Multiplying (18) by $a_{\alpha\beta}$, one can further find

$$a = \frac{1}{2}\, e^{\alpha\gamma}\, e^{\beta\delta}\, a_{\alpha\beta}\, a_{\gamma\delta}\,. \tag{19}$$

3.3 Surface Differentiation

The derivative encountered in (2) is called the *parametric derivative*. It is denoted by a comma. Taking another parametric derivative gives

$$\boldsymbol{a}_{\alpha,\beta} = \frac{\partial \boldsymbol{a}_\alpha}{\partial \xi^\beta} = \boldsymbol{x}_{,\alpha\beta}\,. \tag{20}$$

Generally, vector $\boldsymbol{a}_{\alpha,\beta}$ has both in-plane and out-of-plane components. But only the latter is needed in order to describe surface curvature. This motivates the introduction of another derivative, the so-called *co-variant derivative*. It is denoted by a semicolon. For basis vectors $\boldsymbol{a}_\alpha$ and $\boldsymbol{a}^\alpha$ it is defined by

$$\boldsymbol{a}_{\alpha;\beta} := (\boldsymbol{n} \otimes \boldsymbol{n})\, \boldsymbol{a}_{\alpha,\beta} \tag{21}$$

and

$$\boldsymbol{a}^\alpha_{;\beta} := (\boldsymbol{n} \otimes \boldsymbol{n})\, \boldsymbol{a}^\alpha_{,\beta}\,. \tag{22}$$

Using Eqs. (9) and (8), leads to

$$\boldsymbol{a}_{\alpha;\beta} = \boldsymbol{a}_{\alpha,\beta} - \Gamma^\gamma_{\alpha\beta}\, \boldsymbol{a}_\gamma \tag{23}$$

and

$$\boldsymbol{a}^\alpha_{;\beta} := \boldsymbol{a}^\alpha_{,\beta} + \Gamma^\alpha_{\beta\gamma}\, \boldsymbol{a}^\gamma\,. \tag{24}$$

where

$$\Gamma^\gamma_{\alpha\beta} := \boldsymbol{a}^\gamma \cdot \boldsymbol{a}_{\alpha,\beta} \tag{25}$$

are the so-called *Christoffel symbols*. For scalars $\phi \in \mathbb{R}$ and general vectors $\boldsymbol{v} \in \mathbb{R}^3$ (that are independent of the surface parameterization), such as the normal vector $\boldsymbol{n}$, the covariant derivative is defined to be equal to the parametric derivative. From (7) thus follow $\boldsymbol{n}_{;\alpha} = \boldsymbol{n}_{,\alpha}$, $v_{;\alpha} = v_{,\alpha}$, $(v^\alpha \boldsymbol{a}_\alpha)_{;\beta} = (v^\alpha \boldsymbol{a}_\alpha)_{,\beta}$, $(v_\alpha \boldsymbol{a}^\alpha)_{;\beta} = (v_\alpha \boldsymbol{a}^\alpha)_{,\beta}$, and further

$$\begin{aligned} v_{\alpha;\beta} &= v_{\alpha,\beta} - \Gamma^\gamma_{\alpha\beta}\, v_\gamma\,, \\ v^\alpha_{;\beta} &= v^\alpha_{,\beta} + \Gamma^\alpha_{\beta\gamma}\, v^\gamma\,. \end{aligned} \tag{26}$$

In classical physics, the gradient, divergence, and Laplacian are important differential operators. They can now be defined on the surface $\mathcal{S}$. The surface gradient of a scalar function ϕ is defined through the regular gradient $\nabla\phi$ as

$$\nabla_{\!s}\phi := \nabla\phi \cdot \boldsymbol{i}\,, \tag{27}$$

Inserting (8), gives $\nabla_{\!s}\phi = \phi_{,\alpha}\,\boldsymbol{a}^{\alpha}$. Likewise, the surface gradient for a vector function $\boldsymbol{v}$ is defined as

$$\nabla_{\!s}\boldsymbol{v} := \nabla\boldsymbol{v} \cdot \boldsymbol{i}\,, \tag{28}$$

such that $\nabla_{\!s}\boldsymbol{v} = \boldsymbol{v}_{,\alpha} \otimes \boldsymbol{a}^{\alpha}$. The surface divergence follows from the gradient as

$$\mathrm{div}_{\mathrm{s}}\boldsymbol{v} := \mathrm{tr}\,\nabla_{\!s}\boldsymbol{v}\,. \tag{29}$$

i.e. $\mathrm{div}_{\mathrm{s}}\boldsymbol{v} = \boldsymbol{v}_{,\alpha} \cdot \boldsymbol{a}^{\alpha}$. The surface Laplacian of a scalar ϕ is then defined by

$$\Delta_{\mathrm{s}}\phi := \mathrm{div}_{\mathrm{s}}\nabla_{\!s}\phi\,, \tag{30}$$

which leads to $\Delta_{\mathrm{s}}\phi = \phi_{;\alpha\beta}\,a^{\alpha\beta}$. In the above expressions $\phi_{;\alpha} = \phi_{,\alpha}$ and $\boldsymbol{v}_{;\alpha} = \boldsymbol{v}_{,\alpha}$. However, $\phi_{;\alpha\beta} \neq \phi_{,\alpha\beta}$. Instead

$$\phi_{;\alpha\beta} = \phi_{,\alpha\beta} - \Gamma^{\gamma}_{\alpha\beta}\,\phi_{,\gamma}\,. \tag{31}$$

3.4 Surface Curvature

The surface curvature is characterized by the normal component of $\boldsymbol{a}_{\alpha,\beta}$, i.e., by the four numbers

$$b_{\alpha\beta} := \boldsymbol{n} \cdot \boldsymbol{a}_{\alpha,\beta} = \boldsymbol{n} \cdot \boldsymbol{a}_{\alpha;\beta}\,. \tag{32}$$

They are known as the covariant components of the *curvature tensor* $\boldsymbol{b} = b_{\alpha\beta}\,\boldsymbol{a}^{\alpha} \otimes \boldsymbol{a}^{\beta}$. The curvature tensor is a surface tensor like $\boldsymbol{i}$ and $\boldsymbol{c}_{\mathrm{s}}$. It is symmetric and has the mixed components $b^{\alpha}_{\beta} = a^{\alpha\gamma}\,b_{\gamma\beta}$ and the contra-variant components $b^{\alpha\beta} = b^{\alpha}_{\gamma}\,a^{\gamma\beta}$. It appears in the formulas of Gauss,

$$\boldsymbol{a}_{\alpha;\beta} = b_{\alpha\beta}\,\boldsymbol{n}\,, \tag{33}$$

and Weingarten,

$$\boldsymbol{n}_{,\alpha} = -b^{\beta}_{\alpha}\,\boldsymbol{a}_{\beta}\,. \tag{34}$$

Its two invariants

$$H := \frac{1}{2}\,\mathrm{tr}_{\mathrm{s}}\,\boldsymbol{b} \tag{35}$$

and

$$\kappa := \mathrm{det}_{\mathrm{s}}\,\boldsymbol{b} \tag{36}$$

are known as the mean curvature and Gaussian curvature of surface $\mathcal{S}$. According to Sect. 3.2, those can also be written as $H = \frac{1}{2}\,b^\alpha_\alpha = \frac{1}{2}\,a^{\alpha\beta}\,b_{\alpha\beta}$ and $\kappa = b/a$, where $b := \det[b_{\alpha\beta}]$ and $a := \det[a_{\alpha\beta}]$. The eigenvalues of $\boldsymbol{b}$,

$$\kappa_{1/2} = H \pm \sqrt{H^2 - \kappa}\,, \tag{37}$$

are the principal curvatures of $\mathcal{S}$. Note that $2H = \kappa_1 + \kappa_2$ and $\kappa = \kappa_1\,\kappa_2$. Using (7) and Weingarten's formula, the surface divergence of vector $\boldsymbol{v}$ can also be written as

$$\mathrm{div}_{\mathrm{s}}\boldsymbol{v} = v^\alpha_{;\alpha} - 2H\,v\,. \tag{38}$$

3.5 Surface Cayley–Hamilton

According to the surface Cayley–Hamilton theorem, a tensor $\boldsymbol{c}$ satisfies the identity

$$c^\gamma_\gamma\,a^{\alpha\beta} - c^{\alpha\beta} = \tilde{c}^{\alpha\beta}\,, \tag{39}$$

where $\tilde{c}^{\alpha\beta} := \dfrac{c}{a}c^{\alpha\beta}_{\mathrm{inv}}$ are the contra-variant components of the adjugate tensor of $\boldsymbol{c}$. For the curvature tensor in particular, the Cayley–Hamilton-theorem becomes

$$2H\,a^{\alpha\beta} - b^{\alpha\beta} = \kappa\,b^{\alpha\beta}_{\mathrm{inv}}\,. \tag{40}$$

Multiplying this by b^γ_β gives

$$b^{\alpha\gamma}\,b^\beta_\gamma = 2H\,b^{\alpha\beta} - \kappa\,a^{\alpha\beta}\,. \tag{41}$$

Lowering indices with $a_{\alpha\beta}$, then gives

$$b^\gamma_\alpha\,b_{\gamma\beta} = 2H\,b_{\alpha\beta} - \kappa\,a_{\alpha\beta}\,. \tag{42}$$

4 Surface Kinematics

This section discusses the kinematics of deforming surfaces and examines its consequences. Important kinematical objects are the surface strain tensor (Sect. 4.1), the surface velocity gradient (Sect. 4.2) and the area-incompressibility constraint (Sect. 4.3). For the subsequent developments, all kinematical objects need to be varied (Sect. 4.4) and linearized (Sect. 4.5).

4.1 Surface Deformation

In order to describe the deformation of surface $\mathcal{S}$, a reference configuration, denoted $\mathcal{S}_0$, is introduced. This could for example be a flat plane. But that is not a requirement. The only requirement for $\mathcal{S}_0$ is that it is fixed in time. The reference surface $\mathcal{S}_0$ can be described in the same form as $\mathcal{S}$. Therefore all the quantities introduced in Sect. 3 can be re-defined for $\mathcal{S}_0$. This is done by using upper-case letters, or adding subscript '0.' Surface $\mathcal{S}_0$ is thus described by the mapping $\boldsymbol{X} = \boldsymbol{X}(\xi^\alpha)$ and the tangent vectors $\boldsymbol{A}_\alpha = \boldsymbol{X}_{,\alpha}$, see Fig. 2. Further objects that follow in that fashion are $A_{\alpha\beta}$, $A^{\alpha\beta}$, $\boldsymbol{N}$, $\boldsymbol{A}_{\alpha,\beta}$ and so forth. In particular,

$$\boldsymbol{I} := \boldsymbol{A}_\alpha \otimes \boldsymbol{A}^\alpha = \boldsymbol{A}^\alpha \otimes \boldsymbol{A}_\alpha \tag{43}$$

denotes the surface identity tensor on $\mathcal{S}_0$, such that $\mathbf{1} = \boldsymbol{I} + \boldsymbol{N} \otimes \boldsymbol{N}$.

The mapping between $\mathcal{S}_0$ and $\mathcal{S}$, denoted $\boldsymbol{x} = \varphi(\boldsymbol{X})$, is characterized by the surface deformation gradient

$$\boldsymbol{F} := \boldsymbol{a}_\alpha \otimes \boldsymbol{A}^\alpha \,. \tag{44}$$

and the surface stretch

$$J := \frac{J_a}{J_A} = \frac{\sqrt{\det[a_{\alpha\beta}]}}{\sqrt{\det[A_{\alpha\beta}]}} \,. \tag{45}$$

They relate differential line and area elements according to $\mathrm{d}\boldsymbol{x} = \boldsymbol{F}\,\mathrm{d}\boldsymbol{X}$ and $\mathrm{d}a = J\,\mathrm{d}A$. If the number of surface particles is conserved during deformation, as will be considered here,[2] then

$$\rho\,\mathrm{d}a = \rho_0\,\mathrm{d}A \,, \tag{46}$$

such that

$$J = \frac{\rho_0}{\rho} \,, \tag{47}$$

where ρ and ρ_0 are the surface densities at $\boldsymbol{x} \in \mathcal{S}$ and $\boldsymbol{X} \in \mathcal{S}_0$, respectively.

Two important objects for describing in-plane deformation, are the left and right surface Cauchy–Green tensors, given by

$$\begin{aligned} \boldsymbol{C} &:= \boldsymbol{F}^T\boldsymbol{F} = a_{\alpha\beta}\,\boldsymbol{A}^\alpha \otimes \boldsymbol{A}^\beta \,, \\ \boldsymbol{B} &:= \boldsymbol{F}\boldsymbol{F}^T = A^{\alpha\beta}\,\boldsymbol{a}_\alpha \otimes \boldsymbol{a}_\beta \,. \end{aligned} \tag{48}$$

$\boldsymbol{C}$ is a surface tensor on $\mathcal{S}_0$, while $\boldsymbol{B}$ is a surface tensor on $\mathcal{S}$. Their trace $I_1 := \mathrm{tr}\,\boldsymbol{C} = \boldsymbol{I} : \boldsymbol{C} = \mathrm{tr}\,\boldsymbol{B} = \boldsymbol{i} : \boldsymbol{B}$ is equal to

$$I_1 = A^{\alpha\beta} a_{\alpha\beta} \,. \tag{49}$$

[2]For an extension to changing mass, e.g., due to protein binding, see Sahu et al. (2017).

From $\boldsymbol{C}$ follows the surface Green–Lagrange strain tensor

$$\boldsymbol{E} := \big(\boldsymbol{C} - \boldsymbol{I}\big)/2\,. \tag{50}$$

Its surface components are

$$E_{\alpha\beta} := \big(a_{\alpha\beta} - A_{\alpha\beta}\big)/2\,, \tag{51}$$

such that $\boldsymbol{E} = E_{\alpha\beta}\,\boldsymbol{A}^{\alpha} \otimes \boldsymbol{A}^{\beta}$. Likewise, the relative curvature tensor $\boldsymbol{K} = K_{\alpha\beta}\,\boldsymbol{A}^{\alpha} \otimes \boldsymbol{A}^{\beta}$, with the components

$$K_{\alpha\beta} := b_{\alpha\beta} - B_{\alpha\beta}\,, \tag{52}$$

is defined. It is an important object for describing bending.

4.2 *Surface Motion*

In general, the deformation of the surface is time-dependent. The consequences of this on the surface description and kinematics are discussed here. The velocity of a surface particle (e.g., a lipid molecule) at $\boldsymbol{x} \in \mathcal{S}$, is $\boldsymbol{v} = \dot{\boldsymbol{x}}$, where the notation

$$\dot{(...)} := \frac{D...}{Dt} := \frac{\partial ...}{\partial t}\bigg|_{\boldsymbol{X}=\text{fixed}} \tag{53}$$

denotes the so-called *material time derivative*. The time derivative of the tangent vectors and their parametric derivatives then follow as $\dot{\boldsymbol{a}}_{\alpha} = \dot{\boldsymbol{x}}_{,\alpha} = \boldsymbol{v}_{,\alpha}$ and $\dot{\boldsymbol{a}}_{\alpha,\beta} = \dot{\boldsymbol{x}}_{,\alpha\beta} = \boldsymbol{v}_{,\alpha\beta}$. This then leads to

$$\dot{a}_{\alpha\beta} = \boldsymbol{a}_{\alpha} \cdot \dot{\boldsymbol{a}}_{\beta} + \dot{\boldsymbol{a}}_{\alpha} \cdot \boldsymbol{a}_{\beta} \tag{54}$$

and

$$\dot{b}_{\alpha\beta} = \boldsymbol{a}_{\alpha,\beta} \cdot \dot{\boldsymbol{n}} + \boldsymbol{n} \cdot \dot{\boldsymbol{a}}_{\alpha,\beta}\,. \tag{55}$$

Taking a time derivative of $\boldsymbol{n} \cdot \boldsymbol{n} = 1$ and $\boldsymbol{n} \cdot \boldsymbol{a}_{\alpha} = 0$, one can find

$$\dot{\boldsymbol{n}} = -(\boldsymbol{a}^{\alpha} \otimes \boldsymbol{n})\,\dot{\boldsymbol{a}}_{\alpha} = -\boldsymbol{a}^{\alpha}(\boldsymbol{n} \cdot \dot{\boldsymbol{a}}_{\alpha})\,, \tag{56}$$

such that

$$\dot{b}_{\alpha\beta} = \big(\dot{\boldsymbol{a}}_{\alpha,\beta} - \Gamma^{\gamma}_{\alpha\beta}\,\dot{\boldsymbol{a}}_{\gamma}\big) \cdot \boldsymbol{n}\,. \tag{57}$$

Taking a time derivative of (4) and $\boldsymbol{n} \cdot \boldsymbol{a}^{\alpha} = 0$, one can find

$$\dot{\boldsymbol{a}}^{\alpha} = \big(a^{\alpha\beta}\,\boldsymbol{n} \otimes \boldsymbol{n} - \boldsymbol{a}^{\beta} \otimes \boldsymbol{a}^{\alpha}\big)\,\dot{\boldsymbol{a}}_{\beta}\,. \tag{58}$$

From (19) follows

$$\dot{a} = a\, a^{\alpha\beta}\, \dot{a}_{\alpha\beta}\,, \tag{59}$$

and therefore

$$\dot{J} = \frac{\partial J}{\partial a_{\alpha\beta}}\, \dot{a}_{\alpha\beta} = \frac{J}{2}\, a^{\alpha\beta}\, \dot{a}_{\alpha\beta}\,. \tag{60}$$

From (18) follows

$$\dot{a}^{\alpha\beta} = a^{\alpha\beta\gamma\delta}\, \dot{a}_{\gamma\delta}\,, \tag{61}$$

with

$$a^{\alpha\beta\gamma\delta} := \frac{\partial a^{\alpha\beta}}{\partial a_{\gamma\delta}} = \frac{1}{2a}\left(e^{\alpha\gamma}e^{\beta\delta} + e^{\alpha\delta}e^{\beta\gamma}\right) - a^{\alpha\beta}a^{\gamma\delta}\,. \tag{62}$$

A component-wise comparison shows that

$$a^{\alpha\beta\gamma\delta} = -\frac{1}{2}\left(a^{\alpha\gamma}a^{\beta\delta} + a^{\alpha\delta}a^{\beta\gamma}\right), \tag{63}$$

i.e., $a^{\alpha\beta\gamma\delta}$ corresponds to the contra-variant components of a fourth-order identity tensor: Contracting $a^{\alpha\beta\gamma\delta}$ with any symmetric tensor with components $c_{\gamma\delta}$, yields

$$a^{\alpha\beta\gamma\delta}\, c_{\gamma\delta} = -c^{\alpha\beta}\,. \tag{64}$$

It is noted that $a^{\alpha\beta\gamma\delta}$ has major and minor symmetries. Given $a^{\alpha\beta\gamma\delta}$, Eq. (61) turns into

$$\dot{a}^{\alpha\beta} = -a^{\alpha\gamma}\, a^{\beta\delta}\, \dot{a}_{\gamma\delta}\,. \tag{65}$$

An important object for fluids is the symmetric surface velocity gradient

$$\boldsymbol{d} := \left(\boldsymbol{v}_{,\alpha} \otimes \boldsymbol{a}^{\alpha} + \boldsymbol{a}^{\alpha} \otimes \boldsymbol{v}_{,\alpha}\right)/2\,. \tag{66}$$

Its covariant and contra-variant components, according to (11), simply are $d_{\alpha\beta} = \dot{a}_{\alpha\beta}/2$ and $d^{\alpha\beta} = -\dot{a}^{\alpha\beta}/2$. In terms of the velocity components $v_\alpha := \boldsymbol{v}\cdot\boldsymbol{a}_\alpha$ and $v := \boldsymbol{v}\cdot\boldsymbol{n}$, also $d^{\alpha\beta} = a^{\alpha\gamma}a^{\beta\delta}(v_{\gamma;\delta} + v_{\delta;\gamma})/2 - v\, b^{\alpha\beta}$ holds.

The time derivative of the mean curvature yields

$$\dot{H} = \frac{1}{2}\,\dot{a}^{\alpha\beta}\, b_{\alpha\beta} + \frac{1}{2}\, a^{\alpha\beta}\, \dot{b}_{\alpha\beta}\,. \tag{67}$$

Using Eqs. (61) and (64) gives

$$\dot{H} = \frac{\partial H}{\partial a_{\alpha\beta}}\, \dot{a}_{\alpha\beta} + \frac{\partial H}{\partial b_{\alpha\beta}}\, \dot{b}_{\alpha\beta}\,, \tag{68}$$

with

$$\begin{aligned}\frac{\partial H}{\partial a_{\alpha\beta}} &= -\frac{1}{2}b^{\alpha\beta}\,,\\ \frac{\partial H}{\partial b_{\alpha\beta}} &= \frac{1}{2}a^{\alpha\beta}\,.\end{aligned}\tag{69}$$

Analogously, the change of the Gaussian curvature is

$$\dot{\kappa} = \frac{\partial \kappa}{\partial a_{\alpha\beta}}\,\dot{a}_{\alpha\beta} + \frac{\partial \kappa}{\partial b_{\alpha\beta}}\,\dot{b}_{\alpha\beta}\,,\tag{70}$$

with

$$\begin{aligned}\frac{\partial \kappa}{\partial a_{\alpha\beta}} &= -\kappa\, a^{\alpha\beta}\,,\\ \frac{\partial \kappa}{\partial b_{\alpha\beta}} &= \kappa\, b^{\alpha\beta}_{\text{inv}} = \tilde{b}^{\alpha\beta}\,.\end{aligned}\tag{71}$$

e.g., see Sauer and Duong (2017).

The last object of interest is $\dot{b}^{\alpha\beta}$. Taking the time derivative of $b^{\alpha\beta} = b_{\gamma\delta}\, a^{\gamma\alpha}\, a^{\delta\beta}$ yields

$$\dot{b}^{\alpha\beta} = \frac{\partial b^{\alpha\beta}}{\partial a_{\gamma\delta}}\,\dot{a}_{\gamma\delta} + \frac{\partial b^{\alpha\beta}}{\partial b_{\gamma\delta}}\,\dot{b}_{\gamma\delta}\,,\tag{72}$$

with

$$\begin{aligned}\frac{\partial b^{\alpha\beta}}{\partial a_{\gamma\delta}} &= b^{\alpha\beta\gamma\delta}\,,\\ \frac{\partial b^{\alpha\beta}}{\partial b_{\gamma\delta}} &= -a^{\alpha\beta\gamma\delta}\,,\end{aligned}\tag{73}$$

and

$$b^{\alpha\beta\gamma\delta} := -\frac{1}{2}\Big(a^{\alpha\gamma}\, b^{\beta\delta} + b^{\alpha\gamma}\, a^{\beta\delta} + a^{\alpha\delta}\, b^{\beta\gamma} + b^{\alpha\delta}\, a^{\beta\gamma}\Big)\tag{74}$$

(Sauer and Duong 2017). From a component-wise comparison, it can be shown that $b^{\alpha\beta\gamma\delta}$ is also equal to

$$b^{\alpha\beta\gamma\delta} = 2H\left(a^{\alpha\beta}\, a^{\gamma\delta} + a^{\alpha\beta\gamma\delta}\right) - \left(a^{\alpha\beta}\, b^{\gamma\delta} + b^{\alpha\beta}\, a^{\gamma\delta}\right).\tag{75}$$

4.3 Surface Incompressibility

An important constraint on the surface motion is surface- (or area-) incompressibility. During such motion

$$g := J - 1 = 0 \;\; \forall t\,,\tag{76}$$

such that $\dot{J} = 0$. From (60) and (54) follows that area-incompressibility implies

$$\dot{\boldsymbol{a}}_\alpha \cdot \boldsymbol{a}^\alpha = 0\,, \tag{77}$$

which is equivalent to

$$\mathrm{div}_\mathrm{s} \boldsymbol{v} = 0\,. \tag{78}$$

4.4 Surface Variation

In order to derive the weak form, which is essential for the finite element method, the variation of several kinematical quantities is required. Therefore, a variation of position $\boldsymbol{x} \in \mathcal{S}$ by the amount $\delta\boldsymbol{x}$ is considered, and the effect on various kinematical quantities is examined. The variation of the tangent vectors and its parametric derivative are $\delta\boldsymbol{a}_\alpha = \delta\boldsymbol{x}_{,\alpha}$ and $\delta\boldsymbol{a}_{\alpha,\beta} = \delta\boldsymbol{x}_{,\alpha\beta}$. Since the variation follows the laws of differentiation, $\delta(...)$ has the same format as $(\dot{...})$, and one can immediately extract the expressions for $\delta a_{\alpha\beta}$, $\delta b_{\alpha\beta}$, $\delta\boldsymbol{n}$, $\delta\boldsymbol{a}^\alpha$, δJ, δH, $\delta\kappa$, $\delta a^{\alpha\beta}$ and $\delta b^{\alpha\beta}$ from the preceding section. In particular,

$$\delta a_{\alpha\beta} = \boldsymbol{a}_\alpha \cdot \delta\boldsymbol{a}_\beta + \delta\boldsymbol{a}_\alpha \cdot \boldsymbol{a}_\beta \tag{79}$$

$$\delta b_{\alpha\beta} = \boldsymbol{a}_{\alpha,\beta} \cdot \delta\boldsymbol{n} + \boldsymbol{n} \cdot \delta\boldsymbol{a}_{\alpha,\beta} \tag{80}$$

or

$$\delta b_{\alpha\beta} = \left(\delta\boldsymbol{a}_{\alpha,\beta} - \Gamma^\gamma_{\alpha\beta}\, \delta\boldsymbol{a}_\gamma\right) \cdot \boldsymbol{n} \tag{81}$$

and

$$\delta\boldsymbol{n} = -(\boldsymbol{a}^\alpha \otimes \boldsymbol{n})\, \delta\boldsymbol{a}_\alpha\,. \tag{82}$$

4.5 Surface Linearization

In order to employ Newton's method, as is considered for the solution of the resulting finite element equations, the weak form needs to be linearized w.r.t. configuration $\boldsymbol{x}$. Therefore, an increment $\Delta\boldsymbol{x}$ is considered and its effect on the system is examined. The change of $\boldsymbol{a}_\alpha$ and $\boldsymbol{a}_{\alpha,\beta}$, due to $\Delta\boldsymbol{x}$, thus is $\Delta\boldsymbol{a}_\alpha = \Delta\boldsymbol{x}_{,\alpha}$ and $\Delta\boldsymbol{a}_{\alpha,\beta} = \Delta\boldsymbol{x}_{,\alpha\beta}$. Since the linearization follows the laws of differentiation, $\Delta(...)$ has the same format as $(\dot{...})$, and one can immediately extract the expressions for $\Delta a_{\alpha\beta}$, $\Delta b_{\alpha\beta}$, $\Delta\boldsymbol{n}$, $\Delta\boldsymbol{a}^\alpha$, ΔJ, ΔH, $\Delta\kappa$, $\Delta a^{\alpha\beta}$ and $\Delta b^{\alpha\beta}$ from Sect. 4.2. Since linearization follows after variation, the variations that still depend on $\boldsymbol{x}$ (instead of just depending on $\delta\boldsymbol{x}$), also need to be linearized. Linearizing (79) and (80), gives

$$\begin{aligned} \Delta\delta a_{\alpha\beta} &= \delta\boldsymbol{a}_\alpha \cdot \Delta\boldsymbol{a}_\beta + \delta\boldsymbol{a}_\beta \cdot \Delta\boldsymbol{a}_\alpha\,, \\ \Delta\delta b_{\alpha\beta} &= \delta\boldsymbol{a}_{\alpha,\beta} \cdot \Delta\boldsymbol{n} + \delta\boldsymbol{n} \cdot \Delta\boldsymbol{a}_{\alpha,\beta} + \boldsymbol{a}_{\alpha,\beta} \cdot \Delta\delta\boldsymbol{n}\,. \end{aligned} \tag{83}$$

From (82) and (58) follows

$$\begin{aligned} \Delta\delta\boldsymbol{n} = {} & (\delta\boldsymbol{a}_\alpha \cdot \boldsymbol{n})(\boldsymbol{n} \cdot \Delta\boldsymbol{a}_\beta)\, a^{\alpha\beta}\, \boldsymbol{n} \\ & + (\delta\boldsymbol{a}_\alpha \cdot \boldsymbol{n})(\boldsymbol{a}^\alpha \cdot \Delta\boldsymbol{a}_\beta)\, \boldsymbol{a}^\beta + (\delta\boldsymbol{a}_\alpha \cdot \boldsymbol{a}^\beta)(\boldsymbol{n} \cdot \Delta\boldsymbol{a}_\beta)\, \boldsymbol{a}^\alpha\,, \end{aligned} \tag{84}$$

such that

$$\boldsymbol{a}_{\alpha,\beta} \cdot \Delta\delta\boldsymbol{n} = \delta\boldsymbol{a}_\gamma \cdot \big(\Gamma^\gamma_{\alpha\beta}\, \boldsymbol{a}^\delta \otimes \boldsymbol{n} + \Gamma^\delta_{\alpha\beta}\, \boldsymbol{n} \otimes \boldsymbol{a}^\gamma - a^{\gamma\delta}\, b_{\alpha\beta}\, \boldsymbol{n} \otimes \boldsymbol{n}\big)\, \Delta\boldsymbol{a}_\delta\,. \tag{85}$$

Inserting (85) into (83) and using (82), then gives

$$\begin{aligned} \Delta\delta b_{\alpha\beta} = {} & -\delta\boldsymbol{a}_\gamma \cdot (\boldsymbol{n} \otimes \boldsymbol{a}^\gamma)\, \Delta\boldsymbol{a}_{\alpha,\beta} - \delta\boldsymbol{a}_{\alpha,\beta} \cdot (\boldsymbol{a}^\gamma \otimes \boldsymbol{n})\, \Delta\boldsymbol{a}_\gamma \\ & + \delta\boldsymbol{a}_\gamma \cdot \big(\Gamma^\gamma_{\alpha\beta}\, \boldsymbol{a}^\delta \otimes \boldsymbol{n} + \Gamma^\delta_{\alpha\beta}\, \boldsymbol{n} \otimes \boldsymbol{a}^\gamma - a^{\gamma\delta}\, b_{\alpha\beta}\, \boldsymbol{n} \otimes \boldsymbol{n}\big)\, \Delta\boldsymbol{a}_\delta\,. \end{aligned} \tag{86}$$

Note that all these expressions are symmetric w.r.t. linearization and variation.

5 Surface Balance

This section presents the mechanical balance laws for shells. The sectional forces and sectional moments are introduced (Sect. 5.1), and then linear momentum (Sect. 5.2), angular momentum (Sect. 5.3) and mechanical power (Sect. 5.5) are discussed. Section 5.4 discusses boundary moments. The presentation follows Sauer and Duong (2017).

5.1 *Sectional Forces and Moments*

Consider an infinitesimal surface element $\mathrm{d}a \subset \mathcal{S}$, located at $\boldsymbol{x}$ and aligned along $\boldsymbol{a}_1$ and $\boldsymbol{a}_2$ as is shown in Fig. 3. On the cut surfaces the distributed[3] sectional force and moment components $N^{\alpha\beta}$, S^α and $M^{\alpha\beta}$ are defined as shown. The sectional forces are collected in the stress tensor

$$\boldsymbol{\sigma} := N^{\alpha\beta}\, \boldsymbol{a}_\alpha \otimes \boldsymbol{a}_\beta + S^\alpha\, \boldsymbol{a}_\alpha \otimes \boldsymbol{n}\,, \tag{87}$$

such that the traction vector on the cut normal to $\boldsymbol{\nu}$ is given through Cauchy's formula

$$\boldsymbol{T} := \boldsymbol{\sigma}^{\mathrm{T}}\boldsymbol{\nu}\,. \tag{88}$$

[3]Per current length of the cut face.

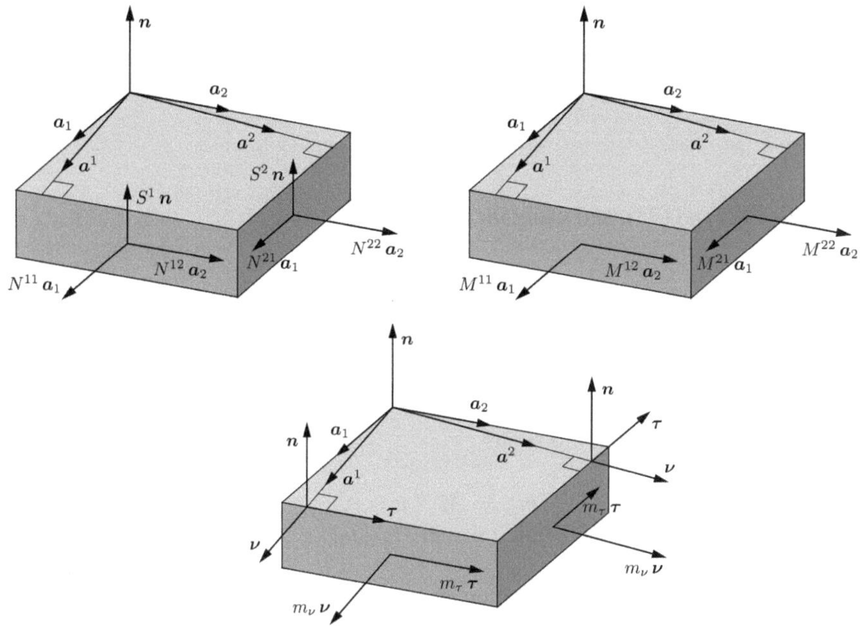

Fig. 3 Sectional forces and moments (Sauer and Duong 2017): Components of the traction and moment vectors $\boldsymbol{T}^1$, $\boldsymbol{T}^2$, $\boldsymbol{M}^1$ and $\boldsymbol{M}^2$ defined on the faces normal to $\boldsymbol{a}^1$ and $\boldsymbol{a}^2$ (*top*). Components of the physical moment vector $\boldsymbol{m}$ acting on the same faces (*bottom*)

With $\boldsymbol{\nu} = \nu_\alpha\, \boldsymbol{a}^\alpha$ one can write $\boldsymbol{T} = \boldsymbol{T}^\alpha\, \nu_\alpha$, where

$$\boldsymbol{T}^\alpha := \boldsymbol{\sigma}^{\mathrm{T}}\, \boldsymbol{a}^\alpha = N^{\alpha\beta}\, \boldsymbol{a}_\beta + S^\alpha\, \boldsymbol{n}\,, \tag{89}$$

are then the tractions defined on the face normal to $\boldsymbol{a}^\alpha$, see Fig. 3.

The distributed section moments are collected in the moment tensor

$$\boldsymbol{\mu} := -M^{\alpha\beta}\, \boldsymbol{a}_\alpha \otimes \boldsymbol{a}_\beta\,, \tag{90}$$

such that one can define the distributed moment vector

$$\boldsymbol{M} := \boldsymbol{\mu}^{\mathrm{T}}\, \boldsymbol{\nu} \tag{91}$$

on the cut normal to $\boldsymbol{\nu}$. Similar to before, one can write

$$\boldsymbol{M} = \boldsymbol{M}^\alpha\, \nu_\alpha\,, \tag{92}$$

with

$$\boldsymbol{M}^\alpha := \boldsymbol{\mu}^{\mathrm{T}}\, \boldsymbol{a}^\alpha = -M^{\alpha\beta}\, \boldsymbol{a}_\beta\,. \tag{93}$$

The components of $-\boldsymbol{M}^\alpha$ are shown in the top right inset of Fig. 3. Vector $\boldsymbol{M}$ can be associated with a force couple (Sahu et al. 2017). The moment vector physically acting on the element is given by the quantity

$$\boldsymbol{m} := \boldsymbol{n} \times \boldsymbol{M} \,. \tag{94}$$

Inserting (92) and (93), and using the identity

$$\boldsymbol{a}_\beta \times \boldsymbol{n} = \tau_\beta\, \boldsymbol{\nu} - \nu_\beta\, \boldsymbol{\tau} \,, \tag{95}$$

gives

$$\boldsymbol{m} = m_\nu\, \boldsymbol{\nu} + m_\tau\, \boldsymbol{\tau} \tag{96}$$

with the local Cartesian components

$$\begin{aligned} m_\nu &:= M^{\alpha\beta}\, \nu_\alpha\, \tau_\beta \,, \\ m_\tau &:= -M^{\alpha\beta}\, \nu_\alpha\, \nu_\beta \,. \end{aligned} \tag{97}$$

The vector $\boldsymbol{M}$ can then also be written as

$$\boldsymbol{M} = m_\tau\, \boldsymbol{\nu} - m_\nu\, \boldsymbol{\tau} \,. \tag{98}$$

The bottom inset of Fig. 3 shows the vector $\boldsymbol{m}$ acting on faces $\boldsymbol{a}^\alpha$.

5.2 Balance of Linear Momentum

Consider a part of the surface $\mathcal{S}$, denoted $\mathcal{R}$ that is assumed to have a smooth boundary $\partial\mathcal{R}$. The 'body' force (per current surface area) acting on $\mathcal{R}$ is denoted by $\boldsymbol{f}$. For every such surface part, the change of its linear momentum is equal to the external forces acting on it, i.e.,

$$\frac{D}{Dt}\int_{\mathcal{R}} \rho\, \boldsymbol{v}\, \mathrm{d}a = \int_{\mathcal{R}} \boldsymbol{f}\, \mathrm{d}a + \int_{\partial\mathcal{R}} \boldsymbol{T}\, \mathrm{d}s \quad \forall\, \mathcal{R} \subset \mathcal{S} \,. \tag{99}$$

Here, D/Dt denotes the material time derivative introduced in (53), and $\boldsymbol{v}$ is the material velocity at $\boldsymbol{x}$. From the local conservation of mass (46) and the surface divergence theorem

$$\int_{\partial\mathcal{R}} \boldsymbol{T}^\alpha\, \nu_\alpha\, \mathrm{d}s = \int_{\mathcal{R}} \boldsymbol{T}^\alpha_{;\alpha}\, \mathrm{d}a \,, \tag{100}$$

immediately follows the local form of (99),

$$\boldsymbol{T}^\alpha_{;\alpha} + \boldsymbol{f} = \rho\, \dot{\boldsymbol{v}} \quad \forall\, \boldsymbol{x} \in \mathcal{S} \,, \tag{101}$$

which is the strong form equilibrium equation at $\boldsymbol{x} \in \mathcal{S}$. If desired, it can be decomposed into in-plane and out-of-plane contributions (Jenkins 1977; Sauer and Duong 2017).

5.3 Balance of Angular Momentum

For every surface part $\mathcal{R} \subset \mathcal{S}$, the change of angular momentum is equal to the moment of the external forces, i.e.,

$$\frac{D}{Dt}\int_{\mathcal{R}} \rho\, \boldsymbol{x} \times \boldsymbol{v}\, \mathrm{d}a = \int_{\mathcal{R}} \boldsymbol{x} \times \boldsymbol{f}\, \mathrm{d}a + \int_{\partial\mathcal{R}} \boldsymbol{x} \times \boldsymbol{T}\, \mathrm{d}s + \int_{\partial\mathcal{R}} \boldsymbol{m}\, \mathrm{d}s \quad \forall \mathcal{R} \subset \mathcal{S}\,. \tag{102}$$

Sauer and Duong (2017) show that this is satisfied if and only if

$$\sigma^{\alpha\beta} := N^{\alpha\beta} - b^{\beta}_{\gamma}\, M^{\gamma\alpha} \tag{103}$$

is symmetric and

$$S^{\alpha} = -M^{\beta\alpha}_{\;\;;\beta}\,. \tag{104}$$

The last equation expresses the well-known Kirchhoff–Love result that the out-of-plane shear component follows as the derivative of the bending moments. It turns out that apart from $\sigma^{\alpha\beta}$ also $M^{\alpha\beta}$ is symmetric, see Sect. 6.2. According to relation (103), the in-plane stress component

$$N^{\alpha\beta} = \sigma^{\alpha\beta} + b^{\beta}_{\gamma}\, M^{\gamma\alpha} \tag{105}$$

is influenced by bending, and consequently $N^{\alpha\beta}$ is generally nonsymmetric.

5.4 Boundary Conditions

At the boundary of the surface, $\partial\mathcal{S}$, the boundary conditions

$$\begin{aligned} \boldsymbol{x} &= \bar{\boldsymbol{\varphi}} && \text{on } \partial_x\mathcal{S}\,, \\ \boldsymbol{t} &= \bar{\boldsymbol{t}} && \text{on } \partial_t\mathcal{S}\,, \\ m_{\tau} &= \bar{m}_{\tau} && \text{on } \partial_m\mathcal{S} \end{aligned} \tag{106}$$

can be prescribed. Here, m_{τ} is the bending moment component parallel to boundary $\partial\mathcal{S}$. For Kirchhoff–Love shells, bending moments perpendicular to boundary $\partial\mathcal{S}$, denoted m_{ν}, affect the boundary traction. Therefore, the effective traction

$$\boldsymbol{t} := \boldsymbol{T} - (m_{\nu}\boldsymbol{n})' \tag{107}$$

is introduced, e.g., see Sauer and Duong (2017). In the following examples, $m_\nu = 0$ is considered.

5.5 *Mechanical Power Balance*

The mechanical power balance follows from equilibrium. Contracting the equilibrium equation (101) with the velocity $\boldsymbol{v}$ and integrating over $\mathcal{R} \subset \mathcal{S}$, gives

$$\int_{\mathcal{R}} \boldsymbol{v} \cdot \big(\boldsymbol{T}^\alpha_{;\alpha} + \boldsymbol{f} - \rho\,\dot{\boldsymbol{v}}\big)\,\mathrm{d}a = 0 \quad \forall \mathcal{R} \subset \mathcal{S}\,. \tag{108}$$

In here, the last term corresponds to the change of the kinetic energy

$$K := \frac{1}{2}\int_{\mathcal{R}} \rho\,\boldsymbol{v}\cdot\boldsymbol{v}\,\mathrm{d}a\,, \tag{109}$$

which, due to mass conservation, is given by

$$\dot{K} := \int_{\mathcal{R}} \rho\,\boldsymbol{v}\cdot\dot{\boldsymbol{v}}\,\mathrm{d}a\,. \tag{110}$$

Applying the surface divergence theorem to the first term, rearranging terms and applying the surface divergence theorem again, leads to the mechanical power balance (Sauer and Duong 2017)

$$\dot{K} + P_{\mathrm{int}} = P_{\mathrm{ext}} \quad \forall \mathcal{R} \subset \mathcal{S}\,, \tag{111}$$

where

$$P_{\mathrm{int}} = \frac{1}{2}\int_{\mathcal{R}} \sigma^{\alpha\beta}\,\dot{a}_{\alpha\beta}\,\mathrm{d}a + \int_{\mathcal{R}} M^{\alpha\beta}\,\dot{b}_{\alpha\beta}\,\mathrm{d}a \tag{112}$$

is the internal stress power of $\mathcal{R}$ and

$$P_{\mathrm{ext}} = \int_{\mathcal{R}} \boldsymbol{v}\cdot\boldsymbol{f}\,\mathrm{d}a + \int_{\partial\mathcal{R}} \boldsymbol{v}\cdot\boldsymbol{T}\,\mathrm{d}s + \int_{\partial\mathcal{R}} \dot{\boldsymbol{n}}\cdot\boldsymbol{M}\,\mathrm{d}s \tag{113}$$

is the power of the external forces acting on $\mathcal{R}$ and $\partial\mathcal{R}$. Using definition (107), P_{ext} can be rewritten into (Sauer and Duong 2017)

$$P_{\mathrm{ext}} = \int_{\mathcal{R}} \boldsymbol{v}\cdot\boldsymbol{f}\,\mathrm{d}a + \int_{\partial\mathcal{R}} \big(\boldsymbol{v}\cdot\boldsymbol{t} + \dot{\boldsymbol{n}}\cdot m_\tau\,\boldsymbol{\nu}\big)\,\mathrm{d}s + \big[\boldsymbol{v}\cdot m_\nu\,\boldsymbol{n}\big]\,, \tag{114}$$

where the last term denotes the power of the point loads $m_\nu\,\boldsymbol{n}$ that are present at corners of boundary $\partial\mathcal{R}$. For smooth boundaries, or for $m_\nu = 0$, the last term vanishes.

The derivation of the weak form of Eq. (101), considered in Sect. 8, is analogous to the derivation of the mechanical power balance.

6 Surface Constitution

This section discusses the constitutive framework of lipid bilayers, accounting for elastic bending, (near-) area-incompressibility, and viscous shear. The framework follows from the dissipation inequality (Sect. 6.1) using classical thermodynamical arguments (Sect. 6.2). For later use, linearization (Sect. 6.3) and stability (Sect. 6.4) are also discussed briefly.

6.1 Dissipation Inequality

The local power density $\sigma^{\alpha\beta}\,\dot{a}_{\alpha\beta}/2 + M^{\alpha\beta}\,\dot{b}_{\alpha\beta}$, appearing within (112), also appears in the mechanical dissipation inequality

$$\mathcal{D} := \frac{1}{2}\sigma^{\alpha\beta}\,\dot{a}_{\alpha\beta} + M^{\alpha\beta}\,\dot{b}_{\alpha\beta} - \rho\dot{T}s - \rho\dot{\psi} \geq 0\,, \tag{115}$$

where T is the temperature, s is specific entropy, and ψ is the specific Helmholtz free energy (per unit mass). Equation (115) is a consequence of the second law of thermodynamics for surfaces, e.g., see Sahu et al. (2017). Under isothermal conditions, considered here, the $\rho\dot{T}s$ term vanishes. The dissipation $\mathcal{D}$ has units of power per current area. Multiplying by J, $\mathcal{D}$ can be related to the reference area. Introducing

$$\begin{aligned} \tau^{\alpha\beta} &:= J\sigma^{\alpha\beta}\,, \\ M_0^{\alpha\beta} &:= J M^{\alpha\beta}\,, \end{aligned} \tag{116}$$

the isothermal dissipation inequality can thus be written as

$$\mathcal{D}_0 := \frac{1}{2}\tau^{\alpha\beta}\,\dot{a}_{\alpha\beta} + M_0^{\alpha\beta}\,\dot{b}_{\alpha\beta} - \dot{\Psi}_0 \geq 0\,, \tag{117}$$

where $\Psi_0 := \rho_0\psi$ is the Helmholtz free energy per reference area. Here, (47) and mass conservation have been used.

6.2 Constrained Visco-Elasticity

The free energy Ψ_0 is a function of the deformation, which, for thin shells, is fully characterized by $a_{\alpha\beta}$ and $b_{\alpha\beta}$. In order to account for constraints on $a_{\alpha\beta}$, such as area-incompressibility, Ψ_0 is expressed as

$$\Psi_0 = \Psi_{0\mathrm{x}} + \Psi_{0\mathrm{g}}\,, \tag{118}$$

where

$$\Psi_{0\mathrm{x}} = \Psi_{0\mathrm{x}}(a_{\alpha\beta}, b_{\alpha\beta}) \tag{119}$$

denotes the contribution from deformation, and

$$\Psi_{0\mathrm{g}} = q\,g(a_{\alpha\beta}) \tag{120}$$

denotes the contribution associated with a constraint $g = 0$. q denotes the Lagrange multiplier associated with the constraint. Applying chain rule then yields

$$\dot{\Psi}_0 = \frac{\partial \Psi_0}{\partial a_{\alpha\beta}}\,\dot{a}_{\alpha\beta} + \frac{\partial \Psi_0}{\partial b_{\alpha\beta}}\,\dot{b}_{\alpha\beta} + g\,\dot{q}\,, \tag{121}$$

so that (117) yields

$$\left(\frac{1}{2}\tau^{\alpha\beta} - \frac{\partial \Psi_0}{\partial a^{\alpha\beta}}\right)\dot{a}_{\alpha\beta} + \left(M_0^{\alpha\beta} - \frac{\partial \Psi_0}{\partial b^{\alpha\beta}}\right)\dot{b}_{\alpha\beta} - g\,\dot{q} \geq 0\,. \tag{122}$$

The surface stress $\sigma^{\alpha\beta}$ is considered to contain elastic and viscous contributions in the form

$$\sigma^{\alpha\beta} = \sigma^{\alpha\beta}_{\mathrm{elas}} + \sigma^{\alpha\beta}_{\mathrm{visc}}\,. \tag{123}$$

The elastic contribution is independent of the rate $\dot{a}_{\alpha\beta}$, while the viscous contribution depends on the rate $\dot{a}_{\alpha\beta}$ such that $\dot{a}_{\alpha\beta} \to 0$ implies $\sigma^{\alpha\beta}_{\mathrm{visc}} \to 0$. The moment $M^{\alpha\beta}$ is considered to be purely elastic.

Since (122) applies to all thermodynamic processes (with general $\dot{a}_{\alpha\beta}$, $\dot{b}_{\alpha\beta}$ and $\dot{q}$), the classical argument by Coleman and Noll (1964) (based on considering a set of special $\dot{a}_{\alpha\beta}$, $\dot{b}_{\alpha\beta}$ and $\dot{q}$) leads to the constitutive equations

$$\begin{aligned} \sigma^{\alpha\beta}_{\mathrm{elas}} &= \frac{2}{J}\,\frac{\partial \Psi_0}{\partial a_{\alpha\beta}}\,, \\ M^{\alpha\beta} &= \frac{1}{J}\,\frac{\partial \Psi_0}{\partial b_{\alpha\beta}}\,, \\ g &= 0\,, \\ \sigma^{\alpha\beta}_{\mathrm{visc}}\,\dot{a}_{\alpha\beta} &\geq 0\,. \end{aligned} \tag{124}$$

The first two relations correspond to classical hyperelasticity, the third is just the constraint, and the fourth implies that viscous stresses are dissipative. A simple expression that satisfies this[4] is

$$\sigma^{\alpha\beta}_{\text{visc}} = -\eta\, \dot{a}^{\alpha\beta} . \tag{125}$$

where $\eta \geq 0$ is a constant. Comparing to 3D fluids, η can be identified as the dynamic surface viscosity. An extension considering more general viscous stresses, as well as thermal fields and changing mass is provided by Sahu et al. (2017).

For the later developments, the variation of Ψ_0 is required. Similar to (121), this can be written as

$$\delta\Psi_0 = \delta_{\text{x}}\Psi_0 + g\,\delta q , \tag{126}$$

with

$$\delta_{\text{x}}\Psi_0 := \frac{\partial \Psi_0}{\partial a_{\alpha\beta}}\,\delta a_{\alpha\beta} + \frac{\partial \Psi_0}{\partial b_{\alpha\beta}}\,\delta b_{\alpha\beta} . \tag{127}$$

From (124) follows

$$\delta_{\text{x}}\Psi_0 = \tfrac{1}{2}\,\tau^{\alpha\beta}\,\delta a_{\alpha\beta} + M_0^{\alpha\beta}\,\delta b_{\alpha\beta} . \tag{128}$$

If no constraint is present q and δq are zero.

6.3 *Linearization of $\delta\Psi_0$*

Linearizing (126), gives

$$\Delta\delta\Psi_0 = \Delta_{\text{x}}\delta_{\text{x}}\Psi_0 + \delta g\,\Delta q + \delta q\,\Delta g , \tag{129}$$

with

$$\delta g = \frac{\partial g}{\partial a_{\alpha\beta}}\delta a_{\alpha\beta} , \qquad \Delta g = \frac{\partial g}{\partial a_{\alpha\beta}}\Delta a_{\alpha\beta} , \tag{130}$$

and

$$\begin{aligned}\Delta_{\text{x}}\delta_{\text{x}}\Psi_0 = \delta a_{\alpha\beta}\frac{\partial^2\Psi_0}{\partial a_{\alpha\beta}\,\partial a_{\gamma\delta}}\Delta a_{\gamma\delta} + \delta a_{\alpha\beta}\frac{\partial^2\Psi_0}{\partial a_{\alpha\beta}\,\partial b_{\gamma\delta}}\Delta b_{\gamma\delta} + \frac{\partial\Psi_0}{\partial a_{\alpha\beta}}\Delta\delta a_{\alpha\beta} \\ + \delta b_{\alpha\beta}\frac{\partial^2\Psi_0}{\partial b_{\alpha\beta}\,\partial a_{\gamma\delta}}\Delta a_{\gamma\delta} + \delta b_{\alpha\beta}\frac{\partial^2\Psi_0}{\partial b_{\alpha\beta}\,\partial b_{\gamma\delta}}\Delta b_{\gamma\delta} + \frac{\partial\Psi_0}{\partial b_{\alpha\beta}}\Delta\delta b_{\alpha\beta} .\end{aligned} \tag{131}$$

Introducing the material tangents

[4] Since $\sigma^{\alpha\beta}_{\text{visc}}\,\dot{a}_{\alpha\beta} = 4\eta\,\boldsymbol{d} : \boldsymbol{d} = 4\eta\|\boldsymbol{d}\|^2 > 0$ due to (65) and (66).

$$\begin{aligned} c^{\alpha\beta\gamma\delta} &:= 4\frac{\partial^2\Psi_0}{\partial a_{\alpha\beta}\,\partial a_{\gamma\delta}} = 2\frac{\partial\tau^{\alpha\beta}}{\partial a_{\gamma\delta}}\,, \\ d^{\alpha\beta\gamma\delta} &:= 2\frac{\partial^2\Psi_0}{\partial a_{\alpha\beta}\,\partial b_{\gamma\delta}} = \frac{\partial\tau^{\alpha\beta}}{\partial b_{\gamma\delta}}\,, \\ e^{\alpha\beta\gamma\delta} &:= 2\frac{\partial^2\Psi_0}{\partial b_{\alpha\beta}\,\partial a_{\gamma\delta}} = 2\frac{\partial M_0^{\alpha\beta}}{\partial a_{\gamma\delta}}\,, \\ f^{\alpha\beta\gamma\delta} &:= \frac{\partial^2\Psi_0}{\partial b_{\alpha\beta}\,\partial b_{\gamma\delta}} = \frac{\partial M_0^{\alpha\beta}}{\partial b_{\gamma\delta}}\,, \end{aligned} \tag{132}$$

gives

$$\begin{aligned} \Delta_{\mathrm{x}}\delta_{\mathrm{x}}\Psi_0 = {} & c^{\alpha\beta\gamma\delta}\,\tfrac{1}{2}\delta a_{\alpha\beta}\,\tfrac{1}{2}\Delta a_{\gamma\delta} + d^{\alpha\beta\gamma\delta}\,\tfrac{1}{2}\delta a_{\alpha\beta}\,\Delta b_{\gamma\delta} + \tau^{\alpha\beta}\,\tfrac{1}{2}\Delta\delta a_{\alpha\beta} \\ & + e^{\alpha\beta\gamma\delta}\,\delta b_{\alpha\beta}\,\tfrac{1}{2}\Delta a_{\gamma\delta} \;+ f^{\alpha\beta\gamma\delta}\,\delta b_{\alpha\beta}\,\Delta b_{\gamma\delta} \;+ M_0^{\alpha\beta}\,\Delta\delta b_{\alpha\beta}\,. \end{aligned} \tag{133}$$

Note that $c^{\alpha\beta\gamma\delta}$ and $f^{\alpha\beta\gamma\delta}$ posses both minor and major symmetries; $d^{\alpha\beta\gamma\delta}$ and $e^{\alpha\beta\gamma\delta}$ posses only minor symmetries, but additionally satisfy

$$d^{\alpha\beta\gamma\delta} = e^{\gamma\delta\alpha\beta}\,. \tag{134}$$

Due to the symmetries of c, d, and e, and due to Eqs. (79) and (81), one finds

$$\begin{aligned} c^{\alpha\beta\gamma\delta}\,\tfrac{1}{2}\delta a_{\alpha\beta}\,\tfrac{1}{2}\Delta a_{\gamma\delta} &= \delta\boldsymbol{a}_\alpha\cdot\boldsymbol{a}_\beta\,c^{\alpha\beta\gamma\delta}\,\boldsymbol{a}_\gamma\cdot\Delta\boldsymbol{a}_\delta\,, \\ d^{\alpha\beta\gamma\delta}\,\tfrac{1}{2}\delta a_{\alpha\beta}\,\Delta b_{\gamma\delta} &= \delta\boldsymbol{a}_\alpha\cdot\boldsymbol{a}_\beta\,d^{\alpha\beta\gamma\delta}\,\boldsymbol{n}\cdot\Delta\tilde{\boldsymbol{a}}_{\alpha,\beta}\,, \\ e^{\alpha\beta\gamma\delta}\,\delta b_{\alpha\beta}\,\tfrac{1}{2}\Delta a_{\gamma\delta} &= \delta\tilde{\boldsymbol{a}}_{\alpha,\beta}\cdot\boldsymbol{n}\,e^{\alpha\beta\gamma\delta}\,\boldsymbol{a}_\gamma\cdot\Delta\boldsymbol{a}_\delta\,, \\ f^{\alpha\beta\gamma\delta}\,\delta b_{\alpha\beta}\,\Delta b_{\gamma\delta} &= \delta\tilde{\boldsymbol{a}}_{\alpha,\beta}\cdot\boldsymbol{n}\,f^{\alpha\beta\gamma\delta}\,\boldsymbol{n}\cdot\Delta\tilde{\boldsymbol{a}}_{\alpha,\beta}\,, \end{aligned} \tag{135}$$

where

$$\begin{aligned} \delta\tilde{\boldsymbol{a}}_{\alpha,\beta} &:= \delta\boldsymbol{a}_{\alpha,\beta} - \Gamma^{\epsilon}_{\alpha\beta}\,\delta\boldsymbol{a}_\epsilon\,, \\ \Delta\tilde{\boldsymbol{a}}_{\alpha,\beta} &:= \Delta\boldsymbol{a}_{\alpha,\beta} - \Gamma^{\epsilon}_{\alpha\beta}\,\Delta\boldsymbol{a}_\epsilon\,. \end{aligned} \tag{136}$$

Expressions for $\Delta\delta a_{\alpha\beta}$ and $\Delta\delta b_{\alpha\beta}$ are given in (83) and (86).

6.4 Material Stability

For many material models, the four tangent matrices introduced in (132) can be written in the format

$$\hat{c}^{\alpha\beta\gamma\delta} = \hat{c}_{aa}\,a^{\alpha\beta}\,a^{\gamma\delta} + \hat{c}_a\,a^{\alpha\beta\gamma\delta} + \hat{c}_{ab}\,a^{\alpha\beta}\,b^{\gamma\delta} + \hat{c}_{ba}\,b^{\alpha\beta}\,a^{\gamma\delta} + \hat{c}_{bb}\,b^{\alpha\beta}\,b^{\gamma\delta}\,, \tag{137}$$

with suitable definitions of coefficients $\hat{c}_{aa}$, $\hat{c}_a$, $\hat{c}_{ab}$, $\hat{c}_{ba}$ and $\hat{c}_{bb}$. Sauer and Duong (2017) show that material stability requires

$$2\hat{c}_{aa} - \hat{c}_a > 0 \quad \& \quad \hat{c}_a < 0\,. \tag{138}$$

7 The Helfrich Energy

In order to fully characterize the constitutive behavior, the Helmholtz free energy Ψ_0 needs to be specified. The bending behavior of lipid bilayers is commonly described by the bending model of Helfrich (1973)

$$w = k\,(H - H_0)^2 + k_g \kappa\,. \tag{139}$$

Here k is the bending modulus, k_g is the Gaussian modulus and H_0 denotes the so-called spontaneous curvature that can be used to model the presence of certain proteins embedded within the lipid bilayer.

This section presents the Helfrich energy for the cases of area-compressibility (Sect. 7.1) and area-incompressibility (Sect. 7.2), and discussed its properties (Sect. 7.3) and tangent matrices (Sect. 7.4). Section 7.5 discusses the relation between the models of Helfrich and Canham. The presentation follows Sauer et al. (2017) and Sauer and Duong (2017).

7.1 *Area-Compressible Lipid Bilayer*

The Helfrich energy is an energy density per current surface area. Multiplying it by J and adding a quadratic energy term for the surface area change, gives the Helmholtz free energy

$$\Psi_0 = J\,w + \frac{K}{2}(J-1)^2\,, \tag{140}$$

where K is the surface bulk modulus. A quadratic energy term is suitable for small area changes. For lipid bilayers, typically $|J - 1| < 4\%$ before rupture occurs. According to (123)–(125) and (105), the stress and moment components then become

$$\begin{aligned}
\sigma^{\alpha\beta} &= \big(K\,(J-1) + k\,\Delta H^2 - k_g\,\kappa\big)\,a^{\alpha\beta} - 2\,k\,\Delta H\,b^{\alpha\beta} - \eta\,\dot{a}^{\alpha\beta}\,,\\
M^{\alpha\beta} &= \big(k\,\Delta H + 2\,k_g H\big)\,a^{\alpha\beta} - k_g\,b^{\alpha\beta}\,,\\
N^{\alpha\beta} &= \big(K(J-1) + k\,\Delta H^2\big)\,a^{\alpha\beta} - k\,\Delta H\,b^{\alpha\beta} - \eta\,\dot{a}^{\alpha\beta}\,,
\end{aligned} \tag{141}$$

where $\Delta H := H - H_0$.

7.2 Area-Incompressible Lipid Bilayer

Since K is usually very large for lipid-bilayers, one may as well consider the surface to be fully area-incompressible. Using the Lagrange multiplier approach, one now has

$$\Psi_0 = J\,w + q\,g\,, \tag{142}$$

where the incompressibility constraint (76) is enforced by the Lagrange multiplier q. q is an independent variable that needs to be accounted for in the solution procedure (see Sect. 11). Physically, q corresponds to a surface tension. The stress and moment components now become

$$\begin{aligned} \sigma^{\alpha\beta} &= \big(q + k\,\Delta H^2 - k_g\,\kappa\big)\,a^{\alpha\beta} - 2\,k\,\Delta H\,b^{\alpha\beta} - \eta\,\dot{a}^{\alpha\beta}\,, \\ M^{\alpha\beta} &= \big(k\,\Delta H + 2\,k_g H\big)\,a^{\alpha\beta} - k_g\,b^{\alpha\beta}\,, \\ N^{\alpha\beta} &= \big(q + k\,\Delta H^2\big)\,a^{\alpha\beta} - k\,\Delta H\,b^{\alpha\beta} - \eta\,\dot{a}^{\alpha\beta}\,. \end{aligned} \tag{143}$$

They are identical to (141) for $q = K g$.

As K becomes larger and larger, both models approach the same solution. So from a physical point of view it may not make a big difference which model is used. Computationally, model (140) is easier to handle but can become inaccurate for large K, as is shown in Sauer et al. (2017). In analytical approaches, often (142) is preferred as it usually simplifies the solution. Examples for (142) are found in Baesu et al. (2004) and Agrawal and Steigmann (2009); (140) is considered in the original work of Helfrich (1973).

7.3 Model Properties

In both preceding models, the membrane part only provides bulk stiffness, but lacks shear stiffness. For quasi-static computations the model can thus become unstable and should be stabilized, as is discussed in Sect. 12. Interestingly, the bending part of the Helfrich model can contribute an in-plane shear stiffness, which is shown in the following.

To this end, the surface tension

$$\gamma := \tfrac{1}{2}\boldsymbol{\sigma} : \boldsymbol{i} = \tfrac{1}{2} N^{\alpha}_{\alpha}\,, \tag{144}$$

is first introduced. For both (141) and (143) one finds

$$\gamma = q - k\,H_0\,\Delta H\,, \tag{145}$$

where $q = K g$ in the former case. It can be seen that for $H_0 \neq 0$, the bending part contributes to the surface tension. This dependency has also been noted by Lipowsky

(2013) and Rangamani et al. (2014). The surface tension is therefore not given by the membrane part alone. For the compressible case, the effective bulk modulus can then be determined from

$$K_{\mathrm{eff}} := \frac{\partial \gamma}{\partial J}\,, \tag{146}$$

i.e. as the change of γ w.r.t. J. One finds

$$K_{\mathrm{eff}} = K + k\,H_0\,H/J\,, \tag{147}$$

since $\partial H/\partial J = -H/J$. Likewise, the effective shear modulus can be defined from

$$\mu_{\mathrm{eff}} := J\,a_{\alpha\gamma}\,\frac{\partial N^{\alpha\beta}_{\mathrm{dev}}}{2\,\partial a_{\gamma\delta}}\,a_{\beta\delta}\,, \tag{148}$$

i.e., as the change of the deviatoric stress w.r.t. the deviatoric deformation (characterized by $a_{\gamma\delta}/J$). The deviatoric in-plane stress is given by

$$N^{\alpha\beta}_{\mathrm{dev}} := N^{\alpha\beta} - \gamma\,a^{\alpha\beta}\,. \tag{149}$$

One finds

$$N^{\alpha\beta}_{\mathrm{dev}} = k\,\Delta H\,\big(H\,a^{\alpha\beta} - b^{\alpha\beta}\big) \tag{150}$$

for both (141) and (143). Evaluating (148) thus gives

$$\mu_{\mathrm{eff}} = Jk\,\big(3H^2 - 2H H_0 - \kappa\big)/2\,. \tag{151}$$

The model therefore provides stabilizing shear stiffness if $3H^2 > 2H H_0 + \kappa$. Since this is not always the case (e.g., for flat surface regions), additional shear stabilization should be provided for quasi-static computations. This is discussed in Sect. 12. The value of μ_{eff} is discussed further in the examples of Sect. 13. It is shown that μ_{eff} can sufficiently stabilize the problem such that no additional shear stabilization is needed. It is also shown that μ_{eff} does not necessarily need to be positive to avoid instabilities. Geometric stiffening, arising in large deformations, can also stabilize the surface.

7.4 Material Tangent

In the following, the material tangents of Eq. (132) are evaluated and assessed. This is done by examining the contributions to (141) and (143) piecewise.

Area-compressibility For the area-compressible case, the elastic membrane stress is characterized by

$$\tau^{\alpha\beta} = K J\,(J-1)\,a^{\alpha\beta}\,. \tag{152}$$

From (132) thus follows

$$c^{\alpha\beta\gamma\delta} = K J\,(2J-1)\,a^{\alpha\beta}a^{\gamma\delta} + 2K J\,(J-1)\,a^{\alpha\beta\gamma\delta}\,. \tag{153}$$

Since $c_a = 2KJ(J-1) \geq 1$ for $J \geq 1$, this model does not satisfy criteria (138) and therefore is unstable by itself.

Area-incompressibility For the area-incompressible case, the elastic membrane stress is characterized by

$$\tau^{\alpha\beta} = -q J\, a^{\alpha\beta}\,, \tag{154}$$

so that

$$c^{\alpha\beta\gamma\delta} = -q J\, a^{\alpha\beta}a^{\gamma\delta} - 2q J\, a^{\alpha\beta\gamma\delta}\,. \tag{155}$$

Since $2c_{aa} - c_a = 0$, this model does not satisfy criteria (138) and therefore is unstable by itself.

Bending part The bending contribution, characterized by

$$\begin{aligned} \tau^{\alpha\beta} &= J\left(k\,\Delta H^2 - k_g\,\kappa\right) a^{\alpha\beta} - 2k\, J\, \Delta H\, b^{\alpha\beta}\,, \\ M_0^{\alpha\beta} &= J\left(k\,\Delta H + 2k_g\, H\right) a^{\alpha\beta} - k_g\, J\, b^{\alpha\beta}\,, \end{aligned} \tag{156}$$

leads to

$$\begin{aligned} c^{\alpha\beta\gamma\delta} &= c_{aa}\, a^{\alpha\beta}\, a^{\gamma\delta} + c_a\, a^{\alpha\beta\gamma\delta} + c_{bb}\, b^{\alpha\beta}\, b^{\gamma\delta} + c_{ab}\left(a^{\alpha\beta}\, b^{\gamma\delta} + b^{\alpha\beta}\, a^{\gamma\delta}\right), \\ d^{\alpha\beta\gamma\delta} &= d_{aa}\, a^{\alpha\beta}\, a^{\gamma\delta} + d_a\, a^{\alpha\beta\gamma\delta} + d_{ab}\, a^{\alpha\beta}\, b^{\gamma\delta} + d_{ba}\, b^{\alpha\beta}\, a^{\gamma\delta} = e^{\gamma\delta\alpha\beta}\,, \\ f^{\alpha\beta\gamma\delta} &= f_{aa}\, a^{\alpha\beta}\, a^{\gamma\delta} + f_a\, a^{\alpha\beta\gamma\delta}\,, \end{aligned} \tag{157}$$

with

$$\begin{aligned} c_{aa} &= J\left(k\,\Delta H\,(\Delta H - 8H) + k_g\,\kappa\right), \\ c_a &= 2J\left(k\,\Delta H\,(\Delta H - 4H) - k_g\,\kappa\right), \\ c_{bb} &= 2k\, J\,, \\ c_{ab} &= c_{ba} = 2k\, J\, \Delta H\,, \\ d_{aa} &= J\left(k\,\Delta H - 2k_g\, H\right), \\ d_a &= 2J\, k\, \Delta H\,, \\ d_{ab} &= J\, k_g\,, \\ d_{ba} &= -J\, k\,, \\ f_{aa} &= J\,(k/2 + k_g)\,, \\ f_a &= J\, k_g\,. \end{aligned} \tag{158}$$

The stability can be assessed by examining the bending tangent $f^{\alpha\beta\gamma\delta}$. According to (138), it is easy to see that stability requires

$$0 < -k_g < k\,. \tag{159}$$

7.5 The Canham Model

A special case of the Helfrich model is the bending model of Canham (1970). It can be expressed as

$$\Psi_0 = J\,w\,, \quad w := \frac{c}{2}\left(\kappa_1^2 + \kappa_2^2\right). \tag{160}$$

Here, w can also be written as $w = c\, b^\alpha_\beta\, b^\beta_\alpha/2$ or $w = c\,(2H^2 - \kappa)$, so that the Canham model follows from the Helfrich model with $k = 2c$, $k_g = -c$ and $H_0 = 0$. Since this satisfies (159), the model is stable in bending. In particular, the Canham model gives

$$\sigma^{\alpha\beta} = c\left(2H^2 + \kappa\right) a^{\alpha\beta} - 4c\,H\,b^{\alpha\beta} - \eta\,\dot{a}^{\alpha\beta} \tag{161}$$

and

$$M^{\alpha\beta} = c\,b^{\alpha\beta}\,. \tag{162}$$

8 Weak Form

This section presents the weak form of the thin shell equation (101), considering the area-compressible case (Sect. 8.1) and the area-incompressible case (Sect. 8.2). The decomposition into in-plane and out-of-plane contributions (Sect. 8.3) and the linearization (Sect. 8.4) follow. The presentation follows Sauer and Duong (2017) and Sauer et al. (2017).

8.1 Unconstrained System

The weak form of equilibrium equation (101) can be derived analogously to the mechanical power balance in Sect. 5.5 by simply replacing the velocity $\boldsymbol{v}$ with the admissible variation $\delta\boldsymbol{x} \in \mathcal{V}$. Immediately one obtains

$$G_{\text{in}} + G_{\text{int}} - G_{\text{ext}} = 0 \quad \forall\,\delta\boldsymbol{x} \in \mathcal{V}\,, \tag{163}$$

with

$$\begin{aligned} G_{\text{in}} &= \int_{\mathcal{S}_0} \delta\boldsymbol{x}\cdot\rho_0\,\dot{\boldsymbol{v}}\,\mathrm{d}A\,, \\ G_{\text{int}} &= \int_{\mathcal{S}_0} \delta_{\text{x}}\Psi_0\,\mathrm{d}A = \int_{\mathcal{S}_0} \frac{1}{2}\,\delta a_{\alpha\beta}\,\tau^{\alpha\beta}\,\mathrm{d}A + \int_{\mathcal{S}_0} \delta b_{\alpha\beta}\,M_0^{\alpha\beta}\,\mathrm{d}A\,, \\ G_{\text{ext}} &= \int_{\mathcal{S}} \delta\boldsymbol{x}\cdot\boldsymbol{f}\,\mathrm{d}a + \int_{\partial\mathcal{S}} \delta\boldsymbol{x}\cdot\boldsymbol{T}\,\mathrm{d}s + \int_{\partial\mathcal{S}} \delta\boldsymbol{n}\cdot\boldsymbol{M}\,\mathrm{d}s\,, \end{aligned} \tag{164}$$

according to Eqs. (110)–(113). As noted in (123), stress $\tau^{\alpha\beta} = J\sigma^{\alpha\beta}$, and hence also G_{int}, has elastic and viscous contributions. Due to Eq. (128), the elastic part of G_{int} can also be obtained as the variation of

$$\Pi_{\mathrm{int}} = \int_{\mathcal{S}_0} \Psi_0 \, \mathrm{d}A \tag{165}$$

w.r.t. $\boldsymbol{x}$, i.e., $G_{\mathrm{int,el}} = \delta_{\mathrm{x}} \Pi_{\mathrm{int}}$. Thus, if G_{ext} is also derivable from a potential, the quasi-static weak form $G_{\mathrm{int}} - G_{\mathrm{ext}} = 0 \;\forall\, \delta\boldsymbol{x} \in \mathcal{V}$ is the result of the principle of stationary potential energy.

8.2 Constrained System

For the constrained problem, the constraint $g = 0$ needs to be included. The weak form of that is simply

$$G_g = \int_{\mathcal{S}_0} \delta q \, g \, \mathrm{d}A = 0 \quad \forall\, \delta q \in \mathcal{Q}\,, \tag{166}$$

where $\delta q \in \mathcal{Q}$ is a suitably chosen variation of the Lagrange multiplier q. The weak form problem statement is then given by solving the two equations

$$\begin{aligned} G_{\mathrm{in}} + G_{\mathrm{int}} - G_{\mathrm{ext}} &= 0 \quad \forall\, \delta\boldsymbol{x} \in \mathcal{V}\,, \\ G_g &= 0 \quad \forall\, \delta q \in \mathcal{Q}\,, \end{aligned} \tag{167}$$

for $\boldsymbol{x}$ and q. Due to Eq. (126), one can find $G_{\mathrm{int,el}} + G_g = \delta\Pi_{\mathrm{int}}$, such that the static version of weak form (167), for suitable G_{ext}, is still the result of the principle of stationary potential energy.

8.3 Decomposition

As noted in Sauer et al. (2014), the weak form can be decomposed into in-plane and out-of-plane contributions. Denoting the in-plane and out-of-plane components of $\delta\boldsymbol{x}$ by w_α and w, such that $\delta\boldsymbol{x} := w_\alpha\, \boldsymbol{a}^\alpha + w\, \boldsymbol{n}$, one finds that

$$\delta a_{\alpha\beta} = w_{\alpha;\beta} + w_{\beta;\alpha} - 2w\, b_{\alpha\beta}\,. \tag{168}$$

Thus, the first part of G_{int} can be split into in-plane and out-of-plane contributions as

$$\int_{\mathcal{S}} \frac{1}{2}\, \delta a_{\alpha\beta}\, \sigma^{\alpha\beta}\, \mathrm{d}a = G_\sigma^{\mathrm{in}} + G_\sigma^{\mathrm{out}}\,, \tag{169}$$

with

$$G_\sigma^{\text{in}} = \int_{\mathcal{S}} w_{\alpha;\beta}\, \sigma^{\alpha\beta}\, \mathrm{d}a \tag{170}$$

and

$$G_\sigma^{\text{out}} = -\int_{\mathcal{S}} w\, b_{\alpha\beta}\, \sigma^{\alpha\beta}\, \mathrm{d}a\,. \tag{171}$$

In principle – although not needed here – the second part of G_{int} can also be split into in-plane and out-of-plane contributions (Sauer and Duong 2017).

8.4 Linearization

In the following, the linearization of the quasi-static case is discussed, where inertia and viscosity are absent. Inertia is linearly dependent on acceleration and thus easy to linearize. Viscosity can be conveniently treated within the framework of the implicit Euler time discretization scheme discussed in Sect. 10.4 and linearized in Sect. 10.5. The quasi-static case of weak form (167) can be written in the combined form

$$\delta\Pi_{\text{int}} - G_{\text{ext}} = 0 \quad \forall\, \delta\boldsymbol{x} \in \mathcal{V}\ \&\ \delta q \in \mathcal{Q}\,, \tag{172}$$

where $\delta\Pi_{\text{int}} = G_{\text{int}} + G_{\text{g}}$. Linearizing the internal virtual work gives, according to (129),

$$\Delta\delta\Pi_{\text{int}} = \int_{\mathcal{S}_0} \Delta_{\boldsymbol{x}}\delta_{\boldsymbol{x}}\Psi_0\, \mathrm{d}A + \int_{\mathcal{S}_0} \delta g\, \Delta q\, \mathrm{d}A + \int_{\mathcal{S}_0} \delta q\, \Delta g\, \mathrm{d}A\,, \tag{173}$$

where $\Delta_{\boldsymbol{x}}\delta_{\boldsymbol{x}}\Psi_0$ is given by (133). In order to linearize G_{ext}, dead loading for $\boldsymbol{f}$, $\boldsymbol{t}$, and $\boldsymbol{M}$ is considered. The case of live pressure loading is given in Sauer et al. (2014). For dead loading, Sauer and Duong (2017) show that

$$\Delta G_{\text{ext}} = \int_{\partial\mathcal{S}} m_\tau\, \delta\boldsymbol{a}_\alpha \cdot \big(\nu^\beta\, \boldsymbol{n} \otimes \boldsymbol{a}^\alpha + \nu^\alpha\, \boldsymbol{a}^\beta \otimes \boldsymbol{n}\big)\, \Delta\boldsymbol{a}_\beta\, \mathrm{d}s\,, \tag{174}$$

which is symmetric w.r.t. variation and linearization.

9 Analytical Solutions

This section presents two analytical solutions that describe simple bilayer deformations. They are useful for the verification of numerical results. Considered are pure bending and stretching of a flat sheet (Sect. 9.1), and the inflation of a sphere (Sect. 9.2).

9.1 Pure Bending and Stretching of a Flat Sheet

The first example considers the pure bending and stretching of a flat sheet. It is taken from Sauer and Duong (2017) and Sauer et al. (2017). The sheet has the dimension $S \times L$ and is parameterized by the coordinates $\xi^1 \in [0, S]$ and $\xi^2 \in [0, L]$. The sheet is deformed into a curved sheet with dimension $s \times \ell$ by applying the homogeneous curvature κ_1 and the homogeneous stretches $\lambda_1 = s/S$ and $\lambda_2 = \ell/L$ as is shown in Fig. 4. The deformed sheet thus forms a circular arc with radius $r = 1/\kappa_1$. The parameters S, L, κ_1, λ_1, and λ_2 are considered given, unless specified otherwise. According to the figure, the surface in its initial configuration can be described by

$$\boldsymbol{X}(\xi^1, \xi^2) = \xi^1\, \boldsymbol{e}_1 + \xi^2\, \boldsymbol{e}_2\,, \tag{175}$$

while its current surface can be described by

$$\boldsymbol{x}(\xi^1, \xi^2) = r\sin\theta\, \boldsymbol{e}_1 + \lambda_2\, \xi^2\, \boldsymbol{e}_2 + r\,(1-\cos\theta)\, \boldsymbol{e}_3\,, \tag{176}$$

with $\theta := \kappa_1\lambda_1\xi^1$ and $r := 1/\kappa_1$. The rotation at the end thus is $\Theta = \kappa_1\lambda_1 S$. From these relations follow the initial tangent vectors

$$\begin{aligned} \boldsymbol{A}_1 &= \frac{\partial \boldsymbol{X}}{\partial \xi^1} = \boldsymbol{e}_1\,, \\ \boldsymbol{A}_2 &= \frac{\partial \boldsymbol{X}}{\partial \xi^2} = \boldsymbol{e}_2\,, \end{aligned} \tag{177}$$

the current tangent vectors

$$\begin{aligned} \boldsymbol{a}_1 &= \frac{\partial \boldsymbol{x}}{\partial \xi^1} = \lambda_1\left(\cos\theta\, \boldsymbol{e}_1 + \sin\theta\, \boldsymbol{e}_3\right), \\ \boldsymbol{a}_2 &= \frac{\partial \boldsymbol{x}}{\partial \xi^2} = \lambda_2\, \boldsymbol{e}_2\,, \end{aligned} \tag{178}$$

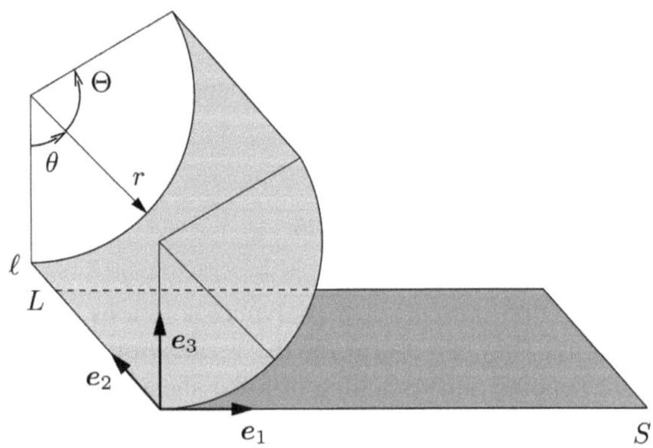

Fig. 4 Pure bending and stretching of a sheet (Sauer and Duong 2017): Deformation of a flat sheet into a curved sheet with constant radius

and the current surface normal

$$\boldsymbol{n} = -\sin\theta\,\boldsymbol{e}_1 + \cos\theta\,\boldsymbol{e}_3\,. \tag{179}$$

This results in the kinematic quantities

$$[A_{\alpha\beta}] = \begin{bmatrix} 1 & 0 \\ 0 & 1 \end{bmatrix}, \quad [A^{\alpha\beta}] = \begin{bmatrix} 1 & 0 \\ 0 & 1 \end{bmatrix}, \tag{180}$$

$$[a_{\alpha\beta}] = \begin{bmatrix} \lambda_1^2 & 0 \\ 0 & \lambda_2^2 \end{bmatrix}, \quad [a^{\alpha\beta}] = \begin{bmatrix} \lambda_1^{-2} & 0 \\ 0 & \lambda_2^{-2} \end{bmatrix}, \quad J = \lambda_1\lambda_2\,, \tag{181}$$

and

$$\begin{aligned} [b_{\alpha\beta}] &= \begin{bmatrix} \kappa_1\lambda_1^2 & 0 \\ 0 & 0 \end{bmatrix}, \quad [b^\alpha_\beta] = \begin{bmatrix} \kappa_1 & 0 \\ 0 & 0 \end{bmatrix}, \\ [b^{\alpha\beta}] &= \begin{bmatrix} \kappa_1\lambda_1^{-2} & 0 \\ 0 & 0 \end{bmatrix}, \quad H = \frac{\kappa_1}{2}, \quad \kappa = 0\,. \end{aligned} \tag{182}$$

With this, the in-plane stress components become

$$\begin{aligned} N_1^1 &= q - k\,H^2\,, \\ N_2^2 &= q + k\,H^2\,, \end{aligned} \tag{183}$$

both for the area-incompressible model of (143) and the area-compressible model of (141) with $q = K(J-1)$.

Now consider a cut at θ that is perpendicular to the normal

$$\boldsymbol{\nu} = \boldsymbol{a}_1/\lambda_1\,, \tag{184}$$

such that

$$\nu_1 = \boldsymbol{a}_1 \cdot \boldsymbol{\nu} = \lambda_1 \quad \text{and} \quad \nu_2 = \boldsymbol{a}_2 \cdot \boldsymbol{\nu} = 0\,. \tag{185}$$

The distributed bending moment acting on the cut is given by $M = M^{\alpha\beta}\nu_\alpha\nu_\beta$. Both models, (141) and (143), lead to the simple linear relation

$$M = k\,H\,, \tag{186}$$

between the prescribed curvature and the resulting bending moment. At $\theta = 0$ and $\theta = \Theta$, M corresponds to the boundary moment (per current length of the support). Measured per reference length, the boundary moment is $M_0 = \lambda_2\,M$.

If the boundaries at $\xi^1 = 0$ and $\xi^1 = S$ are considered stress-free, $N_1^1 = 0$, so that

$$q = k\,H^2, \tag{187}$$

and consequently the support reaction (per current length) along $\xi^2 = 0$ and $\xi^2 = L$ is $N := N_2^2 = 2k\,H^2$. Per reference length this becomes $N_0 = \lambda_1 N$.

For the area-incompressible model of (142), one has $\lambda_1 = 1/\lambda_2$, such that the sheet is in a state of pure shear. For the area-compressible case according to model (140), one can determine λ_1 from (187) with $J = \lambda_1\lambda_2$, giving

$$\lambda_1 = \frac{1}{\lambda_2}\Big[\frac{k}{K}H^2 + 1\Big]. \tag{188}$$

9.2 *Inflation of a Sphere*

The second example considers the inflation of a spherical vesicle. It is taken from Sauer et al. (2017). Since the surface area increases during inflation, the area-incompressible model (140) has to be considered. For this model, the in-plane traction component, given in (141), is

$$N^{\alpha\beta} = N_a\,a^{\alpha\beta} + N_b\,b^{\alpha\beta}\,, \tag{189}$$

with

$$\begin{aligned} N_a &:= k\,\Delta H^2 + K\,(J-1)\,, \\ N_b &:= -k\,\Delta H\,. \end{aligned} \tag{190}$$

The initial radius of the sphere is denoted by R, the initial volume is denoted by $V_0 = 4\pi R^3/3$. The vesicle remains spherical during inflation. The current radius during inflation is denoted by r, the current volume by $V = 4\pi r^3/3$. Considering the surface parameterization

$$\boldsymbol{x}(\phi,\theta) = \begin{bmatrix} r\,\cos\phi\,\sin\theta \\ r\,\sin\phi\,\sin\theta \\ -r\,\cos\theta \end{bmatrix}, \tag{191}$$

one finds

$$[a^{\alpha\beta}] = \frac{1}{r^2}\begin{bmatrix} 1/\sin^2\theta & 0 \\ 0 & 1 \end{bmatrix}, \tag{192}$$

$b^{\alpha\beta} = -a^{\alpha\beta}/r$ and $H = -1/r$. The traction vector $\boldsymbol{T} = \nu_\alpha \boldsymbol{T}^\alpha$ on a cut $\perp \boldsymbol{\nu}$ thus becomes

$$\boldsymbol{T} = (N_a - N_b/r)\,\boldsymbol{\nu} + S^\alpha \nu_\alpha \boldsymbol{n} \tag{193}$$

according to (89). The in-plane component $T_\nu := N_a - N_b/r$ must equilibrate the current pressure according to the well-known relation

$$p = \frac{2T_\nu}{r}\,. \tag{194}$$

One can thus establish the analytical pressure–volume relation

$$\bar{p}(\bar{V}) = 2\bar{H}_0\,\bar{V}^{-\frac{2}{3}} - 2\bar{H}_0^2\,\bar{V}^{-\frac{1}{3}} + 2\bar{K}\left(\bar{V}^{\frac{1}{3}} - \bar{V}^{-\frac{1}{3}}\right), \tag{195}$$

normalized according to the definitions $\bar{p} := pR^3/k$, $\bar{V} := V/V_0$, $\bar{H}_0 := H_0R$ and $\bar{K} := KR^2/k$.

Part III: Computational Formulation

Part III discusses the computational formulation based on the theory described in Part II. The finite element equations are presented for the shell PDE (Sect. 10) and the incompressibility constraint (Sect. 11). Stabilization is addressed (Sect. 12) and several numerical examples are presented (Sect. 13). Part III follows the developments in Duong et al. (2017) and Sauer et al. (2017).

10 Rotation-Free Shell FE

The shell theory presented in Part II results in a fourth order, nonlinear partial differential equation (PDE), which involves displacement degrees of freedom, but no rotations. In order to solve its weak form, a C^1-continuous finite element discretization is required.[5] Such a discretization is provided by isogeometric finite elements. In Duong et al. (2017) a new isogeometric FE formulation is presented for thin shells. The formulation is suitable for a wide range of materials, and it accounts for large deformations and rotations as Fig. 5 demonstrates. This section presents the formulation (Sects. 10.1–10.3) and discussed how to treat surface viscosity (Sect. 10.4), C^1-continuity (Sect. 10.6) and patch boundaries (Sect. 10.7). Linearization is addressed in Sect. 10.5.

10.1 FE Approximation

The surface geometry of the reference and current configuration (see Fig. 2) is discretized into n_{el} finite elements Ω_e, $e = 1, \ldots, n_{\mathrm{el}}$. Within each element, the surface is approximated by the finite element interpolations

$$\boldsymbol{X}^h = \mathbf{N}\,\mathbf{X}_e \tag{196}$$

and

$$\boldsymbol{x}^h = \mathbf{N}\,\mathbf{x}_e\,, \tag{197}$$

[5]Strictly, G^1-continuity (i.e., continuity in $\boldsymbol{n}$ but not necessary in $\boldsymbol{a}_\alpha$) is sufficient.

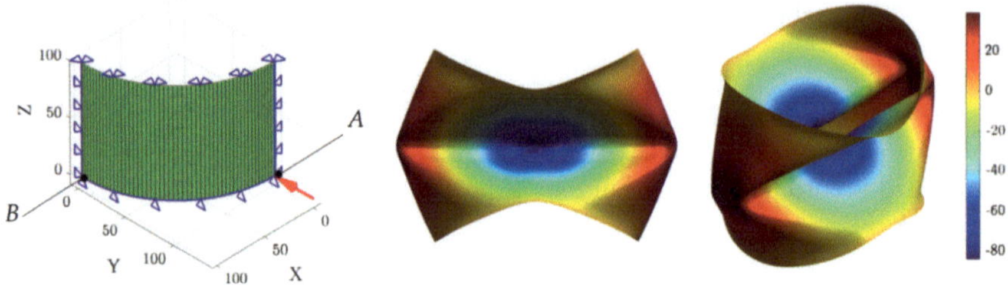

Fig. 5 Pinching of a cylindrical shell (Duong et al. 2017)

where $\mathbf{N} := [N_1\mathbf{1},\ \ldots,\ N_{n_e}\mathbf{1}]$ is a $(3 \times 3n_e)$ array containing the n_e nodal shape functions $N_I = N_I(\xi^1, \xi^2)$ of element Ω^e defined in parameter space $\mathcal{P}$. $\mathbf{X}_e := [\boldsymbol{X}_1^{\mathrm{T}},\ \ldots.,\ \boldsymbol{X}_{n_e}^{\mathrm{T}}]^{\mathrm{T}}$ and $\mathbf{x}_e := [\boldsymbol{x}_1^{\mathrm{T}},\ \ldots.,\ \boldsymbol{x}_{n_e}^{\mathrm{T}}]^{\mathrm{T}}$ contain the n_e nodal position vectors of Ω^e. The tangent vectors of the surface are thus approximated by

$$\boldsymbol{A}_\alpha^h = \frac{\partial \boldsymbol{X}^h}{\partial \xi^\alpha} = \mathbf{N}_{,\alpha}\,\mathbf{X}_e \tag{198}$$

and

$$\boldsymbol{a}_\alpha^h = \frac{\partial \boldsymbol{x}^h}{\partial \xi^\alpha} = \mathbf{N}_{,\alpha}\,\mathbf{x}_e\,. \tag{199}$$

Likewise, the the tangent derivative $\boldsymbol{a}_{\alpha,\beta}$ and the variations $\delta\boldsymbol{x}$ and $\delta\boldsymbol{a}_\alpha$ are approximated by

$$\boldsymbol{a}_{\alpha,\beta}^h = \mathbf{N}_{,\alpha\beta}\,\mathbf{x}_e\,, \tag{200}$$

$$\delta\boldsymbol{x}^h = \mathbf{N}\,\delta\mathbf{x}_e \tag{201}$$

and

$$\delta\boldsymbol{a}_\alpha^h = \mathbf{N}_{,\alpha}\,\delta\mathbf{x}_e\,. \tag{202}$$

According to (6), the surface normals $\boldsymbol{N}$ and $\boldsymbol{n}$ are thus approximated by

$$\boldsymbol{N}^h = \frac{\boldsymbol{A}_1^h \times \boldsymbol{A}_2^h}{\|\boldsymbol{A}_1^h \times \boldsymbol{A}_2^h\|} \tag{203}$$

and

$$\boldsymbol{n}^h = \frac{\boldsymbol{a}_1^h \times \boldsymbol{a}_2^h}{\|\boldsymbol{a}_1^h \times \boldsymbol{a}_2^h\|}\,. \tag{204}$$

With these approximations, all the kinematical quantities of Sect. 4, like $a_{\alpha\beta}$, $a^{\alpha\beta}$, $\boldsymbol{a}^{\alpha}$ and $b_{\alpha\beta}$, can be approximated.

10.2 Discretization of Kinematical Variations

Based on the above expressions, all the variations appearing within weak form (163) can be evaluated. According to (79) and (81), the discretization of $\delta a_{\alpha\beta}$ and $\delta b_{\alpha\beta}$ follow as

$$\begin{aligned} \delta a^h_{\alpha\beta} &= \delta \mathbf{x}_e^{\mathrm{T}} \left[\mathbf{N}^{\mathrm{T}}_{,\alpha} \mathbf{N}_{,\beta} + \mathbf{N}^{\mathrm{T}}_{,\beta} \mathbf{N}_{,\alpha} \right] \mathbf{x}_e \,, \\ \delta b^h_{\alpha\beta} &= \delta \mathbf{x}_e^{\mathrm{T}} \, \mathbf{N}^{\mathrm{T}}_{;\alpha\beta} \, \boldsymbol{n}^h \,, \end{aligned} \tag{205}$$

where

$$\mathbf{N}_{;\alpha\beta} := \mathbf{N}_{,\alpha\beta} - \Gamma^{\gamma}_{\alpha\beta} \mathbf{N}_{,\gamma} \tag{206}$$

has been introduced. In the same fashion, the increments $\Delta a_{\alpha\beta}$ and $\Delta b_{\alpha\beta}$ are discretized by

$$\begin{aligned} \Delta a^h_{\alpha\beta} &= \Delta \mathbf{x}_e^{\mathrm{T}} \left[\mathbf{N}^{\mathrm{T}}_{,\alpha} \mathbf{N}_{,\beta} + \mathbf{N}^{\mathrm{T}}_{,\beta} \mathbf{N}_{,\alpha} \right] \mathbf{x}_e \,, \\ \Delta b^h_{\alpha\beta} &= \Delta \mathbf{x}_e^{\mathrm{T}} \, \mathbf{N}^{\mathrm{T}}_{;\alpha\beta} \, \boldsymbol{n}^h \,. \end{aligned} \tag{207}$$

For the increments of $\delta a_{\alpha\beta}$ and $\delta b_{\alpha\beta}$, given in (83) and (86), the approximations

$$\begin{aligned} \Delta\delta a^h_{\alpha\beta} &= \delta \mathbf{x}_e^{\mathrm{T}} \left[\mathbf{N}^{\mathrm{T}}_{,\alpha} \mathbf{N}_{,\beta} + \mathbf{N}^{\mathrm{T}}_{,\beta} \mathbf{N}_{,\alpha} \right] \Delta \mathbf{x}_e \\ \Delta\delta b^h_{\alpha\beta} &= -\,\delta \mathbf{x}_e^{\mathrm{T}} \Big[\mathbf{N}^{\mathrm{T}}_{,\gamma} \, (\boldsymbol{n} \otimes \boldsymbol{a}^{\gamma}) \, \mathbf{N}_{;\alpha\beta} + \mathbf{N}^{\mathrm{T}}_{;\alpha\beta} \, (\boldsymbol{a}^{\gamma} \otimes \boldsymbol{n}) \, \mathbf{N}_{,\gamma} \\ &\qquad + \mathbf{N}^{\mathrm{T}}_{,\gamma} \, a^{\gamma\delta} \, b_{\alpha\beta} \, (\boldsymbol{n} \otimes \boldsymbol{n}) \, \mathbf{N}_{,\delta} \Big] \Delta \mathbf{x}_e \end{aligned} \tag{208}$$

then follow. Here, superscript h has been omitted from $\boldsymbol{n}$, $\boldsymbol{a}^{\gamma}$, $a^{\gamma\delta}$, and $b_{\alpha\beta}$ for simplicity. For the rest of the paper, all quantities are understood to be discrete even without explicit use of superscript h.

10.3 Discretized Weak Form

In the discrete system the weak form of Sect. 8 takes the form

$$G = \sum_{e=1}^{n_{\mathrm{el}}} G^e = \sum_{e=1}^{n_{\mathrm{el}}} \left(G^e_{\mathrm{in}} + G^e_{\mathrm{int}} + G^e_{\mathrm{c}} - G^e_{\mathrm{ext}} \right) , \tag{209}$$

where $G^e_{\bullet}$ are the elemental contributions to the expressions in (164). Inserting the above interpolations into (209) leads to

$$G^e = \delta\mathbf{x}_e^{\mathrm{T}}\,\mathbf{f}^e\,, \tag{210}$$

with

$$\mathbf{f}^e := \mathbf{f}^e_{\mathrm{in}} + \mathbf{f}^e_{\mathrm{int}} + \mathbf{f}^e_{\mathrm{c}} - \mathbf{f}^e_{\mathrm{ext}}\,. \tag{211}$$

The first term,

$$\mathbf{f}^e_{\mathrm{in}} := -\int_{\Omega^e} \rho\,\mathbf{N}^{\mathrm{T}}\,\mathbf{N}\,\mathrm{d}a\,\dot{\mathbf{v}}_e\,, \tag{212}$$

defines the inertia forces acting on the nodes of element Ω^e. The second term, $\mathbf{f}^e_{\mathrm{int}} := \mathbf{f}^e_{\sigma} + \mathbf{f}^e_{M}$, defines the internal forces of element Ω^e caused by the membrane stress $\sigma^{\alpha\beta}$ and the bending moment $M^{\alpha\beta}$. The two contributions are given by

$$\mathbf{f}^e_{\sigma} := \int_{\Omega^e} \sigma^{\alpha\beta}\,\mathbf{N}^{\mathrm{T}}_{,\alpha}\,\boldsymbol{a}_\beta\,\mathrm{d}a\,, \tag{213}$$

and

$$\mathbf{f}^e_{M} := \int_{\Omega^e} M^{\alpha\beta}\,\mathbf{N}^{\mathrm{T}}_{;\alpha\beta}\,\boldsymbol{n}\,\mathrm{d}a\,. \tag{214}$$

Following decomposition (169), $\mathbf{f}^e_{\sigma}$ can be split into the in-plane and out-of-plane contributions (Sauer et al. 2014)

$$\begin{aligned} \mathbf{f}^e_{\sigma\mathrm{in}} &:= \mathbf{f}^e_{\sigma} - \mathbf{f}^e_{\sigma\mathrm{out}}\,, \\ \mathbf{f}^e_{\sigma\mathrm{out}} &:= -\int_{\Omega^e} \sigma^{\alpha\beta}\,b_{\alpha\beta}\,\mathbf{N}^{\mathrm{T}}\,\boldsymbol{n}\,\mathrm{d}a\,. \end{aligned} \tag{215}$$

The third term,

$$\mathbf{f}^e_{\mathrm{c}} = \int_{\Omega^e} \mathbf{N}^{\mathrm{T}}\,p_{\mathrm{c}}\,\boldsymbol{n}\,\mathrm{d}a\,, \tag{216}$$

defines the FE contact forces due to the contact pressure p_{c}. The last term, $\mathbf{f}^e_{\mathrm{ext}} := \mathbf{f}^e_{f} + \mathbf{f}^e_{t} + \mathbf{f}^e_{m}$, defines the FE forces due to the external loads $\boldsymbol{f}$, $\boldsymbol{t}$ and m_τ. The three pieces are given by

$$\begin{aligned} \mathbf{f}^e_{f} &:= \int_{\Omega^e} \mathbf{N}^{\mathrm{T}}\,\boldsymbol{f}\,\mathrm{d}a\,, \\ \mathbf{f}^e_{t} &:= \int_{\partial_t\Omega^e} \mathbf{N}^{\mathrm{T}}\,\boldsymbol{t}\,\mathrm{d}s\,, \\ \mathbf{f}^e_{m} &:= \int_{\partial_m\Omega^e} \mathbf{N}^{\mathrm{T}}_{,\alpha}\,\nu^\alpha\,m_\tau\,\boldsymbol{n}\,\mathrm{d}s\,. \end{aligned} \tag{217}$$

In the examples considered here, the external forces are zero.

The discretized system is in equilibrium if all nodal forces sum up to zero (see Sect. 11.2 for details). This force balance is a second-order system of ordinary differential equations due to the inertia term. If inertia is neglected, as is considered in the remainder of this paper, the force balance is a first-order system of ODEs due to the viscosity term. The temporal discretization of the viscosity term is discussed next.

10.4 Temporal Discretization of the Viscosity Term

In order to solve the time-dependent problem, time is discretized into a set of n_t steps and the solution is advanced from step t_n to t_{n+1}. The viscosity dependant stress $\sigma^{\alpha\beta}_{\text{visc}} = -\eta\,\dot{a}^{\alpha\beta}$ can be discretized at t_{n+1} by the first-order rate approximation

$$\dot{a}^{\alpha\beta}_{n+1} \approx \frac{1}{\Delta t_{n+1}}\left(a^{\alpha\beta}_{n+1} - a^{\alpha\beta}_{n}\right), \tag{218}$$

where $\bullet_n := \bullet(t_n)$ and $\Delta t_{n+1} := t_{n+1} - t_n$. At the new step t_{n+1}, the problem is then solved implicitly for the current nodal positions $\boldsymbol{x}_I(t_{n+1})$, given the previous positions $\boldsymbol{x}_I(t_n)$. The reference configuration is taken as the initial configuration at time $t_0 = 0$, i.e. $\boldsymbol{X}_I = \boldsymbol{x}_I(t_0)$. This temporal discretization approach corresponds to the implicit Euler scheme.

10.5 Linearization

The resulting nonlinear equations at the current time step are solved with the Newton–Raphson method. This requires the linearization of the discretized weak form. The most important contribution is the linearization of the internal virtual work. The linearization of inertia and the external forces is not required for the later examples, and they are therefore omitted here. The interested reader can find them in Duong et al. (2017). The linearization of the contact forces can be found in the contact literature, e.g., see Sauer and De Lorenzis (2013, 2015).

According to (173) and (133), the linearization of G^e_{int} (in the absence of the incompressibility constraint) leads to

$$\begin{aligned}\Delta G^e_{\text{int}} = \int_{\Omega^e_0} \Big(& c^{\alpha\beta\gamma\delta}\,\tfrac{1}{2}\delta a_{\alpha\beta}\,\tfrac{1}{2}\Delta a_{\gamma\delta} && + d^{\alpha\beta\gamma\delta}\,\tfrac{1}{2}\delta a_{\alpha\beta}\,\Delta b_{\gamma\delta} \\ & + e^{\alpha\beta\gamma\delta}\,\delta b_{\alpha\beta}\,\tfrac{1}{2}\Delta a_{\gamma\delta} && + f^{\alpha\beta\gamma\delta}\,\delta b_{\alpha\beta}\,\Delta b^h_{\gamma\delta} \\ & + J\sigma^{\alpha\beta}\,\tfrac{1}{2}\Delta\delta a_{\alpha\beta} && + J M^{\alpha\beta}\,\Delta\delta b_{\alpha\beta} \\ & - \frac{J\eta}{4\Delta t}\delta a_{\alpha\beta}\Big[\big(a^{\alpha\beta} - a^{\alpha\beta}_n\big)a^{\gamma\delta} + 2a^{\alpha\beta\gamma\delta}\Big]\Delta a_{\alpha\beta}\Big)\,\mathrm{d}A\,,\end{aligned} \tag{219}$$

where subscript $n+1$ has been omitted. The tangent matrices $c^{\alpha\beta\gamma\delta}$, $d^{\alpha\beta\gamma\delta}$, $e^{\alpha\beta\gamma\delta}$ and $f^{\alpha\beta\gamma\delta}$ have been given in Sect. 7.4. The last term arises from the viscosity approximation of (218). In can be absorbed into $c^{\alpha\beta\gamma\delta}$ if one replaces

$$c^{\alpha\beta\gamma\delta} \leftarrow c^{\alpha\beta\gamma\delta} - \frac{J\eta}{\Delta t}\Big[\big(a^{\alpha\beta} - a_n^{\alpha\beta}\big)a^{\gamma\delta} + 2a^{\alpha\beta\gamma\delta}\Big]. \tag{220}$$

Using (135), and exploiting the minor symmetries in the tangent matrices, one finds

$$\begin{aligned}
c^{\alpha\beta\gamma\delta}\,\tfrac{1}{2}\delta a_{\alpha\beta}\,\tfrac{1}{2}\Delta a_{\gamma\delta} &= c^{\alpha\beta\gamma\delta}\,\delta\mathbf{x}_e^{\mathrm{T}}\,\mathbf{N}_{,\alpha}^{\mathrm{T}}\,(\boldsymbol{a}_\beta\otimes\boldsymbol{a}_\gamma)\,\mathbf{N}_{,\delta}\,\Delta\mathbf{x}_e\,,\\
d^{\alpha\beta\gamma\delta}\,\tfrac{1}{2}\delta a_{\alpha\beta}\,\Delta b_{\gamma\delta} &= d^{\alpha\beta\gamma\delta}\,\delta\mathbf{x}_e^{\mathrm{T}}\,\mathbf{N}_{,\alpha}^{\mathrm{T}}\,(\boldsymbol{a}_\beta\otimes\boldsymbol{n})\,\mathbf{N}_{;\gamma\delta}\,\Delta\mathbf{x}_e\,,\\
e^{\alpha\beta\gamma\delta}\,\delta b_{\alpha\beta}\,\tfrac{1}{2}\Delta a_{\gamma\delta} &= e^{\alpha\beta\gamma\delta}\,\delta\mathbf{x}_e^{\mathrm{T}}\,\mathbf{N}_{;\alpha\beta}^{\mathrm{T}}\,(\boldsymbol{n}\otimes\boldsymbol{a}_\gamma)\,\mathbf{N}_{,\delta}\,\Delta\mathbf{x}_e\,,\\
f^{\alpha\beta\gamma\delta}\,\delta b_{\alpha\beta}\,\Delta b_{\gamma\delta} &= f^{\alpha\beta\gamma\delta}\,\delta\mathbf{x}_e^{\mathrm{T}}\,\mathbf{N}_{;\alpha\beta}^{\mathrm{T}}\,(\boldsymbol{n}\otimes\boldsymbol{n})\,\mathbf{N}_{;\gamma\delta}\,\Delta\mathbf{x}_e\,,
\end{aligned} \tag{221}$$

such that

$$\Delta G_{\mathrm{int}}^e = \delta\mathbf{x}_e^{\mathrm{T}}\Big[\mathbf{k}_{\sigma\sigma}^e + \mathbf{k}_{\sigma M}^e + \mathbf{k}_{M\sigma}^e + \mathbf{k}_{MM}^e + \mathbf{k}_\sigma^e + \mathbf{k}_M^e\Big]\Delta\mathbf{x}_e\,, \tag{222}$$

with

$$\begin{aligned}
\mathbf{k}_{\sigma\sigma}^e &:= \int_{\Omega_0^e} c^{\alpha\beta\gamma\delta}\,\mathbf{N}_{,\alpha}^{\mathrm{T}}\,(\boldsymbol{a}_\beta\otimes\boldsymbol{a}_\gamma)\,\mathbf{N}_{,\delta}\,\mathrm{d}A\,,\\
\mathbf{k}_{\sigma M}^e &:= \int_{\Omega_0^e} d^{\alpha\beta\gamma\delta}\,\mathbf{N}_{,\alpha}^{\mathrm{T}}\,(\boldsymbol{a}_\beta\otimes\boldsymbol{n})\,\mathbf{N}_{;\gamma\delta}\,\mathrm{d}A\,,\\
\mathbf{k}_{M\sigma}^e &:= \int_{\Omega_0^e} e^{\alpha\beta\gamma\delta}\,\mathbf{N}_{;\alpha\beta}^{\mathrm{T}}(\boldsymbol{n}\otimes\boldsymbol{a}_\gamma)\,\mathbf{N}_{,\delta}\,\mathrm{d}A\,,\\
\mathbf{k}_{MM}^e &:= \int_{\Omega_0^e} f^{\alpha\beta\gamma\delta}\,\mathbf{N}_{;\alpha\beta}^{\mathrm{T}}\,(\boldsymbol{n}\otimes\boldsymbol{n})\,\mathbf{N}_{;\gamma\delta}\,\mathrm{d}A\,,
\end{aligned} \tag{223}$$

and

$$\begin{aligned}
\mathbf{k}_\sigma^e &= \int_{\Omega^e} \mathbf{N}_{,\alpha}^{\mathrm{T}}\,\sigma^{\alpha\beta}\,\mathbf{N}_{,\beta}\,\mathrm{d}a\,,\\
\mathbf{k}_M^e &= \mathbf{k}_{M1}^e + \mathbf{k}_{M2}^e + (\mathbf{k}_{M2}^e)^{\mathrm{T}}\,,
\end{aligned} \tag{224}$$

and

$$\begin{aligned}
\mathbf{k}_{M1}^e &:= -\int_{\Omega^e} b_{\alpha\beta}\,M^{\alpha\beta}\,a^{\gamma\delta}\,\mathbf{N}_{,\gamma}^{\mathrm{T}}\,(\boldsymbol{n}\otimes\boldsymbol{n})\,\mathbf{N}_{,\delta}\,\mathrm{d}a\,,\\
\mathbf{k}_{M2}^e &:= -\int_{\Omega^e} M^{\alpha\beta}\,\mathbf{N}_{,\gamma}^{\mathrm{T}}\,(\boldsymbol{n}\otimes\boldsymbol{a}^\gamma)\,\mathbf{N}_{;\alpha\beta}\,\mathrm{d}a\,.
\end{aligned} \tag{225}$$

The first four $\mathbf{k}^e$ are the material tangent matrices of element Ω^e. In order for those to be positive definite, stability criterion (138) needs to be satisfied. $\mathbf{k}_\sigma^e$ and $\mathbf{k}_M^e$ are the geometric tangent matrices of element Ω^e.

10.6 C1-Continuous Shape Functions

As noted before, the FE shape function have to be at least C^1-continuous everywhere in the domain, including element boundaries. This property is provided by the shape functions used in isogeometric analysis (Hughes et al. 2005; Cottrell et al. 2009). An example are NURBS (Nonuniform rational B-splines). Thanks to the Bézier extraction operator $\mathbf{C}^e$ introduced by Borden et al. (2011), the usual finite element structure can be used for NURBS basis functions. The NURBS shape function of node (= control point) A is given by

$$N_A(\xi^\alpha) = \frac{w_A\,\hat{N}_A^e(\xi^\alpha)}{\sum_{A=1}^{n_e} w_A\,\hat{N}_A^e(\xi^\alpha)}\,. \tag{226}$$

Here, n_e is the number of control points defining element Ω^e, w_A is a weight, and $\hat{N}_A^e$ is the B-spline basis function expressed in terms of Bernstein polynomials according to

$$\hat{\mathbf{N}}^e(\xi^\alpha) = \mathbf{C}_1^e\,\mathbf{B}(\xi^1) \otimes \mathbf{C}_2^e\,\mathbf{B}(\xi^2)\,, \tag{227}$$

with $\hat{N}_A^e$ being the corresponding entries of matrix $\hat{\mathbf{N}}^e$. Further details can be found in Borden et al. (2011). Figure 6 shows the basis function $\hat{N}_A^e$ for a one-dimensional example with five control points. The tensor-based structure of (227) provides a simple extension to two dimensions, as long as the surface $\mathcal{S}$ can be globally defined from a rectangular parameter domain. If this is not the case, alternatives exists. One possibility is to use T-spline basis functions (Scott et al. 2011). Another option is to construct the surface from multiple NURBS patches (e.g., see the example in Fig. 16). In this case, the relative rotation between neighboring patches has to be suppressed. This is discussed in the following section. It is also possible to apply local refinement to the patches (Johannessen et al. 2014; Zimmermann and Sauer 2017).

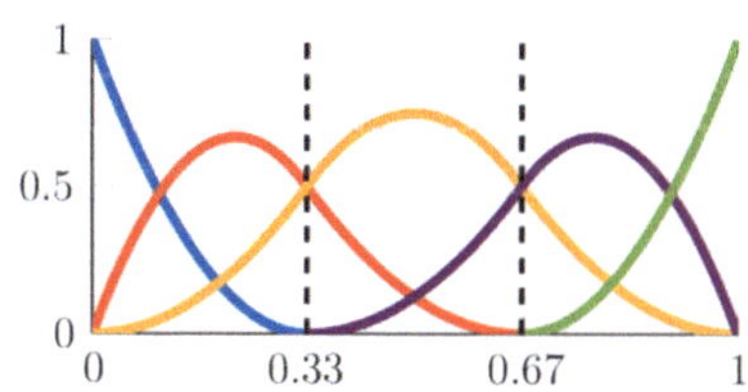

Fig. 6 The B-spline basis functions for a patch of three elements and five control points (Corbett 2016)

10.7 Patch Interfaces

To constrain rotations between patches, the constraint potential

$$\Pi_{\mathrm{n}} = \int_{\mathcal{L}_0} \frac{\epsilon}{2}\,(\boldsymbol{n} - \tilde{\boldsymbol{n}}) \cdot (\boldsymbol{n} - \tilde{\boldsymbol{n}})\,\mathrm{d}S \tag{228}$$

is added to the formulation. $\mathcal{L}_0$ denotes the patch interface in the reference configuration, ϵ is a penalty parameter, and $\boldsymbol{n}$ and $\tilde{\boldsymbol{n}}$ are the normal vectors on the two sides of the patch interface. The variation, linearization, and FE discretization of (228) is discussed in Duong et al. (2017). Careful implementation of the approach leads to no loss in accuracy compared to single patches as Fig. 7 shows.

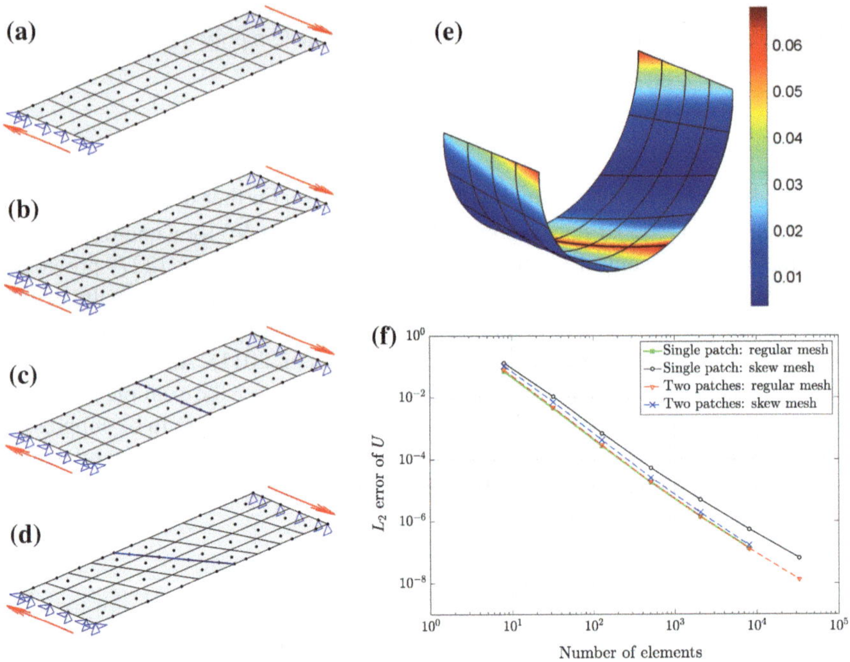

Fig. 7 Pure bending of a flat sheet (cf. Sect. 9) considering: **a** single patch with regular mesh, **b** single patch with skew mesh, **c** two patches with regular mesh, and **d** two patches with skew mesh. **e** Deformed configuration coloured by the relative error in mean curvature H. **f** L_2 error of the solution w.r.t. mesh refinement (Duong et al. 2017)

11 Mixed Finite Elements

For the area-incompressible case of Eq. (142), the Lagrange multiplier q and the corresponding weak form equation have to be discretized as well. This is discussed in the following (Sect. 11.1) using LBB-conforming finite elements (Sect. 11.3). The resulting solution procedure is presented in Sect. 11.2 using the normalization scheme of Sect. 11.4

11.1 Discretization of the Area Constraint

The Lagrange multiplier is approximated by the interpolation

$$q^h = \mathbf{L}\,\mathbf{q}_e\,, \tag{229}$$

analogously to the deformation in Eq. (197). Here $\mathbf{L} := [L_1, \ ..., \ L_{m_e}]$ is a $(1 \times m_e)$ array containing the m_e nodal shape functions $L_I = L_I(\xi^\alpha)$ of surface element Ω^e, and $\mathbf{q}_e := [q_1, \ ...., \ q_{m_e}]^{\mathrm{T}}$ contains the m_e nodal Lagrange multipliers of the element. It follows that

$$\delta q^h = \mathbf{L}\,\delta\mathbf{q}_e\,, \tag{230}$$

such that weak form (166) becomes

$$G_g = \sum_{e=1}^{n_{\mathrm{el}}} G_g^e\,, \tag{231}$$

where

$$G_g^e = \delta\mathbf{q}_e^{\mathrm{T}}\,\mathbf{g}^e\,, \tag{232}$$

with

$$\mathbf{g}^e := \int_{\Omega_0^e} \mathbf{L}^{\mathrm{T}}\,g\,\mathrm{d}A\,. \tag{233}$$

11.2 Solution Procedure

The mixed problem is characterized by the two unknown fields $\boldsymbol{x}$ and q, or their discrete counterparts $\mathbf{x}$ and $\mathbf{q}$. The combined weak form of the discrete problem is given by

$$\delta\mathbf{x}^{\mathrm{T}}\,\mathbf{f}(\mathbf{x},\mathbf{q}) + \delta\mathbf{q}^{\mathrm{T}}\,\mathbf{g}(\mathbf{x}) = 0\,, \quad \forall\,\delta\mathbf{x} \in \mathcal{V}^h \ \&\ \delta\mathbf{q} \in \mathcal{Q}^h\,, \tag{234}$$

which follows from adding the elemental contributions of G^e and G_g^e given in (210) and (232). Here $\mathbf{x}$, $\mathbf{q}$, $\delta\mathbf{x}$, and $\delta\mathbf{q}$ are global vectors containing all nodal deformations, Lagrange multipliers and their variations. $\mathcal{V}^h$ and $\mathcal{Q}^h$ are the discrete counterparts to spaces $\mathcal{V}$ and $\mathcal{Q}$. The global vectors $\mathbf{f}$ and $\mathbf{g}$ are assembled from the elemental contributions $\mathbf{f}^e_{\mathrm{int}}$, $\mathbf{f}^e_{\mathrm{c}}$, $\mathbf{f}^e_{\mathrm{ext}}$, and $\mathbf{g}^e$ by adding corresponding entries. Equation (234) is satisfied if $\mathbf{f}=\mathbf{0}$ and $\mathbf{g}=\mathbf{0}$ at nodes where no Dirichlet BC apply. These two nonlinear equations are then solved with Newton's method for the unknowns $\mathbf{x}$ and $\mathbf{q}$.

If no constraint is present (like in model (140)), the parts containing $\mathbf{q}$ and $\mathbf{g}$ are simply skipped.

11.3 LBB Condition

For mixed FE problems, the discretization of $\boldsymbol{x}$ and q cannot be chosen independently. Instead, $\boldsymbol{x}^h$ and q^h should satisfy the LBB condition[6]

$$\inf_{q^h\in\mathcal{Q}^h}\ \sup_{\boldsymbol{v}^h\in\mathcal{V}^h}\frac{\int_{\mathcal{S}} q^h\,\mathrm{div}_{\mathrm{s}}\,\boldsymbol{v}^h\,\mathrm{d}A}{\|\boldsymbol{v}^h\|_{H^1}\ \|q^h\|_{L^2}}=\gamma^h\geq\gamma>0 \tag{235}$$

(Babuška 1973; Bathe 1996). For the presented shell discretization, the inf-sup value γ^h corresponds to the smallest eigenvalue of

$$\mathbf{G}\,\boldsymbol{\phi}=\lambda\,\mathbf{S}\,\boldsymbol{\phi}\,, \tag{236}$$

Bathe (2001), where

$$\begin{aligned}
\mathbf{G}\ &:=\mathbf{k}_{\mathrm{g}}^{\mathrm{T}}\,\mathbf{T}^{-1}\,\mathbf{k}_{\mathrm{g}}\,,\\
\mathbf{k}_{\mathrm{g}}\ &:=\int_{\mathcal{S}^h}\mathbf{L}^{\mathrm{T}}\boldsymbol{a}^{\alpha}\cdot\mathbf{N}_{,\alpha}\,\mathrm{d}a\,,\\
\mathbf{T}\ &:=\int_{\mathcal{S}^h}\mathbf{L}^{\mathrm{T}}\mathbf{L}\,\mathrm{d}a\,,\\
\mathbf{S}\ &:=\int_{\mathcal{S}^h}\Big(\mathbf{N}^{\mathrm{T}}\mathbf{N}+a^{\alpha\beta}\,\mathbf{N}^{\mathrm{T}}_{,\alpha}\mathbf{N}_{,\beta}\Big)\,\mathrm{d}a\,,
\end{aligned} \tag{237}$$

and $(\lambda,\,\boldsymbol{\phi})$ denotes the eigenvalue/eigenvector pair. Eigenvalue problem (236) is defined on the entire system, and hence, the integrals are taken over the entire surface $\mathcal{S}^h$ and the arrays $\mathbf{N}$ and $\mathbf{L}$ now extended to all nodes. The LBB condition can be satisfied if $\boldsymbol{x}$ is interpolated by C^1-continuous, biquadratic NURBS and q is interpolated by C^0-continuous, bilinear Lagrange shape functions (Loc et al. 2013). This is demonstrated in Fig. 8. If the LBB condition is violated, oscillations in the Lagrange multiplier appear. Such oscillations do not appear in this example. Also the

[6]Named after Ladyzhenskaya, Babuška, and Brezzi.

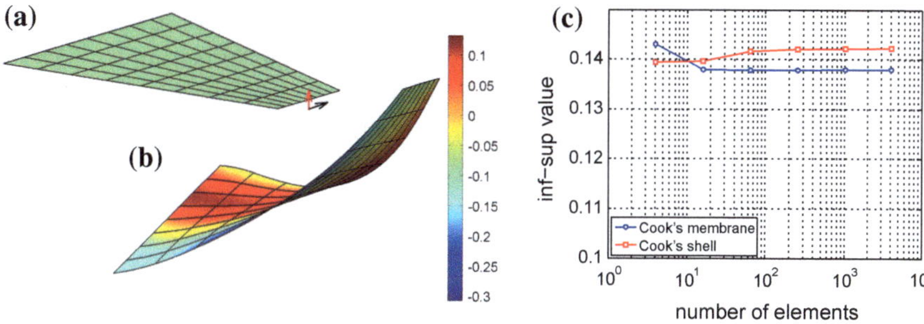

Fig. 8 Cook's membrane test for mixed displacement/pressure FE (here for NURBS-based biquadratic displacement and Lagrange-based bilinear pressure interpolation): initial configuration (**a**), deformed configuration (**b**), inf-sup value (**c**). The left boundary is fully clamped, while on the right boundary a distributed force is applied considering two cases: only an in-plane force ('Cook's membrane') and a force with an out-of-plane component ('Cook's shell'). Area-incompressibility together with shear model 'A-st' with $\bar{\mu} = 5$ is considered. The color in the middle figure shows the Lagrange multiplier. Since the inf-sup value is bounded the formulation is LBB-stable

multi-patch example in Sect. 13.2 does not exhibit such oscillations if it is discretized by the mixed approach described above. However, If a penalty regularization of constraint (76) is used, oscillations appear as the penalty parameter (i.e., the bulk modulus K) is increased (Sauer et al. 2017).

11.4 Normalization

For a numerical implementation, the preceding expressions need to be normalized. For this purpose a length scale L_0, time scale T_0, and force F_0 are chosen, and used to normalize all lengths, times, and forces in the system. Velocities, surface densities, surface pressures, membrane stiffness, and membrane viscosity are then normalized by the scales

$$v_0 := \frac{L_0}{T_0}, \quad \rho_0 := \frac{F_0 T_0^2}{L_0^3}, \quad p_0 := \frac{F_0}{L_0^2}, \quad \mu_0 := \frac{F_0}{L_0}, \quad \eta_0 := \frac{F_0 T_0}{L_0}. \tag{238}$$

Weak form (234) can then be expressed in the normalized form

$$\delta\bar{\mathbf{x}}^{\mathrm{T}}\,\bar{\mathbf{f}}(\bar{\mathbf{x}}, \bar{\mathbf{q}}) + \delta\bar{\mathbf{q}}^{\mathrm{T}}\,\bar{\mathbf{g}}(\bar{\mathbf{x}}) = 0\,, \tag{239}$$

where a bar denotes normalization with the corresponding scale from above, e.g.,

$$\bar{\mathbf{f}}^e_\sigma := \int_{\bar{\Omega}^e} \bar{\sigma}^{\alpha\beta}\,\bar{\mathbf{N}}^{\mathrm{T}}_{,\alpha}\,\boldsymbol{a}_\beta\,\mathrm{d}\bar{a}\,, \tag{240}$$

with $\bar{\sigma}^{\alpha\beta} = \sigma^{\alpha\beta}/\mu_0$, $\bar{\mathbf{N}}_{,\alpha} = \mathbf{N}_{,\alpha}\, L_0$ and $\mathrm{d}\bar{a} = \mathrm{d}a/L_0^2$. By choice, parameter ξ^α is supposed to carry units of length, so that $\boldsymbol{a}_\alpha$ and $a^{\alpha\beta}$ become dimensionless. All the other quantities appearing in (234) are normalized in the same fashion.

If F_0 is defined through $k = F_0 L_0$ (the bending modulus k has the unit [force × length]), the system is effectively normalized by k. The nondimensional material parameters thus are

$$\begin{aligned} \bar{k} &= 1\,, \\ \bar{k}_{\mathrm{g}} &= k_{\mathrm{g}}/k\,, \\ \bar{K} &= K\,L^2/k\,, \\ \bar{\mu} &= \mu\,L^2/k\,, \\ \bar{\epsilon} &= \epsilon\,L/k\,, \end{aligned} \tag{241}$$

while the normalization of stress and moment components become

$$\begin{aligned} \bar{q} &= q\,L^2/k\,, \\ \bar{\sigma}^{\alpha\beta} &= \sigma^{\alpha\beta}\,L^2/k\,, \\ \bar{M}^{\alpha\beta} &= M^{\alpha\beta}\,L/k\,. \end{aligned} \tag{242}$$

12 Lipid Bilayer Stabilization

As noted in Sects. 1, 7.4 and 10.5, the lipid bilayer is unstable for quasi-static computations (i.e., when no inertia and viscosity is considered). There are two principal ways to stabilize the system without modifying and affecting the original problem. They are discussed in the following two sections and then summarized in Sect. 12.3. The presentation is taken from Sauer et al. (2017).

12.1 Adding Stiffness

One way to stabilize the system is to add a stabilization stress $\sigma^{\alpha\beta}_{\mathrm{sta}}$ to $\sigma^{\alpha\beta}$ in order to provide additional stiffness. This stress can be defined from a (convex) shear energy or from numerical viscosity. An elegant and accurate option is to add the stabilization stress only to the in-plane contribution (170) while leaving the out-of-plane contribution (171) unchanged. The advantage of this approach is that the out-of-plane part, responsible for the shape of the bilayer, is not affected by the stabilization, at least not in the continuum limit of the discretization. There are several different ways to define the stabilization stress, and they are grouped into two categories. An overview of all the options is then summarized in Table 2.

In-plane shear and bulk stabilization The first category goes back to Sauer (2014), who used it to stabilize liquid membranes governed by constant surface tension. The stabilization stress for such membranes requires shear and bulk contributions. Those are given for example by the stabilization stress

$$\sigma^{\alpha\beta}_{\text{sta}} = \mu/J\left(A^{\alpha\beta} - a^{\alpha\beta}\right), \tag{243}$$

based on numerical stiffness, and

$$\sigma^{\alpha\beta}_{\text{sta}} = \mu/J\left(a^{\alpha\beta}_{\text{pre}} - a^{\alpha\beta}\right), \tag{244}$$

based on numerical viscosity. Here, $a^{\alpha\beta}_{\text{pre}}$ denotes the value of $a^{\alpha\beta}$ at the preceding computational step. These stabilization stresses are then only included within Eq. (170) and not in Eq. (171), and the resulting two stabilization schemes are denoted 'A' (for (243)) and 'a' (for (244)) following Sauer (2014). This reference shows that scheme 'a' is highly accurate and performs much better than scheme 'A.' It also shows that applying the stabilization stresses (243) and (244) only to the in-plane part is much more accurate than applying it throughout the system (i.e., in both Eqs. (170) and (171)), which we denote as schemes 'A-t' and 'a-t.'

Sole in-plane shear stabilization If the surface tension is not constant, as in the lipid bilayer models introduced above, only shear stabilization is required. A suitable stabilization stress can be derived from the shear energy

$$\Psi_0 = \frac{\mu}{2}\left(\widehat{I}_1 - 2\right), \tag{245}$$

where $\widehat{I}_1 = I_1/J$ (Sauer et al. 2017). Equations (124) and (116) then give

$$\tau^{\alpha\beta}_{\text{sta}} = \frac{\mu}{J}\left(A^{\alpha\beta} - \frac{I_1}{2}\,a^{\alpha\beta}\right). \tag{246}$$

As before, this stress will only be applied to Eq. (170) and not to Eq. (171), even though it has been derived from a potential and should theoretically apply to both terms. Following earlier nomenclature, this scheme is denoted by 'A-s'. Replacing $A^{\alpha\beta}$ by $a^{\alpha\beta}_{\text{pre}}$ in (246) gives

$$\tau^{\alpha\beta}_{\text{sta}} = \frac{\mu}{J^*}\left(a^{\alpha\beta}_{\text{pre}} - \frac{I_1^*}{2}\,a^{\alpha\beta}\right), \tag{247}$$

with $J^* := \sqrt{\det a_{\alpha\beta}/\det a^{\text{pre}}_{\alpha\beta}}$ and $I_1^* := a^{\alpha\beta}_{\text{pre}}\,a_{\alpha\beta}$, which is an alternative shear-stabilization scheme based on the numerical viscosity. It is denoted 'a-s'. If stresses (246) and (247) are applied throughout the system (i.e., to both (170) and (171)), the corresponding schemes are denoted 'A-st' and 'a-st'.

If the shell is (nearly) area-incompressible the two stabilization methods of sections "In-plane shear and bulk stabilization" and "Sole in-plane shear stabilization" can behave identical, as can be seen by the example in Sect. 12.4.

12.2 Normal Projection

The second principal way to stabilize the system consists of a simple projection of the formulation onto the solution space defined by the normal surface direction. This step can be applied directly to the discretized formulation as was proposed by Sauer (2014). According to this, for the discrete system of linear equations for displacement increment $\Delta\mathbf{u}$, which is given by $\mathbf{K}\,\Delta\mathbf{u} = -\mathbf{f}$, the reduced system for increment $\Delta\mathbf{u}_{\text{red}} = \mathbf{P}\,\Delta\mathbf{u}$ is simply obtained as

$$\mathbf{K}_{\text{red}}\,\Delta\mathbf{u}_{\text{red}} = -\mathbf{f}_{\text{red}}\,,\quad \mathbf{K}_{\text{red}} := \mathbf{P}\,\mathbf{K}\,\mathbf{P}^T\,,\quad \mathbf{f}_{\text{red}} := \mathbf{P}\,\mathbf{f}\,, \tag{248}$$

where

$$\mathbf{P} := \begin{bmatrix} \boldsymbol{n}_1^T & \mathbf{0}^T & \cdots & \mathbf{0}^T \\ \mathbf{0}^T & \boldsymbol{n}_2^T & \cdots & \mathbf{0}^T \\ \vdots & \vdots & \ddots & \vdots \\ \mathbf{0}^T & \mathbf{0}^T & \cdots & \boldsymbol{n}_{n_{\text{no}}}^T \end{bmatrix} \tag{249}$$

is a projection matrix defined by the nodal normal vectors $\boldsymbol{n}_I$. Since this method can lead to distorted FE meshes, a mesh update can be performed by applying any of the stabilization techniques discussed above. If this is followed by a projection step at the same load level, a dependency on parameter μ is avoided.

As noted in Sauer et al. (2017), the projection approach does not work very well for surfaces with (near) area-incompressibility, but it does work very well for area-compressible surfaces (Sauer 2014).

12.3 Summary of the Stabilization Schemes

The nine stabilization schemes presented above are summarized in Table 2. They can be grouped into three classes: **A**, **a** and **P**. The schemes of class **A** depend only on μ but require this value to be quite low. The schemes of class **a** also depend on the number of computational steps, n_t. If this number is high, the schemes provide stiffness without adding much stress. The shell is then stabilized without modifying the solution much, even when μ is high. Scheme 'P' depends on the nodal projection vector $\boldsymbol{n}_I$, which is usually taken as the surface normal. The performance of the different stabilization schemes is investigated in the following section.

Table 2 Summary of the stabilization schemes presented in Sects. 12.1 and 12.2 (Sauer et al. 2017)

Class	Scheme	Stab. stress $\sigma^{\alpha\beta}_{\mathrm{sta}}/\mu$	Application of $\sigma^{\alpha\beta}_{\mathrm{sta}}$	Dependence
A	A	$(A^{\alpha\beta} - a^{\alpha\beta})/J$	Only in (170)	Only on μ
	A-t	$(A^{\alpha\beta} - a^{\alpha\beta})/J$	In (170) and (171)	Only on μ
	A-s	$(A^{\alpha\beta} - \frac{1}{2}I_1\, a^{\alpha\beta})/J^2$	Only in (170)	Only on μ
	A-st	$(A^{\alpha\beta} - \frac{1}{2}I_1\, a^{\alpha\beta})/J^2$	In (170) and (171)	Only on μ
a	a	$(a^{\alpha\beta}_{\mathrm{pre}} - a^{\alpha\beta})/J$	Only in (170)	On μ and n_t
	a-t	$(a^{\alpha\beta}_{\mathrm{pre}} - a^{\alpha\beta})/J$	In (170) and (171)	On μ and n_t
	a-s	$(a^{\alpha\beta}_{\mathrm{pre}} - \frac{1}{2}I_1^*\, a^{\alpha\beta})/J^{*2}$	Only in (170)	On μ and n_t
	a-st	$(a^{\alpha\beta}_{\mathrm{pre}} - \frac{1}{2}I_1^*\, a^{\alpha\beta})/J^{*2}$	In (170) and (171)	On μ and n_t
P	P	0	–	On nodal $\boldsymbol{n}_I$

12.4 Performance of the Stabilization Schemes

Two examples are considered in order to examine the performance of the proposed stabilization schemes. They are based on the two analytical examples presented in Sect. 9.

Pure bending and stretching of a flat sheet The first example considers the pure bending and stretching of a flat sheet. The analytical solution for this problem is given in Sect. 9.1. The problem is solved numerically using the computational setup shown in Fig. 9. The FE mesh is discretized by m elements along X. The parameter t is introduced to apply the rotation $\Theta = t\pi/6$ and stretch $\lambda_2 = 1 + t/2$ by increasing t linearly from 0 to 1 in n_t steps, where n_t is chosen as a multiple of m. The mean curvature then follows as $H = \Theta/(2\lambda_1 S)$. Numerically, the rotation is applied according to (228) considering the penalty parameter $\epsilon = 100\, m\, k/L$. Figure 9b shows the FE solution and analytical solution for the bending moment $M_0(t)$ and the stress $N_0(t)$, normalizing M_0 by k/L and N_0 by k/L^2.

Next, the accuracy of the different stability schemes is studied in detail by examining the L_2-error of the solution, defined by

$$L_2 := \sqrt{\frac{1}{SL}\int_{\mathcal{S}_0} \|\boldsymbol{u}_{\mathrm{exact}} - \boldsymbol{u}_{\mathrm{FE}}\|^2\, \mathrm{d}A}\,, \tag{250}$$

and the error in M and N, defined by

$$E_{MN} := \frac{|M_{\text{exact}} - M_{\text{FE}}|}{M_{\text{exact}}} + \frac{|N_{\text{exact}} - N_{\text{FE}}|}{N_{\text{exact}}}, \tag{251}$$

where M_{FE} and N_{FE} are the computed mean values along the respective boundaries. The first error is a measure of the kinematic accuracy, while the second is a measure of the kinetic accuracy. Figure 10 shows the two errors for the area-incompressible model of Eq. (142). Looking at the L_2-error, schemes 'A-t,' 'A-st,' 'a-t,' and 'a-st' perform best. In case of error E_{NM}, schemes 'a' and 'a-s' perform best. Class **A** generally converges with μ, but it may not converge with the number of elements for high values of μ. Interestingly, the L_2-error of scheme 'A-t' and 'A-st' is not affected by μ, as schemes 'A' and 'A-s' are. For sufficiently low μ (for $m = 32$ about $\bar{\mu} < 10^{-3}$), the accuracy of class **A** (both in L_2 and E_{MN}) reaches that of class **a** and then only improves with mesh refinement. Class **A** with low μ may even surpass class **a** with high μ. But generally, class **a** is more accurate and robust (as μ does not need to be very small). There is no clear favorite in class **a** for this test case.

As the plots show, not a single stabilization scheme stands out here and the accuracy depends both on the model and the error measure. In general, all schemes are suitable to solve the problem. If class **A** is used, the value of μ needs to be suitably low. For class **a** even large values for μ can be used. In this example it is even possible to set $\mu = 0$ in the code. This works since the effective shear stiffness according to (151) is positive here, i.e., $\mu_{\text{eff}} = 3\,JkH^2/2 > 0$. For other problems μ_{eff} can be negative, and stabilization is required.

Inflation of a sphere The second example considers the inflation of a spherical cell. Contrary to the previous example, the FE mesh now also contains interfaces between NURBS patches. Since the surface area increases during inflation, potential (140) is considered. Figure 11 shows the computational setup of the problem. The computational domain consists of a quarter sphere discretized with four NURBS patches. The quarter sphere contains $3\text{m}^2/2$ elements where m is the number of elements

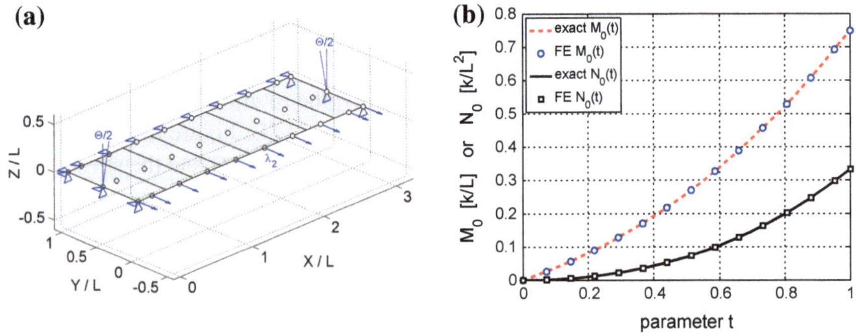

Fig. 9 Pure bending and stretching of a sheet (Sauer et al. 2017): **a** initial FE configuration and boundary conditions (discretized with $m = 8$ elements); **b** distributed boundary moment $M_0(t)$ and normal traction $N_0(t)$ as obtained analytically and computationally for the area-incompressible model (142)

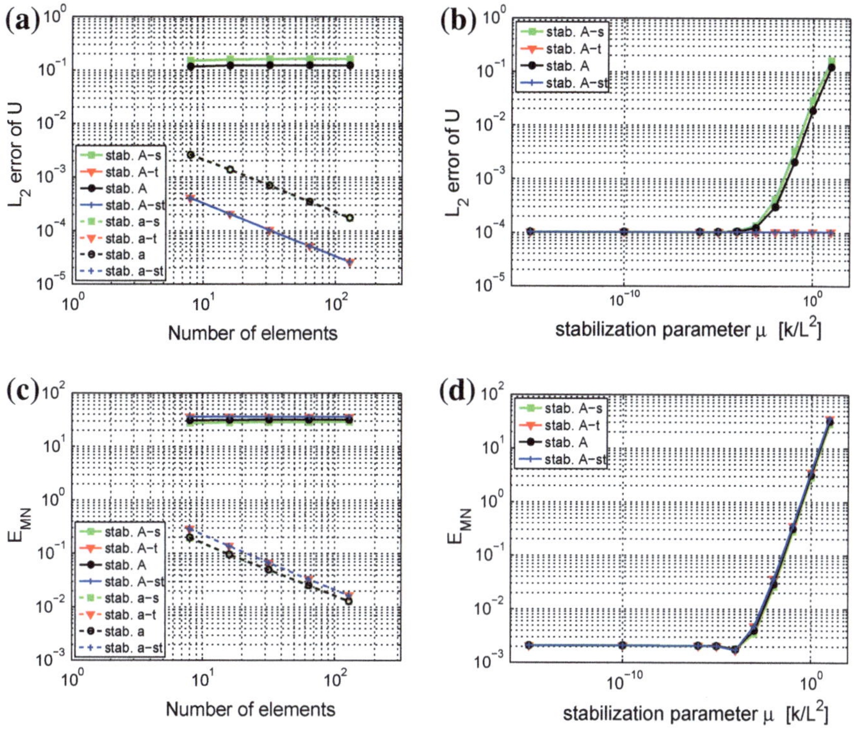

Fig. 10 Pure bending and stretching of a sheet (Sauer et al. 2017): accuracy for the area-incompressible model (142): (**a**) L_2-error versus m considering stabilization classes **A** and **a** with $\bar{\mu} = 10$ and $n_t = 12.5\,m$; (**b**) L_2-error versus μ considering stabilization class **A** with $m = 32$; (**c**)–(**d**) same as (**a**)–(**b**), but now for error E_{MN}. Considered is $\Theta/2 = \pi/3$ and $\lambda_2 = 1.5$

along the equator of the quarter sphere. At the boundaries and at the patch interfaces C^1-continuity is enforced using (228) with $\epsilon = 4000\,mk/R$. The area bulk modulus is taken as $K = 5\,k/R^2$, while k_g is taken as zero. Two cases are considered: $H_0 = 0$ and $H_0 = 1/R$. Figure 12 shows that the computational $p(V)$-data converge to the exact analytical result of (195). Here the pressure error

$$e_p = \frac{|p_{\text{exact}} - p_{\text{FE}}|}{p_{\text{exact}}} \tag{252}$$

is examined for $H_0 = 1/R$ and $\bar{V} = 2$ considering the 9 stabilization schemes of Table 2 with $\bar{\mu} = 0.01$ for class **A** and $\bar{\mu} = 1$ and $n_t = 5\,\text{m}$ for class **a**. For schemes 'A,' 'A-s,' 'A-st,' 'a,' 'a-s,' 'a-st,' and 'P' this error converges nicely (and is indistinguishable in the figure). Only schemes 'A-t' and 'a-t' behave significantly different. They introduce further errors that only converge if μ is decreased or n_t is increased. The reason why all other schemes have equal error, is that here the error is actually determined by the penalty parameter ϵ used within patch constraint (228). The error

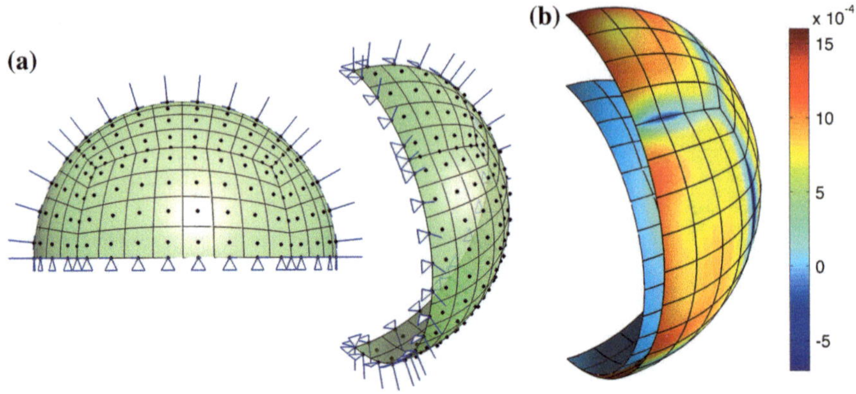

Fig. 11 Sphere inflation (Sauer et al. 2017): **a** initial FE configuration and boundary conditions (for mesh $m = 8$); **b** current FE configuration for an imposed volume of $\bar{V} = 2$ compared to the initial configuration; the colors show the relative error in the surface tension T_ν

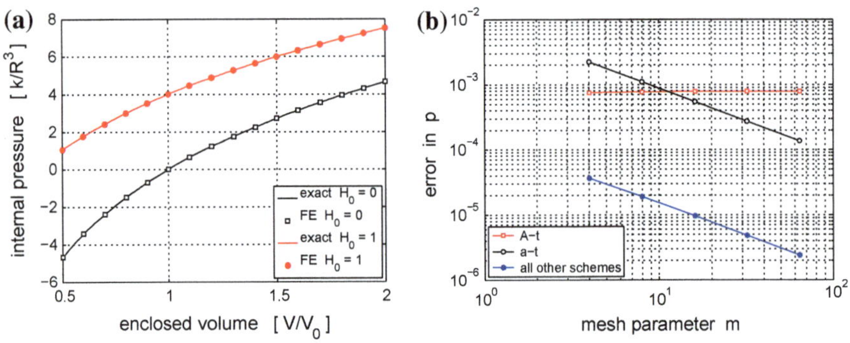

Fig. 12 Sphere inflation (Sauer et al. 2017): **a** pressure-volume relation; **b** FE convergence for the different stabilization schemes

stemming from the stabilization methods (apart from 'A-t' and 'a-t') is insignificant compared to that. It is interesting to note that 'A-st' and 'a-st' perform much better than 'A-t' and 'a-t', even though no shear is present in the analytical solution. 'A-st' and 'a-st' can therefore be considered as the best choices here, since they are the most efficient schemes to implement.

We finally note that for a sphere $\mu_{\text{eff}} = JkH(H - H_0)$, where $H = -1/r$. Thus $\mu_{\text{eff}} > 0$ for $H < H_0$, which is the case here.

13 Numerical Examples

This section presents three numerical examples based on the computational formulation presented in the preceding three sections. The first two examples are taken from Sauer et al. (2017). The third example is new.

13.1 Bilayer Tethering

If a surface point of the bilayer is pulled in normal direction $\boldsymbol{n}$, a thin tether forms, e.g., see Cuvelier et al. (2005). In order to simulate the tether drawing process, the setup of Fig. 13 is considered. The bilayer membrane is modeled as a circular, initially flat disc with initial radius L. The effect of the surrounding membrane is captured by the boundary tension σ (measured w.r.t. the current boundary length). The surface is described by material model (140). L and k are used for normalization. The remaining material parameters are chosen as $k_{\mathrm{g}} = -0.7\,k$ and $K = 20{,}000\,k/L^2$. The cases $\sigma \in \{100,\ 200,\ 400,\ 800\}\,k/L^2$ are considered. Stabilization scheme 'A-s' is used with $\mu = 0.1\,k/L^2$. The bilayer is clamped at the boundary, but free to move in the in-plane direction. The traction $\boldsymbol{t} = \sigma\boldsymbol{\nu}$ is imposed and applied numerically via (217.2). Even though $\boldsymbol{t}$ is constant during deformation, the boundary length $\mathrm{d}s$ appearing in $\mathbf{f}_t^e$ changes and has to be linearized (Sauer 2014). At the center, the displacement u is imposed on the initially flat, circular surface.

Figure 13b also shows one of the chosen finite element discretizations of the initial configuration. Quadratic, NURBS-based, C^1-continuous finite elements are used. A finer discretization is chosen at the center, where the tube is going to form. The chosen NURBS discretization degenerates at the center, such that the C^1-continuity is lost there. It is regained if displacement u is applied not only to the central control point but also to the first ring of control points around the center. This ensures that the tangent plane remains horizontal at the tip. Likewise, a horizontal tangent is enforced at the outer boundary by fixing the height of the outer two rings of control points.

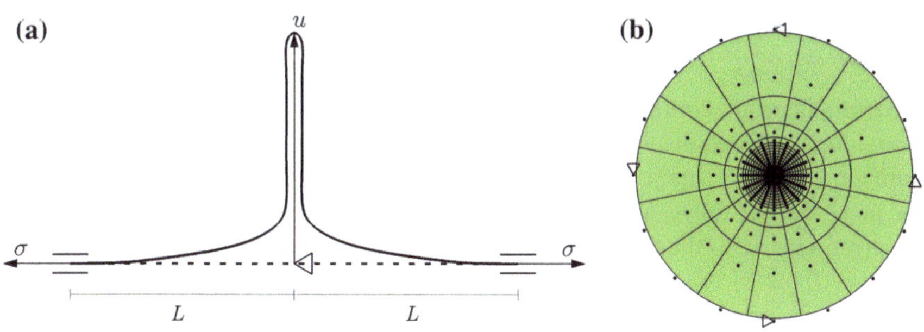

Fig. 13 Bilayer tethering (Sauer et al. 2017): **a** boundary conditions and **b** coarse FE mesh of the initial configuration

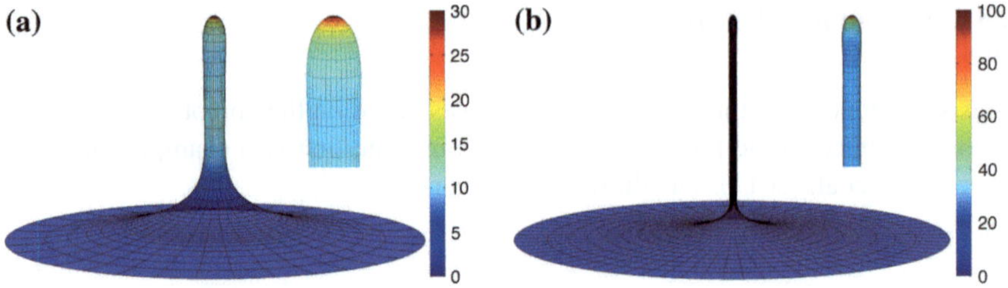

Fig. 14 Bilayer tethering (Sauer et al. 2017): Results for **a** $\sigma = 100\,k/L^2$ and **b** $\sigma = 800\,k/L^2$; the colors show the mean curvature H normalized by L^{-1}

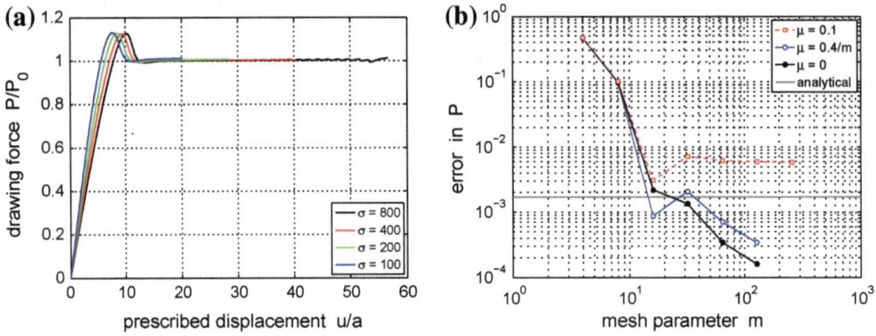

Fig. 15 Bilayer tethering (Sauer et al. 2017): **a** load-displacement curve; **b** FE convergence

Figure 14 shows the deformed surface for $u = L$ with $\bar{\sigma} = 100$ and $\bar{\sigma} = 800$. Further cases are shown in Sauer et al. (2017). The surface tension affects the slenderness of the appearing tube. Derényi et al. (2002) showed from theoretical considerations[7] that the tube radius is

$$a = \frac{1}{2}\sqrt{\frac{k}{\sigma}}\,, \tag{253}$$

while the steady force during tube drawing is

$$P_0 = 2\pi\sqrt{\sigma\,k}\,. \tag{254}$$

These values are confirmed by the computations, as is shown in Fig. 15. The left side shows the force-displacement relation during drawing. Oscillations appear in the numerical solution due to the mesh discretization error. They are more pronounced for more slender tubes, as the black curve in Fig. 15a shows. They disappear upon mesh refinement, as the solution converges. The convergence of P_0 for $\sigma = 200\,k/L^2$ and $u = L$ $(= 28.28a)$ is shown in Fig. 15b by examining the error

[7] Assuming that the tube is sufficiently long and can be idealized by a perfect cylinder.

$$e(P_0^{\mathrm{FE}}) := \frac{|P_0^{\mathrm{ref}} - P_0^{\mathrm{FE}}|}{P_0^{\mathrm{ref}}}, \tag{255}$$

where P_0^{ref} is the FE solution for $m = 256$ and $\mu = 0$. Different values of stability parameter μ are considered. Even the case $\mu = 0$ works, due to the inherent shear stiffness of the Helfrich model given in (151). In all cases, the error reported in Fig. 15b is assessed by comparison to the finest FE mesh. From this one can find that the analytical solution itself has an error of about 0.2%, due to its underlying assumptions.

13.2 Bilayer Budding

The adsorption of proteins can lead to shape change in lipid bilayers (Zimmerberg and Kozlov 2006; McMahon and Gallop 2005; Kozlov et al. 2014; Shi and Baumgart 2015). The lipid membrane deforms whenever its curvature is incompatible with the inherent structure of a protein, giving rise to a spontaneous curvature. In order to study this, a hemi-spherical cell with initial radius R and curvature $H = -1/R$ is considered. The cell surface is clamped at the boundary, but free to expand radially as is shown in Fig. 16. On the top of the cell, within the circular region of radius $0.2R$, a constant spontaneous curvature $\bar{H}_0$ is prescribed in a Lagrangian fashion, such that the proteins causing H_0 move along with the lipid bilayer and no diffusion occurs. Unless otherwise specified, model (140) is used with the material parameters $\bar{k}_g = -0.7$ and $\bar{K} = 10{,}000$, while k and R are used for normalization according to Sect. 11.4 and remain unspecified. Further, stabilization scheme 'A-s' is used with $\bar{\mu} = 0.01$. The FE discretization shown in Fig. 16, consisting of five NURBS patches, is used. Where the patches meet, constraint (228) is added to ensure rotational continuity and moment transfer. Constraint (228) is also used to fix the surface normal at the boundary. The actual FE mesh is much finer than in Fig. 16 and uses 12288 elements (64 times more than in the figure).

In past numerical studies, axisymmetric bud shapes have been reported, e.g., Walani et al. (2015). These shapes should be a natural solution due to the axisymmetry of the problem. However, as is shown below, non-axisymmetric solutions are also possible, and can be energetically favorable, indicating that axisymmetric solutions can become unfavored. This is illustrated by considering the three different test cases listed in Table 3 and discussed in the following:

Case a. (Fig. 17a): Here, the deformation is constrained to remain axisymmetric (i.e., the FE nodes are only allowed to move in radial direction). The resulting deformation at $\bar{H}_0 = -25$ is shown in Fig. 17a.

Case b. (Fig. 17b): Here, the deformation is not constrained to remain axisymmetric. Consequently, a non-axisymmetric bud shape appears. To induce it, H_0 is prescribed within an imperfect circle, i.e., an ellipse with half-axes $a = 0.22R$ and $b = 0.18R$. It is energetically favorable for the bud to evade into an elongated

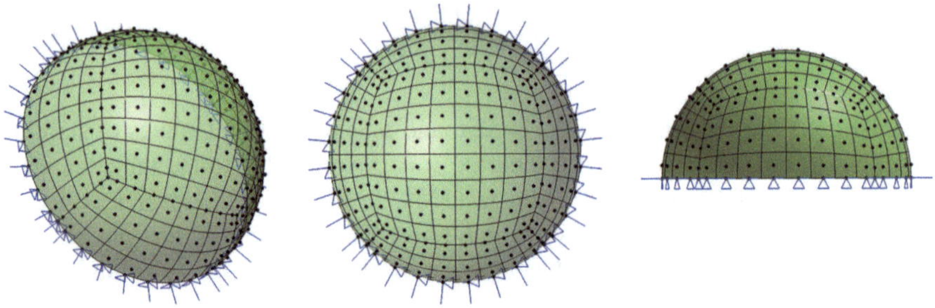

Fig. 16 Bilayer budding (Sauer et al. 2017): Considered setup showing the initial configuration, FE discretization and boundary conditions. The surface normal at the boundary is fixed and the boundary nodes are only free to move in the radial direction

Table 3 Bilayer budding: different physical test cases considered

Case	Bud shape	H_0 region	Stabilization	$\bar{\mu}$	In-plane stress
a	Axisym.	Circle	A-s	0.01	Hydro-static
b	General	Ellipse	A-st	10	Elastic shear
c	General	Ellipse	a-st	1250	Viscous shear

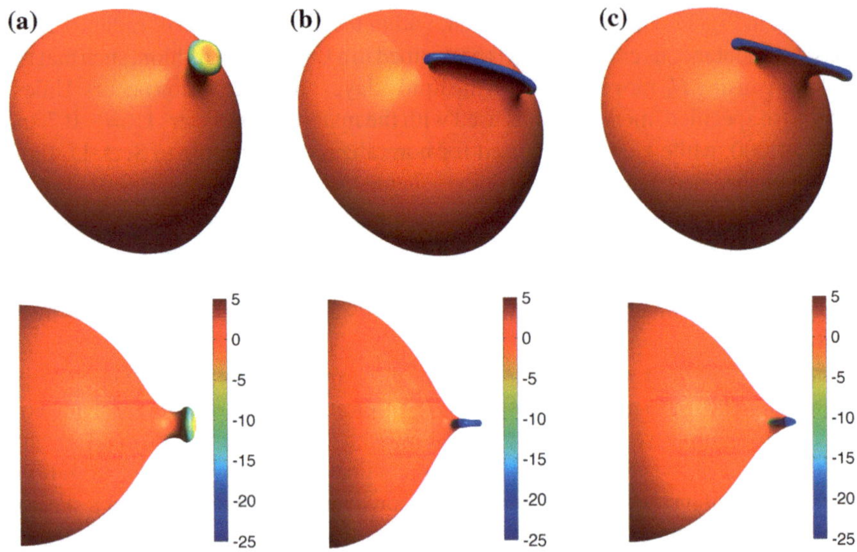

Fig. 17 Bilayer budding (Sauer et al. 2017): **a** axisymmetric case, **b** shear stiff case, and **c** viscous case at $\bar{H}_0 = -25$. The colors show the mean curvature $\bar{H}$

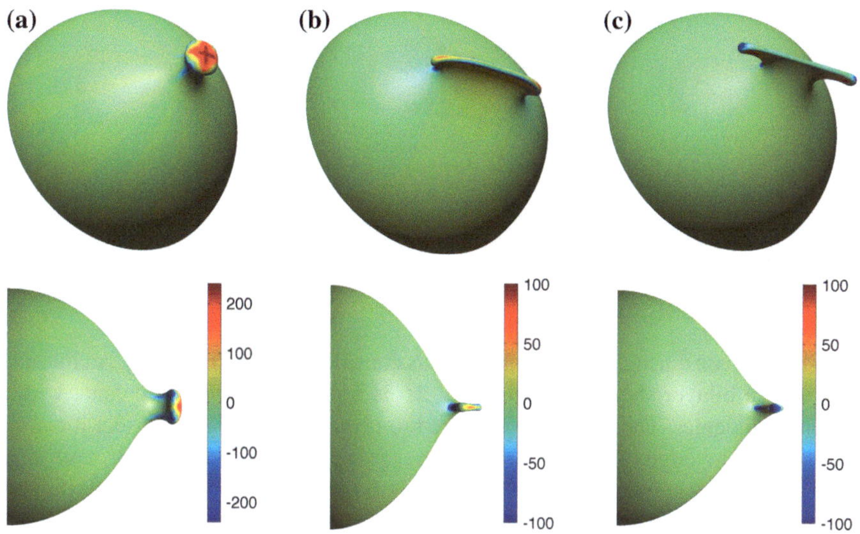

Fig. 18 Bilayer budding (Sauer et al. 2017): **a** axisymmetric case, **b** shear stiff case, and **c** viscous case at $\bar{H}_0 = -25$. The colors show the normalized surface tension $\bar{\gamma}$

plate-like shape (see Fig. 17b). To counter this, shear resistance is provided by elastic shear stresses according to model 'A-st.'[8]

Case c. (Fig. 17c): Here, the deformation is also not constrained to remain axisymmetric. H_0 is again prescribed within an imperfect circle ($a = 0.22R$ and $b = 0.18R$). But now shear resistance is provided through physical viscosity. This is captured through model 'a-st' using the relation $\eta = \mu\,\Delta t$ with $\bar{\mu} = 1250$ and a load stepping increment for H_0 of $\Delta\bar{H}_0 = 0.02$ (such that $\eta = 25\,k/L^3/\dot{H}_0$, where $\dot{H}_0$ is the rate with which the spontaneous curvature is prescribed). As Fig. 17c shows, the bud splits into two separate buds.

In Sauer et al. (2017) movies can be found that animate the bud growth for the three cases. One of the advantages of the proposed finite element formulation is that the surface tension γ can be studied. This is shown in Fig. 18. As seen the surface tension is not a constant. At extreme values of γ, rupture might occur, depending on the strength of the lipid bilayer. Further details on bilayer budding are discussed Sauer et al. (2017).

13.3 Bilayer Indentation

The preceding two examples show that liquid shells, such as lipid bilayers, exhibit out-of-plane deformations that are very different to those observed for solid shells.

[8]The shear stresses are now physical and need to be applied both in-plane and out-of-plane.

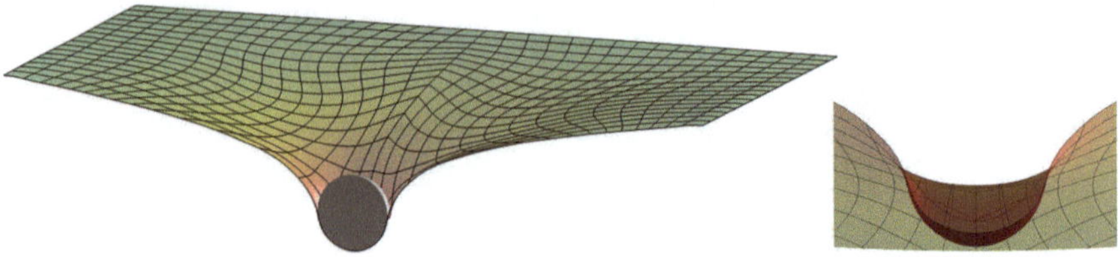

Fig. 19 Indentation of a square bilayer sheet by a spherical indenter. The inset on the right shows that the surface remains smooth during deformation. Only half of the system is shown

This is also seen during indentation, which is considered in the following. An initially flat, square bilayer sheet with size $2L \times 2L$ is brought into normal contact with a spherical indenter. The sheet is considered to be area-extensible such that it can be clamped at the edge. The considered model parameters are $\bar{k}_g = -0.83$ and $\bar{K} = 7.60 \cdot 10^5$, using stabilization scheme 'a-s' with $\bar{\mu} = 12.20$. Contact is described by the penalty method, according to which the contact pressure is given by

$$p_c = \begin{cases} -\epsilon_n\, g_n & \text{if } g_n < 0\,, \\ 0 & \text{else}\,, \end{cases} \tag{256}$$

where ϵ_n is the contact penalty parameter and

$$g_n = (\boldsymbol{x} - \boldsymbol{x}_0) \cdot \boldsymbol{n}_0 - R_0 \tag{257}$$

denotes the normal gap between sphere and bilayer surface. Here, $\boldsymbol{x} \in \mathcal{S}$ denotes a surface point, while R_0, $\boldsymbol{x}_0$, and $\boldsymbol{n}_0$ denote radius, center, and surface normal of the sphere. In the example, $\bar{R}_0 = 0.1$ and $\bar{\epsilon}_n = 7.60 \cdot 10^8$ is used.

Figure 19 shows the deformation of the bilayer sheet for an indentation depth of $0.4L$. The problem is symmetric, and therefore the finite element computations are performed on one quarter of the sheet. Along the symmetry boundaries, constraint (228) is used in order to enforce continuity of $\boldsymbol{n}$. The continuity of $\boldsymbol{n}$ across the symmetry boundary is confirmed by the inset shown on the right.

14 Conclusion

This chapter discusses the computational modeling of lipid bilayers based on thin-shell theory. Various model ingredients are presented in order to address the challenges of this. Those ingredients range from theoretical approaches that provide a general description of balance laws, constitution, kinematics, and weak form based on curvilinear coordinates, to computational methods, such as nonlinear finite element analysis, NURBS-based surface discretizations, LBB-conforming mixed methods, and shear stabilization. A necessary component of this is the consistent linearization of the formulation. The proposed formulation is illustrated by several analytical and numerical examples. The analytical examples are used to examine the behavior of the proposed shear stabilization schemes. The constitutive behavior of the bilayer is based on the Helfrich bending model combined with in-plane viscosity and (near) area-incompressibility. Neglecting inertia leads to a nonlinear PDE that is fourth order in space and first order in time. The corresponding weak form is second order in space and first order in time. This requires globally C^1-continuous surface discretizations. Those are provided by NURBS-based FE shape functions together with rotational constraints at patch interfaces.

The generality of the proposed formulation admits many possible extensions. One is the generalization of the bilayer kinematics to account for tilt and interlayer sliding. Another is the consideration of mass-varying systems (Sahu et al. 2017). Further extension are the development of surface-ALE formulations, and local mesh refinement in the framework of LR-NURBS (Zimmermann and Sauer 2017).

Acknowledgements The author is grateful to the German Research Foundation (DFG) for supporting this research under grants GSC 111 and SA1822/5-1. The author also wishes to thank Kranthi Mandadapu, Thang Duong, and Amaresh Sahu for their valuable comments, and Yannick Omar for his help with the example in Sect. 13.3. Special thanks also go to David Steigmann for organizing the CISM summer school that led to this article.

References

A. Agrawal, D. Steigmann, Modeling protein-mediated morphology in biomembranes. Biomech. Model. Mechanobiol. **8**(5), 371–379 (2009)

M. Arroyo, A. DeSimone, Relaxation dynamics of fluid membranes. Phys. Rev. E **79**, 031915 (2009)

I. Babuška, The finite element method with Lagrangian multipliers. Num. Math. **20**, 179–192 (1973)

E. Baesu, R.E. Rudd, J. Belak, M. McElfresh, Continuum modeling of cell membranes. Int. J. Non-lin. Mech. **39**, 369–377 (2004)

K.-J. Bathe, *Finite Element Procedures* (Prentice-Hall, New Jersey, 1996)

K.-J. Bathe, The inf sup condition and its evaluation for mixed finite element methods. Comput. Struct. **79**, 243–252 (2001)

D.J. Benson, Y. Bazilevs, M.-C. Hsu, T.J.R. Hughes, A large deformation, rotation-free, isogeometric shell. Comput. Methods Appl. Mech. Engrg. **200**(13–16), 1367–1378 (2011)

M.J. Borden, M.A. Scott, J.A. Evans, T.J.R. Hughes, Isogeometric finite element data structures based on Bezier extraction of NURBS. Int. J. Numer. Meth. Engng. **87**, 15–47 (2011)

P.B. Canham, The minimum energy of bending as a possible explanation of the biconcave shape of the human red blood cell. J. Theoret. Biol. **26**, 61–81 (1970)

F. Cirak, M. Ortiz, Fully C^1-conforming subdivision elements for finite element-deformation thin-shell analysis. Int. J. Numer. Meth. Engng **51**, 813–833 (2001)

B.D. Coleman, W. Noll, The thermodynamics of elastic materials with heat conduction and viscosity. Arch. Ration. Mech. Anal. **13**, 167–178 (1964)

C.J. Corbett, *Isogeometric Finite Element Enrichment for Problems Dominated by Surface Effects*. Ph.D. thesis, RWTH Aachen University, Aachen, Germany (2016)

J.A. Cottrell, T.J.R. Hughes, Y. Bazilevs, *Isogeometric Analysis* (Wiley, Chichester, 2009)

D. Cuvelier, I. Derényi, P. Bassereau, P. Nassoy, Coalescence of membrane tethers: experiments, theory, and applications. Biophys. J. **88**, 2714–2726 (2005)

M. Dao, C.T. Lim, S. Suresh, Mechanics of the human red blood cell deformed by optical tweezers. J. Mech. Phys. Solids **51**, 2259–2280 (2003)

I. Derényi, F. Jülicher, J. Prost, Formation and interaction of membrane tubes. Phy. Rev. Lett. **88**(23), 238101 (2002)

Q. Du, X.Q. Wang, Convergence of numerical approximations to a phase field bending elasticity model of membrane deformations. Int. J. Numer. Anal. Model. **4**(3–4), 441–459 (2007)

N.T. Dung, G.N. Wells, Geometrically nonlinear formulation for thin shells without rotation degrees of freedom. Comput. Methods Appl. Mech. Engrg. **197**, 2778–2788 (2008)

T.X. Duong, F. Roohbakhshan, R.A. Sauer, A new rotation-free isogeometric thin shell formulation and a corresponding continuity constraint for patch boundaries. Comput. Methods Appl. Mech. Engrg. **316**, 43–83 (2017)

C.M. Elliott, B. Stinner, Modeling and computation of two phase geometric biomembranes using surface finite elements. J. Comp. Phys. **229**(18), 6585–6612 (2010)

F. Feng, W.S. Klug, Finite element modeling of lipid bilayer membranes. J. Comput. Phys. **220**, 394–408 (2006)

F.G. Flores, C.F. Estrada, A rotation-free thin shell quadrilateral. Comput. Methods Appl. Mech. Engrg. **196**(25–28), 2631–2646 (2007)

R. Gu, X. Wang, M. Gunzburger, Simulating vesicle-substrate adhesion using two phase field functions. J. Comput. Phys. **275**, 626–641 (2014)

W. Helfrich, Elastic properties of lipid bilayers: theory and possible experiments. Z. Naturforsch. **28c**, 693–703 (1973)

T.J.R. Hughes, J.A. Cottrell, Y. Bazilevs, Isogeometric analysis: CAD, finite elements, NURBS, exact geometry and mesh refinement. Comput. Methods Appl. Mech. Engrg. **194**, 4135–4195 (2005)

M. Jarić, U. Seifert, W. Wirtz, M. Wortis, Vesicular instabilities: The prolate-to-oblate transition and other shape instabilities of fluid bilayer membranes. Phys. Rev. E **52**(6), 6623–6634 (1995)

J.T. Jenkins, The equations of mechanical equilibrium of a model membrane. SIAM J. Appl. Math. **32**(4), 755–764 (1977)

Y. Jie, L. Quanhui, L. Jixing, O.-Y. Zhong-Can, Numerical observation of nonaxisymmetric vesicles in fluid membranes. Phys. Rev. E **58**(4), 4730–4736 (1998)

K.A. Johannessen, T. Kvamsdal, T. Dokken, Isogeometric analysis using LRB-splines. Comput. Methods Appl. Mech. Engng. **269**, 471–514 (2014)

O. Kahraman, N. Stoop, M.M. Müller, Fluid membrane vesicles in confinement. New J. Phys. **14**, 095021 (2012)

J. Kiendl, K.-U. Bletzinger, J. Linhard, R. Wüchner, Isogeometric shell analysis with Kirchhoff-Love elements. Comput. Methods Appl. Mech. Engrg. **198**, 3902–3914 (2009)

J. Kiendl, Y. Bazilevs, M.-C. Hsu, R. Wüchner, K.-U. Bletzinger, The bending strip method for isogeometric analysis of Kirchhoff-Love shell structures comprised of multiple patches. Comput. Methods Appl. Mech. Engrg. **199**(37–40), 2403–2416 (2010)

J. Kiendl, M.-C. Hsu, M.C. Wu, A. Reali, Isogeometric Kirchhoff-Love shell formulations for general hyperelastic materials. Comput. Methods Appl. Mech. Engrg. **291**, 280–303 (2015)

T. Kloeppel, W.A. Wall, A novel two-layer, coupled finite element approach for modeling the nonlinear elastic and viscoelastic behavior of human erythrocytes. Biomech. Model. Mechanobiol. **10**(4), 445–459 (2011)

M.M. Kozlov, F. Campelo, N. Liska, L.V. Chernomordik, S.J. Marrink, H.T. McMahon, Mechanisms shaping cell membranes. Curr. Opin. Cell Biol. **29**, 53–60 (2014)

C. Lau, W.E. Brownell, A.A. Spector, Internal forces, tension and energy density in tethered cellular membranes. J. Biomech. **45**(7), 1328–1331 (2012)

H. Li, G. Lykotrafitis, Two-component coarse-grained molecular-dynamics model for the human erythrocyte membrane. Biophys. J. **102**(1), 75–84 (2012)

A. Libai, J.G. Simmonds, *The Nonlinear Theory of Elastic Shells*, 2nd edn. (Cambridge University Press, Cambridge, 1998)

J. Linhard, R. Wüchner, K.-U. Bletzinger, "Upgrading" membranes to shells - The CEG rotation free element and its application in structural anaylsis. Finite Elem. Anal. Des. **44**(1–2), 63–74 (2007)

R. Lipowsky, Spontaneous tabulation of membranes and vesicles reveals membrane tension generated by spontaneous curvature. Faraday Discuss. **161**, 305–331 (2013)

T.V. Loc, T.H. Chien, N.X. Hung, On two-field nurbs-based isogeometric formulation for incompressible media problems. Vietnam J. Mech. **35**, 225–237 (2013)

L. Ma, W.S. Klug, Viscous regularization and r-adaptive meshing for finite element analysis of lipid membrane mechanics. J. Comput. Phys. **227**, 5816–5835 (2008)

H.T. McMahon, J.L. Gallop, Membrane curvature and mechanisms of dynamic cell membrane remodelling. Nature **438**(7068), 590–596 (2005)

P.M. Naghdi, Finite deformation of elastic rods and shells, in *Proceedings of the IUTAM Symposium on Finite Elasticity*, ed. by D.E. Carlson, R.T. Shields (Martinus Nijhoff Publishers, The Hague, 1982), pp. 47–103

N. Nguyen-Thanh, J. Kiendl, H. Nguyen-Xuan, R. Wüchner, K.-U. Bletzinger, Y. Bazilevs, T. Rabczuk, Rotation free isogeometric thin shell analysis using pht-splines. Comput. Methods Appl. Mech. Engrg. **200**(47–48), 3410–3424 (2011)

Z. Peng, R.J. Asaro, Q. Zhu, Multiscale simulation of erythrocyte membranes. Phys. Rev. E **81**, 031904 (2010)

W. Pietraszkiewicz, Geometrically nonlinear theories of thin elastic shells. Adv. Mech. **12**(1), 51–130 (1989)

M. Rahimi, M. Arroyo, Shape dynamics, lipid hydrodynamics, and the complex viscoelasticity of bilayer membranes. Phys. Rev. E **86**, 011932 (2012)

N. Ramakrishnan, P.B.S. Kumar, J.H. Ipsen, Monte carlo simulations of fluid vesicles with in-plane orientational ordering. Phys. Rev. E **81**, 041922 (2010)

P. Rangamani, A. Agrawal, K.K. Mandadapu, G. Oster, D.J. Steigmann, Interaction between surface shape and intra-surface viscous flow on lipid membranes. Biomech. Model. Mechanobiol. **12**(4), 833–845 (2013)

P. Rangamani, K.K. Mandadapu, G. Oster, Protein-induced membrane curvature alters local membrane tension. Biophys. J. **107**(3), 751–762 (2014)

R. Rangarajan, H. Gao, A finite element method to compute three-dimensional equilibrium configurations of fluid membranes: Optimal parameterization, variational formulation and applications. J. Comput. Phys. **297**, 266–294 (2015)

J.E. Rim, P.K. Purohit, W.S. Klug, Mechanical collapse of confined fluid membrane vesicles. Biomech. Model. Mechanobio. **13**(6), 1277–1288 (2014)

A. Rosolen, C. Peco, M. Arroyo, An adaptive meshfree method for phase-field models of biomembranes. Part I: approximation with maximum-entropy basis functions. J. Comput. Phys. **249**, 303–319 (2013)

A. Sahu, R.A. Sauer, K.K. Mandadapu, The irreversible thermodynamics of curved lipid membranes (2017), arXiv:1701.06495

D. Salac, M. Miksis, A level set projection model of lipid vesicles in general flows. J. Comput. Phys. **230**(22), 8192–8215 (2011)

R.A. Sauer, Stabilized finite element formulations for liquid membranes and their application to droplet contact. Int. J. Numer. Meth. Fluids **75**(7), 519–545 (2014)

R.A. Sauer, L. De Lorenzis, A computational contact formulation based on surface potentials. Comput. Methods Appl. Mech. Engrg. **253**, 369–395 (2013)

R.A. Sauer, L. De Lorenzis, An unbiased computational contact formulation for 3D friction. Int. J. Numer. Meth. Engrg. **101**(4), 251–280 (2015)

R.A. Sauer, T.X. Duong, On the theoretical foundations of solid and liquid shells. Math. Mech. Solids. **22**(3), 343–371 (2017)

R.A. Sauer, T.X. Duong, C.J. Corbett, A computational formulation for constrained solid and liquid membranes considering isogeometric finite elements. Comput. Methods Appl. Mech. Engrg. **271**, 48–68 (2014)

R.A. Sauer, T.X. Duong, K.K. Mandadapu, D.J. Steigmann, A stabilized finite element formulation for liquid shells and its application to lipid bilayers. J. Comput. Phys. **330**, 436–466 (2017)

M.A. Scott, M.J. Borden, C.V. Verhoosel, T.W. Sederberg, T.J.R. Hughes, Isogeometric finite element data structures based on Bézier extraction of T-splines. Int. J. Numer. Meth. Engng. **88**(2), 126–156 (2011)

Z. Shi, T. Baumgart, Membrane tension and peripheral protein density mediate membrane shape transitions. Nat. commun. **6**, 5974 (2015)

D. Steigmann, E. Baesu, R.E. Rudd, J. Belak, M. McElfresh, On the variational theory of cell-membrane equilibria. Interfaces Free Bound. **5**, 357–366 (2003)

D.J. Steigmann, Fluid films with curvature elasticity. Arch. Rat. Mech. Anal. **150**, 127–152 (1999)

I.V. Tasso, G.C. Buscaglia, A finite element method for viscous membranes. Comput. Methods Appl. Mech. Engrg. **255**, 226–237 (2013)

N. Walani, J. Torres, A. Agrawal, Endocytic proteins drive vesicle growth via instability in high membrane tension environment. Proc. Natl. Acad. Sci. **112**(12), E1423–E1432 (2015)

J. Zimmerberg, M.M. Kozlov, How proteins produce cellular membrane curvature. Nat. Rev. Mol. Cell Biol. **7**(1), 9–19 (2006)

C. Zimmermann, R.A. Sauer, Adaptive local surface refinement based on LR-NURBS and its application to contact (2017), arXiv:1701.08742

Onsager's Variational Principle in Soft Matter: Introduction and Application to the Dynamics of Adsorption of Proteins onto Fluid Membranes

Marino Arroyo, Nikhil Walani, Alejandro Torres-Sánchez and Dimitri Kaurin

1 Introduction

Lipid bilayers are unique soft materials operating in general in the low Reynolds limit. While their shape is predominantly dominated by curvature elasticity as in a solid shell, their in-plane behavior is that of a largely inextensible viscous fluid. These two behaviors, however, are tightly coupled through the membrane geometry. Indeed, shape transformations necessarily induce lipid flows that bring material from one part of the membrane to another (Evans and Yeung 1994). On the other hand, fluid flows in the presence of curvature generate out-of-plane forces, which modify the shape of the membrane and elicit elastic forces (Rahimi et al. 2013). This mechanical duality provides structural stability and adaptability, allowing membranes to build relatively stable structures that can nevertheless undergo dynamic shape transformations. These transformations are critical for the cell function; they are required in vesicular and cellular trafficking (Sprong et al. 2001; Rustom et al. 2004), cell motility and migration (Arroyo et al. 2012; Yamaguchi et al. 2015), or in the mechano-adaptation of cells to stretch and osmotic stress (Kosmalska et al. 2015).

In addition to this solid–fluid duality, lipid membranes are extremely responsive to chemical stimuli. They transiently respond, for instance, to pH gradients by developing tubules and pearled protrusions (Khalifat et al. 2014, 2008; Fournier et al. 2009). Furthermore, a myriad of proteins interact with lipid bilayers through curvature, either to generate it or to sense it (McMahon and Gallop 2005; Zimmerberg and Kozlov 2006; Sens et al. 2008; Shibata et al. 2009; Antonny 2011). A number of quantitative experiments on synthetic reconstituted systems have examined this interaction, notably using tethers pulled out of vesicles and exposed to curvature-active

M. Arroyo (✉) · N. Walani · A. Torres-Sánchez · D. Kaurin
Universitat Politècnica de Catalunya - BarcelonaTech,
Campus Nord, edifici C2 Jordi Girona 1-3, 08034 Barcelona, Spain
e-mail: Marino.arroyo@upc.edu

D.J. Steigmann (ed.), *The Role of Mechanics in the Study of Lipid Bilayers*,
CISM International Centre for Mechanical Sciences 577,
DOI 10.1007/978-3-319-56348-0_6

proteins delivered from either the bulk solution or a membrane reservoir (Sorre et al. 2009; Heinrich et al. 2010a, b; Sorre et al. 2012). More recently, the interplay between membrane tension and curvature generation by adsorbed curving proteins has been examined, with implications in cell mechanosensing and mechano-adaptation (Sinha et al. 2011; Shi and Baumgart 2015).

While there is a very large body of theoretical and computational literature covering different aspects of bilayer mechanics, current models and simulation techniques fail to capture the dynamical and chemically responsive nature of bilayer membranes. We highlight below some of the requirements of a sufficiently general modeling framework that can quantify and predict the behavior of lipid bilayer membranes:

Capture the out-of-equilibrium response. Indeed, bilayers are highly dynamical, but due to the complexity of the chemical and hydrodynamical effects involved, theory and experiments have focused on equilibrium. For instance, the classical bending model of Helfrich (Helfrich 1973; Lipowsky 1991; Jülicher and Lipowsky 1993; Staykova et al. 2013) has been very successful in understanding equilibrium conformations (Steigmann 1999; Capovilla and Guven 2002; Tu and Ou-Yang 2004; Feng and Klug 2006; Rangarajan and Gao 2015; Sauer et al. 2017), but is insufficient to understand the reconfigurations of membranes when subjected to transient stimuli. To address this challenge, models and simulations coupling membrane hydrodynamics and elasticity (Arroyo and DeSimone 2009; Arroyo et al. 2010; Rahimi and Arroyo 2012; Rahimi et al. 2013; Rangamani et al. 2013; Rodrigues et al. 2013; Barrett et al. 2016) or elasticity and the phase-separation of chemical species (Embar et al. 2013; Elliott and Stinner 2013) are emerging in recent years, but only provide initial steps toward a general dynamical framework.

Capture the bilayer architecture. The classical Helfrich model treats bilayer membranes as simple surfaces. Subsequent refinements in equilibrium, such as the Area difference elasticity (ADE) model (Seifert 1997), acknowledge the bilayer architecture, by which bending compresses one monolayer and stretches the other but, since monolayers can slip relative to each other, this mechanism of elastic energy storage can be released to a certain degree in a nonlocal manner. In real biological membranes, however, the ADE effect is thought to play a minor role because of fast cholesterol translocation between monolayers. Beyond equilibrium, the work of Seifert and Langer (1993) and Evans and Yeung (1994) demonstrated that the bilayer architecture is crucial to understand the dynamics of lipid membranes. In particular, these works highlighted the role of inter-monolayer friction as a "hidden" but significant dissipative effect. When it comes to the interaction of bilayers with proteins, the bilayer architecture is bound to play an important role since proteins can merely scaffold the membrane, shallowly insert into one monolayer, or pierce through the entire bilayer. Elasto-hydrodynamical models capturing the bilayer architecture have been developed under the assumption of linearized perturbations (Seifert and Langer 1993; Fournier et al. 2009; Callan-Jones et al. 2016), or in a fully nonlinear albeit axisymmetric setting (Rahimi and Arroyo 2012). In this chapter, we will not focus on this aspect.

Capture mechanical and chemical nonlinearity. Nonlinearity is essential to understand many soft matter systems such as lipid membranes. On the mechanics side, these systems experience very large deformations that elicit geometric nonlinearity. On the chemical side, bilayers exhibit nonlinear chemical effects as a result of molecular crowding, such as nonlinear adsorption (Sorre et al. 2012) or nonlinear sorting of proteins between vesicles and tubules (Zhu et al. 2012; Aimon et al. 2014; Prévost et al. 2015). Thus, linearized chemo-mechanical models can only provide information about the onset of transitions (Shi and Baumgart 2015), about dilute concentrations of protein on the surface (Góźdź 2011), or about the response under very small perturbations (Callan-Jones et al. 2016).

Consistently treat multiple physics. As argued above, the function of lipid bilayers is mediated by the tight interplay between elasticity, hydrodynamics, molecular diffusion, and chemical reactions. All these phenomena act in concert and depend on each other. For instance, the concentration of adsorbed proteins modifies the preferred curvature of the membrane and conversely curvature modulates the adsorption reaction. Application of a force may change the shape, inducing lipid flows that advect proteins, drive diffusion, and feedback into shape. Although several models have coupled mechanical and chemical phenomena resulting from the interaction between membranes and proteins, these focus on equilibrium (Zhu et al. 2012; Singh et al. 2012; Lipowsky 2013), on the linearized setting (Callan-Jones et al. 2016), or consider simplistic models for the mechanical relaxation dynamics (Liu et al. 2009).

In this chapter, we do not attempt to address all these requirements. Instead, we will focus on the last point. More specifically, our main objective is to introduce an emerging variational modeling framework for the dissipative dynamics of soft matter and biological systems, which provides a systematic and transparent approach to generate complex models coupling multiple physics. This approach is founded on Onsager's variational principle, by which the dynamics result from the interplay between energetic driving forces and dissipative drag forces, each of them deriving from potentials that are the sum of individual contributions for each physical mechanism. Models coupling different physics can be assembled by just adding more terms to the energy and dissipation potentials, and encoding in them the interactions between the different physical mechanisms. In this way, this framework provides a flexible and thermodynamically consistent method to generate complex models. The goal of this chapter is to convey Onsager's variational principle through examples. Some of these examples are directly relevant to bilayer mechanics. In the second part of this chapter, we emphasize models relevant to the adsorption of proteins on membranes. We avoid, however, the general formulation of a complete model for bilayers coupling elasticity, bulk and interfacial hydrodynamics, bulk and interfacial diffusion, and adsorption in a deformable membrane, which requires many pages and extensive use of differential geometry, but does not significantly contribute to our goal here.

While going through this chapter, some readers may find some of the material close to trivial or irrelevant. We encourage them to read further because the beauty and importance of Onsager's principle manifests itself when confronted with systems involving multiple physics. We also note that much of the material presented here is standard textbook material or can be found scattered in the recent and not so recent literature. The unified view of Onsager's principle as a powerful and general approach to model nonlinear dissipative systems, however, is an emerging idea. Furthermore, we present some original applications of this principle to model nonlinear adsorption phenomena, and curvature sensing and generation by proteins. The interested reader may find recent applications of Onsager's principle to lipid membranes elsewhere (Arroyo and DeSimone 2009; Rahimi and Arroyo 2012; Rahimi et al. 2013; Fournier 2015; Callan-Jones et al. 2016).

This chapter is organized as follows. In Sect. 2, we introduce Onsager's variational principle by way of elementary mechanical models. We also revisit common models such as Stokes incompressible flow, linear diffusion, diffusion coupled with hydrodynamics in the presence of a rigid semipermeable membrane, or a linear reaction–diffusion system involving two species. See Peletier (2014) for a related pedagogical work. By deriving all these problems using Onsager's principle, we frame them into a unified framework and provide the elements to build more complex models. In Sect. 3, we focus on the adsorption and diffusion of a chemical species from the bulk onto a surface of fixed shape. This allows us to identify the variational structure behind common linear and nonlinear adsorption models including Langmuir model. Finally, in Sect. 4, we provide a minimal model to examine curvature sensing and generation, by introducing a coupling between protein concentration and spontaneous curvature.

2 Onsager's Variational Principle

2.1 Background

Variational principles underly many mechanical and thermodynamic theories. These principles provide a systematic procedure to generate governing equations, and provide an additional mathematical structure that highlights qualitative properties of the solutions not apparent from the Euler–Lagrange equations. For instance, the principle of minimum potential energy provides information about the stability of equilibria, not accessible from the mere equilibrium equations. Hamilton's principle for the inertial mechanics of particles and continua characterizes variationally trajectories otherwise satisfying "$F = mA$." This variational principle provides a natural framework to Noether's theorem and to derive variational time integrators (Lew et al. 2004).

Toward an analogous framework to model soft matter and biological systems, we introduce here Onsager's variational principle (Onsager 1931a, b), in a terminology introduced by Doi (2011). This variational framework describes the dynamics of dissipative systems and is an extension of the principle of least energy dissipation, first introduced by Rayleigh (1873) (Goldstein 1980). Onsager's relations are generally invoked in the context of linear irreversible thermodynamics (Prigogine 1967). As argued earlier, however, nonlinearity is essential in many soft matter systems. Importantly, as noted by Doi (2011), Onsager's relations emerge from a more general variational principle applicable in fully nonlinear settings. This fact was exploited to derive the geometrically nonlinear equations for an inextensible interfacial fluid with bending rigidity coupled to a bulk viscous fluid (Arroyo and DeSimone 2009), or to derive the governing equations for a phase-field model of membranes coupled to a viscous bulk fluid (Peco et al. 2013). This formalism assumes that inertial forces are negligible (see Öttinger (2005) for an extension), but otherwise encompasses the classes of problems encountered in soft matter and biological physics, tightly coupling chemistry, hydrodynamics, and nonlinear solid mechanics.

Besides soft matter physics, variational principles of the Onsager type were introduced in solid mechanics, in particular invoking time-incremental discretized principles to generate algorithms (Ortiz and Repetto 1999) or to develop mathematical analysis (Mielke 2011a, b, 2012). Along similar lines, Jordan et al. (1998); Otto (2001) identified a variational formulation for diffusion equations as gradient flows of entropy functionals, providing mathematical and physical insight and highlighting the importance of adequately parametrizing the *processes* that modify the state of the system. This led to a further formalization of Onsager's variational principle by Peletier (2014) introducing the so-called process operators, which were independently used by Doi (2011) to model viscoelastic fluids and by Rahimi and Arroyo (2012) to derive the equations of a nonlinear dynamical model for lipid bilayers. More recently, the gradient structure of reaction–diffusion systems has been identified (Mielke 2011a), allowing us to couple such problems with other phenomena through Onsager's principle.

We introduce next Onsager's principle through simple mechanical and chemical models. Our goal is to emphasize that this principle provides a systematic way to derive the governing equations for complex systems starting from elementary energetic and dissipative ingredients, which act as building blocks of the theory. Here, we do not address an additional important benefit of Onsager's principle: the fact that it provides a privileged starting point for time and space discretization of the resulting systems of partial differential equations.

2.2 *Onsager's Principle for Elementary Systems*

We consider a spring of elastic constant k coupled in parallel with a dashpot of drag coefficient η and under the action of a force F (see Fig. 1a). It may seem an overkill to invoke Onsager's principle to describe such an elementary model. However, we shall see that the treatment of more complex systems follows the same rationale, and therefore this and subsequent examples provide a simple physical picture to understand the essential ideas.

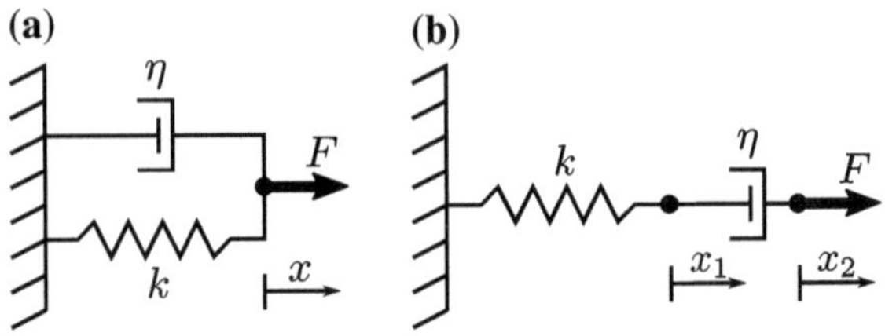

Fig. 1 Diagrams of two elementary mechanical systems. **a** A spring with constant k is in parallel with a dashpot with drag coefficient η and a force F is applied. The system is characterized by the displacement of the point of application of the force from its equilibrium position, x. **b** The spring is now in series with the dashpot and the force is applied to the dashpot; the system in this case is characterized by x_1, the displacement of the spring relative to its equilibrium position, and x_2, the relative displacement of the dashpot with respect to the spring

The state of the system is characterized by the displacement of the spring with respect to its natural elongation, x. The force generated by the spring is

$$F_{\text{cons}} = -kx, \tag{1}$$

where the label "cons" identifies that this force is conservative. The system is also experiencing a viscous force opposing its motion

$$F_{\text{visc}} = -\mu v, \tag{2}$$

where $v = \dot{x}$. If the drag is sufficiently large, inertia can be neglected. Then, balance of forces reads

$$F_{\text{cons}} + F_{\text{visc}} + F = 0, \tag{3}$$

leading to

$$\eta\dot{x} + kx = F. \tag{4}$$

This is an ordinary differential equation that can be easily integrated in time to obtain $x(t)$ given an initial condition. But let us focus on the structure of this equation rather than on its solution; this equation follows from a variational principle. Indeed, on the one hand, the spring and external forces derive from a potential, which includes the stored elastic energy in the spring and the potential for the external force

$$F_{\text{cons}} + F = -\frac{d\mathcal{F}}{dx} \quad \text{where} \quad \mathcal{F}(x) = \frac{k}{2}x^2 - Fx. \tag{5}$$

On the other hand, the viscous force also derives from a potential, usually referred to as the dissipation potential or as the Rayleigh dissipation function, depending on v

$$F_{\text{visc}} = -\frac{\partial \mathcal{D}}{\partial v} \quad \text{where} \quad \mathcal{D}(v) = \frac{\eta}{2} v^2. \tag{6}$$

The rate of change of the energy can be written, using the chain rule, as

$$\frac{d}{dt}\left[\mathcal{F}(x(t))\right] = \frac{d\mathcal{F}}{dx}(x(t))\, \dot{x}(t) = (kx - F)\, v, \tag{7}$$

and therefore $\dot{\mathcal{F}}$ depends on the state of the system x and on the rate of change of the state v. Now, let us define the function

$$\mathcal{R}(x, v) = \dot{\mathcal{F}}(x, v) + \mathcal{D}(v) = (kx - F)\, v + \frac{1}{2}\eta v^2. \tag{8}$$

It is clear that the governing equation for this system (4) follows from $0 = \partial \mathcal{R}/\partial v$. Furthermore, because $\eta > 0$, $\mathcal{R}$ is a convex function of v. Thus, we conclude that the governing equation follows from the variational principle

$$v = \underset{w}{\text{argmin}}\, \mathcal{R}(x, w). \tag{9}$$

This is Onsager's variational principle and the function $\mathcal{R}(x, v)$ is called the *Rayleighian* of the system. The minimization is performed over the rate of change of the state of the system, v, rather than on the state of the system, x, in contrast with the classical equilibrium principle of minimum potential energy. *This is a genuinely dynamical principle establishing a competition between the energy release rate and dissipation* (Onsager 1931a; Doi 2011). Focusing on linear response theory, Onsager showed that this principle holds for general irreversible processes, where *the key assumptions are that (i) dissipation dominates over inertia and (ii) viscous forces are derived from a dissipation potential.* This principle, however, is still valid if $\mathcal{F}$ or $\mathcal{D}$ are general nonharmonic potentials for the spring or for the dashpot.

Before showing the application of Onsager's variational principle to continuous systems, we consider another discrete example consisting of a spring in series with a dashpot loaded with a force (see Fig. 1b). The system is characterized by the displacement of the spring from its equilibrium position, x_1, and by the displacement of the dashpot with respect to the spring, x_2. We denote the rate of change of these coordinates by $v_i = dx_i/dt$. Let us proceed directly following Onsager's variational principle. The energy of this system is just the energy stored by the spring and the potential energy of the load, whose application point is displaced by $x_1 + x_2$

$$\mathcal{F}(x_1, x_2) = \frac{k}{2} x_1^2 - F(x_1 + x_2). \tag{10}$$

The rate of change of the energy is

$$\dot{\mathcal{F}}(x_1; v_1, v_2) = k x_1 v_1 - F(v_1 + v_2), \tag{11}$$

which here happens not to depend on x_2. On the other hand, the dissipation potential can be written in terms of v_2 only

$$\mathcal{D}(v_2) = \frac{\eta}{2} v_2^2. \tag{12}$$

Thus, the Rayleighian is

$$\mathcal{R}(x_1; v_1, v_2) = k x_1 v_1 - F(v_1 + v_2) + \frac{\eta}{2} v_2^2 \tag{13}$$

and Onsager's variational principle states that

$$v_1, v_2 = \underset{w_1, w_2}{\operatorname{argmin}}\ \mathcal{R}(x_1; w_1, w_2). \tag{14}$$

The stationarity necessary conditions for the minimizer, $0 = \partial \mathcal{R}/\partial v_i$, lead to

$$F = k x_1 = \eta v_2, \tag{15}$$

which coincides with the result obtained from direct force balance for this system.

We end this Section with a variation of the model in Fig. 1b, in which we have two dashpots in series with constants η_1 and η_2. This example is intended to make a more subtle point and may be skipped in a first reading. In this case, $\mathcal{F} = -F(x_1 + x_2)$ and we simply add the dissipation potentials of each of the dashpots to form

$$\mathcal{D} = \frac{\eta_1}{2} v_1^2 + \frac{\eta_2}{2} v_2^2. \tag{16}$$

Applying Onsager's principle, we immediately find $v_i = F/\eta_i$. Now, let us define the total displacement of the right end of the system $x = x_1 + x_2$ and its velocity $v = v_1 + v_2$. A simple manipulation shows that the following relation holds

$$v = \frac{\eta_1 + \eta_2}{\eta_1 \eta_2} F, \tag{17}$$

and therefore, the viscous force of the composite system consisting of two dashpots in series derives from the following dissipation potential

$$\hat{\mathcal{D}}(v) = \frac{1}{2} \frac{\eta_1 \eta_2}{\eta_1 + \eta_2} v^2. \tag{18}$$

This potential exhibits a nonadditive structure in that the effective drag coefficient is not the sum of the individual drag coefficients. In contrast, the corresponding dual dissipation potential obtained through a Legendre transform

$$\hat{\mathcal{D}}^*(F) = \min_v \left[Fv - \hat{\mathcal{D}}(v) \right] = \frac{1}{2} \left(\frac{1}{\eta_1} + \frac{1}{\eta_2} \right) F^2 \tag{19}$$

does exhibit an additive structure for this model. This point has been raised in a different but related context (Mielke 2012) to favor a formalism relying on the dual dissipation potential. This simple example shows, however, that the dissipation potential retains an additive structure, see Eq. (16), provided sufficient detail is kept in the formulation describing the rate of change of the system and how these changes dissipate energy.

2.3 *Incompressible Stokes Flow*

We formulate next the governing equations for a Newtonian incompressible fluid in the low Reynolds limit, the familiar Stokes equations, within the framework of Onsager's principle. This formalization will be useful later, when coupling low Re hydrodynamics with different physics. We consider a fluid in a fixed volume Ω with boundary $\partial\Omega$. The motion of material particles in the fluid is characterized by a velocity field $\boldsymbol{v}(\boldsymbol{x})$. The field $\boldsymbol{v}(\boldsymbol{x})$ is the continuous equivalent to v in the previous example. The dissipation potential characterizes the energy dissipated as the fluid deforms. For an incompressible Newtonian fluid, it takes the form

$$\mathcal{D}[\boldsymbol{v}] = \eta \int_\Omega \boldsymbol{d} : \boldsymbol{d} \, dV, \tag{20}$$

where $\boldsymbol{d}$ is the rate-of-deformation tensor $\boldsymbol{d} = \frac{1}{2}\left(\nabla \boldsymbol{v} + (\nabla \boldsymbol{v})^T\right)$ and η is the shear viscosity of the fluid. Recalling that the viscous shear stress is $2\eta \boldsymbol{d}$ for a Newtonian fluid, then $\eta\, \boldsymbol{d} : \boldsymbol{d}$ can be identified as half of the rate of dissipation per unit volume.

We split $\partial\Omega = \Gamma_D \cup \Gamma_N$ into two disjoint subdomains, the Dirichlet boundary Γ_D, where a velocity field is prescribed $\boldsymbol{v}(x) = \bar{\boldsymbol{v}}(\boldsymbol{x})$, and the Neumann boundary Γ_N, where a traction $\boldsymbol{t}(\boldsymbol{x})$ is applied. The traction at the Neumann boundary is supplying power to the system. This power supply can be introduced through a potential of the form

$$\mathcal{P}[\boldsymbol{v}] = -\int_{\Gamma_N} \boldsymbol{t} \cdot \boldsymbol{v} dS. \tag{21}$$

In this problem there is no energetic ingredient, and therefore, the system is oblivious to any variable encoding the state system. Thus the Rayleghian accounting for internal dissipation and power supply through boundary traction is simply $\mathcal{R}[\boldsymbol{v}] = \mathcal{D}[\boldsymbol{v}] + \mathcal{P}[\boldsymbol{v}]$.

Onsager's principle states that the system evolves in such a way that the Rayleghian is minimized with respect to $\boldsymbol{v}$. However, it is important to realize that the velocity field is subjected to constraints. On the one hand, it should satisfy the Dirichlet boundary conditions. On the other hand, since the fluid is incompressible, it should satisfy

$\nabla \cdot \boldsymbol{v} = 0$ in Ω. The variational principle allows us to easily incorporate constraints, for instance using a field of Lagrange multipliers and forming the Lagrangian

$$\begin{aligned} \mathcal{L}[\boldsymbol{v}, p] &= \mathcal{R}[\boldsymbol{v}] - \int_\Omega p \nabla \cdot \boldsymbol{v} \, dV \\ &= \eta \int_\Omega \boldsymbol{d} : \boldsymbol{d} \, dV - \int_{\Gamma_N} \boldsymbol{t} \cdot \boldsymbol{v} dS - \int_\Omega p \nabla \cdot \boldsymbol{v} \, dV, \end{aligned} \tag{22}$$

where p can be interpreted as the pressure in the fluid. Now, the constrained Onsager's principle can be stated as a saddle problem

$$\boldsymbol{v}, p = \underset{q}{\operatorname{argmax}} \; \underset{\boldsymbol{w}}{\operatorname{argmin}} \left\{ \mathcal{L}[\boldsymbol{w}, q] \right\}. \tag{23}$$

The variation of the Lagrangian with respect to the velocity field along $\delta \boldsymbol{v}$ consistent with Dirichlet boundary conditions, i.e., $\delta \boldsymbol{v} = \boldsymbol{0}$ at Γ_D, leads to the stationarity condition

$$2\eta \int_\Omega \boldsymbol{d} : \nabla \delta \boldsymbol{v} \, dV - \int_{\Gamma_N} \boldsymbol{t} \cdot \delta \boldsymbol{v} \, dS - \int_\Omega p \nabla \cdot \delta \boldsymbol{v} \, dV = 0. \tag{24}$$

Variations with respect to p lead to

$$\int_\Omega \delta p \nabla \cdot \boldsymbol{v} \, dV = 0. \tag{25}$$

Eqs. (24) and (25) are the weak form of the problem. The strong form follows after integration by parts and taking into account the arbitrariness in $\delta \boldsymbol{v}$ and δp,

$$\begin{aligned} \nabla \cdot \boldsymbol{\sigma} &= \boldsymbol{0} && \text{in } \Omega, \\ \nabla \cdot \boldsymbol{v} &= 0 && \text{in } \Omega, \\ \boldsymbol{v} &= \bar{\boldsymbol{v}} && \text{on } \Gamma_D, \\ \boldsymbol{\sigma} \cdot \boldsymbol{n} &= \boldsymbol{t} && \text{on } \Gamma_N, \end{aligned} \tag{26}$$

where $\boldsymbol{\sigma} = 2\eta \boldsymbol{d} - p\boldsymbol{I}$ is the stress tensor of the fluid, $\boldsymbol{I}$ is the identity tensor, and $\boldsymbol{n}$ is the unit outward normal to $\partial\Omega$. Thus, the equations characterizing an incompressible Newtonian fluid in the low Re limit can be obtained from Onsager's variational principle. This example also illustrates the treatment of constraints in this formalism.

Note that by replacing $\boldsymbol{v}$ by a displacement field $\boldsymbol{u}$ (now a state variable), $\boldsymbol{d}$ by the linearized strain tensor $\boldsymbol{\varepsilon} = (\nabla \boldsymbol{u} + \nabla \boldsymbol{u}^T)/2$, and η by the shear modulus μ, these equations are those of linear isotropic elasticity for an incompressible material. These equations also follow from Onsager's principle, starting from the free energy

$$\mathcal{F}[\boldsymbol{u}] = \mu \int_\Omega \boldsymbol{\varepsilon} : \boldsymbol{\varepsilon} \, dV - \int_{\Gamma_N} \boldsymbol{t} \cdot \boldsymbol{u} dS. \tag{27}$$

Since we do not have a dissipation source, and noting that $\boldsymbol{v} = \partial_t \boldsymbol{u}$, the constrained Rayleighian becomes

$$\mathcal{L}[\boldsymbol{u}; \boldsymbol{v}, p] = 2\mu \int_{\Omega} \boldsymbol{\varepsilon} : \nabla \boldsymbol{v} \, dV - \int_{\Gamma_N} \boldsymbol{t} \cdot \boldsymbol{v} dS - \int_{\Omega} p \nabla \cdot \boldsymbol{v} \, dV. \tag{28}$$

2.4 *Diffusion of a Solute in a Fluid*

Further building our catalog of models amenable to Onsager's principle, we consider now the diffusion equation. Let Ω be a region of space occupied by a quiescent fluid with a dilute distribution of noninteracting and neutrally buoyant solute molecules. This region is delimited by an impermeable container. We denote by $c(\boldsymbol{x}, t)$ the molar concentration field of this substance at time t. A classical model to describe the time evolution of this field is based on the diffusion equation, $\partial_t c = D \Delta c$, where D is the diffusion coefficient and Δ is the Laplacian, supplemented by appropriate boundary and initial conditions. Furthermore, the Stokes–Einstein equation provides a microscopic expression for the diffusion coefficient as

$$D = \frac{k_B T}{f} \tag{29}$$

where k_B is the Boltzmann constant, T is the absolute temperature, and f is the hydrodynamic drag coefficient, that is, the proportionality coefficient between the drag force experienced by a solute molecule and the speed at which it is moving relative to the fluid. For an incompressible Newtonian fluid at low Re and a spherical solute or radius a,

$$f = 6\pi\eta a \tag{30}$$

where η is the shear viscosity of the fluid (Happel and Brenner 2012).

As discussed by Jordan et al. (1998), a direct calculation shows that the diffusion equation can be formally derived from Onsager's principle using as free energy

$$\mathcal{F}[c] = \frac{D}{2} \int_{\Omega} |\nabla c|^2 dV, \tag{31}$$

characterizing the changes of state of the system simply by $\partial_t c$, and considering as dissipation potential

$$\mathcal{D}[\partial_t c] = \frac{1}{2} \int_{\Omega} (\partial_t c)^2 dV. \tag{32}$$

This approach, however, is not satisfactory for several reasons. First, these potentials do not admit a compelling physical interpretation, nor provide a connection with the microscopic physics. Second, and this cannot be fully appreciated yet, these potentials are not meaningful building blocks that can be combined with elasticity, chemistry or hydrodynamics.

Instead, the main driving force for molecular diffusion is mixing entropy maximization (or minimization of entropic free energy). For a dilute solution at concentration c, it can be motivated from various points of view (Peletier 2014; Pauli and Enz 2000) that the entropy density per unit volume due to mixing between the solute and the solvent is given by $-RTc\log(c/c_0)$, where c_0 is an arbitrary normalization constant and R is the universal gas constant. Therefore, the free energy of the system will be given by the so-called ideal gas mixing entropy

$$\mathcal{F}[c] = RT\int_\Omega c\log\frac{c}{c_0}dV. \tag{33}$$

It is easy to see that c_0 is an arbitrary constant. Indeed, using the properties of the logarithm, the free energy can be written as $\cdots - RT\log c_0\int_\Omega c\,dV$. Because the container is impermeable, conservation of solute molecules implies that this integral is a constant, and therefore c_0 only modifies an additive constant in the energy. If part of the boundary was capable of exchanging solute molecules with a reservoir at fixed chemical potential, then the constant c_0 would not be arbitrary. We leave this as an exercise. An alternative way to express the normalization constant common in the literature is

$$\mathcal{F}[c] = RT\int_\Omega c\,(\log c - 1)\,dV + \int_\Omega c\mu_0 dV, \tag{34}$$

where μ_0 is called standard chemical potential. As we shall see later, the governing equations will not depend on this arbitrary normalization constant.

The state of the system, and hence its free energy, is characterized at time t by the field $c(\cdot,t)$. Then, the free energy functional evaluated at this time-dependent field $\mathcal{F}[c(\cdot,t)]$ generates a function that depends only on time. Its rate of change, noting that the boundary of Ω is impermeable, and therefore, Reynolds transport theorem only involves a bulk term, can then be computed as

$$\frac{d}{dt}\left(\mathcal{F}[c(\cdot,t)]\right) = \int_\Omega \left(\mu_0 + RT\log c\right)\partial_t c\,dV, \tag{35}$$

where

$$\mu(c) = \frac{\delta\mathcal{F}}{\delta c} = \mu_0 + RT\log c \tag{36}$$

is the chemical potential at concentration c, defined as the functional derivative of the free energy with respect to the concentration (here it is simply the partial derivative of the free energy density). The chemical potential $\mu(c)$ locally measures the free energy cost of adding one mole of solutes per unit volume at a given concentration. Therefore, it is natural that gradients in the chemical potential will drive migration of the solutes to reduce the free energy. From this expression we see that μ_0 is the chemical potential at a reference concentration, here $c = 1$.

Now, let us think about the dissipation involved in the diffusive migration of the solutes. Imagining a single solute molecule moving with a velocity $\boldsymbol{w}$ relative to the quiescent fluid, we have seen that the drag force is given by $\boldsymbol{F} = -f\boldsymbol{w}$, and therefore the dissipation potential for this single solute would be $(f/2)|\boldsymbol{w}|^2$. Now, in a unit of volume we have $N_A c$ of such molecules, where N_A is Avogadro's number. Let us think of $\boldsymbol{w}$ as an effective collective velocity of the solutes relative to the quiescent fluid–a *diffusive velocity*–in a given region in space. Then, assuming that the solution is dilute, and therefore the drag on a solute molecule is not affected by the presence of other solute molecules, it is reasonable to write the dissipation potential per unit volume as $(f N_A/2)c|\boldsymbol{w}|^2$ and therefore

$$\mathcal{D}[\boldsymbol{w}] = \frac{f N_A}{2} \int_\Omega c|\boldsymbol{w}|^2 dV. \tag{37}$$

Toward applying Onsager's principle, we can combine Eqs. (35) and (37) to form the Rayleighian. However, we immediately note that $\dot{\mathcal{F}}$ is expressed as a functional of $\partial_t c$, but $\mathcal{D}$ is instead a functional of $\boldsymbol{w}$, and therefore it is not clear what should we minimize with respect to. How to proceed?

The first important observation is that not only $\partial_t c$ but also $\boldsymbol{w}$ characterize the rate of change of the state of the system. Indeed, if solutes move with diffusive velocity $\boldsymbol{w}$, they will rearrange in space and the concentration field will be modified. The second observation is that these two ways of expressing the rate of change of the state are not independent. Indeed, they are related by the continuity equation

$$\partial_t c + \nabla \cdot (c\boldsymbol{w}) - 0 \tag{38}$$

expressing locally the conservation of solute molecules (Landau and Lifshitz 2013). The product $\boldsymbol{j}_D = c\boldsymbol{w}$ is the molar diffusive flux of solute molecules. We will call $\boldsymbol{w}$ the process variable for this system because it describes the rate of change of the system, and allows us to express the dissipation. Plugging this equation into Eq. (35), we can express the rate of change of the energy, after integration by parts, as

$$\begin{aligned} \dot{\mathcal{F}}[c; \boldsymbol{w}] &= -\int_\Omega \mu(c)\nabla \cdot (c\boldsymbol{w})\, dV \\ &= -\int_{\partial\Omega} \mu(c)c\boldsymbol{w} \cdot \boldsymbol{n}\, dS + \int_\Omega c\nabla\mu(c) \cdot \boldsymbol{w}\, dV. \end{aligned} \tag{39}$$

Since we have assumed that the solute molecules cannot cross the boundary of the container, and therefore, $\boldsymbol{j}_D \cdot \boldsymbol{n} = 0$ over $\partial\Omega$, the boundary integral term vanishes and the Rayleighian takes the form

$$\mathcal{R}[c; \boldsymbol{w}] = \int_\Omega c\nabla\mu(c) \cdot \boldsymbol{w}\, dV + \frac{f N_A}{2} \int_\Omega c|\boldsymbol{w}|^2 dV. \tag{40}$$

Recalling Eq. (36) and minimizing this functional with respect to $\boldsymbol{w}$ we obtain the stationarity condition

$$0 = \int_\Omega RT\nabla c \cdot \delta \boldsymbol{w} \, dV + f N_A \int_\Omega c\boldsymbol{w} \cdot \delta \boldsymbol{w} \, dV, \tag{41}$$

which should hold for all admissible variations $\delta \boldsymbol{w}$. This allows us to localize the relation

$$\boldsymbol{j}_D = c\boldsymbol{w} = -\frac{RT}{f N_A} \nabla c = -\frac{k_B T}{f} \nabla c. \tag{42}$$

Thus, not only do we identify Fick's law of diffusion. We also recover the Stokes–Einstein equation for the diffusion coefficient, see Eq. (29). Plugging this expression into the continuity equation, we recover the classical diffusion equation

$$\partial_t c = \frac{k_B T}{f} \Delta c \quad \text{in } \Omega, \tag{43}$$

with homogeneous Neumann boundary conditions $\partial c / \partial n = 0$ in $\partial \Omega$. Thus, we have seen that Fick's law, the Stokes–Einstein equation, and the diffusion equation can be derived using Onsager's principle from physically motivated expressions for the free energy and the dissipation potential. We also see that the resulting governing equations are independent of the normalization constant μ_0.

2.5 Abstract Statement of Onsager's Principle

The previous example has shown that the rate of change of the energy and the dissipation potential may be expressed in terms of different descriptions of the rate of change of the system. $\dot{\mathcal{F}}$ was a functional of $\partial_t c$ while $\mathcal{D}$ was a functional of the diffusive velocity $\boldsymbol{w}$. To place the rate of change of the energy and the dissipation potential on an equal footing in the Rayleighian, we needed a relation between these two quantities (the continuity equation), termed *process operator* in the terminology of Peletier (2014). We follow this reference in this section to formalize an abstract statement of Onsager's principle. The objective of this formal exercise is to conceptualize the procedure and guide our formulation of more complex problems. It remains a nontrivial task, however, to map a particular physical model into this formalism.

In the examples examined so far, we have seen that the main ingredients in Onsager's modeling framework are (1) the state variables, such as x or c, which identify the state of the system, (2) the free energy $\mathcal{F}$, which depends on the state variables, (3) the process variables, such as v, $\boldsymbol{v}$ or $\boldsymbol{w}$, which describe how the system changes its state and generates dissipation, (4) the process operator P, which relates the rate of change of the state variables and the process variables, (5) the dissipation potential $\mathcal{D}$, measuring the energy dissipated by the process variables, and possibly (6) potentials accounting for the externally supplied power $\mathcal{P}$ and (7) constraints such as the incompressibility condition. Constraints may be formulated on the state or on the process variables, but the former can always be linearized and expressed as constraints on the process variables. Collecting all these ingredients, we can abstractly state Onsager's variational principle as follows.

Let us describe a dissipative system through some state variables $X(t)$ evolving in a suitable space (possibly a nonlinear manifold), a free energy $\mathcal{F}(X)$, some process variables V (living in a vector bundle and therefore with a clear notion of 0), a dissipation potential $\mathcal{D}(X; V)$, and a potential for the external power supply $\mathcal{P}(X; V)$. Suppose also that the process variables are linearly constrained by $0 = \mathbb{C}(X)V$ during the time evolution of the system. $\mathcal{F}$ is often a nonlinear function of X, $\mathcal{D}$ may be a nonlinear function of X but is generally quadratic in V, and $\mathcal{P}$ is generally linear in V. However, $\mathcal{D}$ does not need to be quadratic in V in Onsager's formalism as described here. As motivated below, *the thermodynamic requirements we will need on $\mathcal{D}$ are (1) that it is nonnegative, (2) that $\mathcal{D}(X, 0) = 0$ and (3) that it is convex as a function of V*. We will also assume here that the dissipation potential is differentiable. This is not necessarily the case, for instance in rate independent dissipative processes such as dry friction, which can nevertheless be framed in Onsager's principle. The differentiability assumption is justified here because soft and biological matter is generally wet and rate dependent.

To form the Rayleighian, we need to evaluate the rate of change of the energy, which can most of the times be obtained by the chain rule

$$\dot{\mathcal{F}}(X; \partial_t X) = \frac{d}{dt}\left[\mathcal{F}(X(t))\right] = D\mathcal{F}(X) \cdot \partial_t X, \tag{44}$$

where $D\mathcal{F}(X)$ denotes the derivative of the free energy. The situations is slightly complicated when considering free energy integrals over nonmaterial domains (open systems), where Reynolds transport theorem produces an explicit dependence of $\dot{\mathcal{F}}$ on the process variables V. This is the case in the example in Sect. 2.6. This dependence, however, does not complicate the application of Onsager's principle in any way.

In general, the process variable V ($\boldsymbol{w}$ in the previous example) will not be simply the time-derivative of the state variable $\partial_t X$ ($\partial_t c$ in the previous example), although this was the case in the examples of Sect. 2.2. To relate these two descriptions of the evolution of the state of the system, we need a process operator, which we consider here to be linear

$$\partial_t X = P(X)V. \tag{45}$$

This operator will often be either trivial, i.e., $\partial_t X = V$, or a statement of conservation of mass. Importantly, as noted by Otto (2001); Peletier (2014), V often contains redundant information to describe $\partial_t X$, which is however required to properly model dissipation. This is the case in the previous example, where $\partial_t c$ is a scalar field but $\boldsymbol{w}$ is a vector field.

The process operator allows us to express the rate of change of the system in terms of the process variable V, and thus form the Rayleighian as

$$\mathcal{R}(X;V) = D\mathcal{F}(X)\cdot P(X)V + \mathcal{D}(X;V) + \mathcal{P}(X;V). \tag{46}$$

Onsager's variational principle then states that the system evolves such that

$$V = \underset{W}{\text{argmin}}\ \mathcal{R}(X;W) \tag{47}$$

subject to the constraints on W

$$\mathbb{C}(X)W = 0. \tag{48}$$

The constrained dynamics can be equivalently characterized as stationary (saddle) points of the Lagrangian

$$\mathcal{L}(X;V,\Lambda) = D\mathcal{F}(X)\cdot P(X)V + \mathcal{D}(X;V) + \mathcal{P}(X;V) + \Lambda\cdot\mathbb{C}(X)V, \tag{49}$$

where Λ are the Lagrange multipliers. Once V is obtained from this variational principle, we can then integrate $\partial_t X$ in time recalling Eq. (45).

Let us now formally examine an important qualitative property of the resulting dissipative dynamics. For this, we will consider a "homogeneous" system with $\mathcal{P}(X;V) = 0$. The stationarity condition $0 = \delta_\Lambda \mathcal{L}$ simply leads to $0 = \mathbb{C}(X)V$. The stationarity condition $0 = \delta_V \mathcal{L}$ results in the dynamical equilibrium equation

$$0 = D_X\mathcal{F}(X)\cdot P(X) + D_V\mathcal{D}(X;V) + \Lambda\cdot\mathbb{C}(X). \tag{50}$$

Multiplying this equation by the actual V along the dissipative dynamics and rearranging terms, we obtain

$$\underbrace{D_X\mathcal{F}(X)\cdot P(X)V}_{\dot{\mathcal{F}}} = -D_V\mathcal{D}(X;V)V - \Lambda\cdot\underbrace{\mathbb{C}(X)V}_{0}. \tag{51}$$

Now, since $\mathcal{D}$ is convex and differentiable in V and we have required that $\mathcal{D}(X;0) = 0$, we conclude that

$$0 = \mathcal{D}(X;0) \geq \mathcal{D}(X;V) + D_V\mathcal{D}(X;V)(0 - V). \tag{52}$$

Since we have required $\mathcal{D}(X;V) \geq 0$, we conclude from this equation that

$$0 \geq -\mathcal{D}(X;V) \geq -D_V\mathcal{D}(X;V)V. \tag{53}$$

This equation, together with Eq. (51), shows that during the dynamics

$$\dot{\mathcal{F}} \leq 0, \tag{54}$$

and $D_V\mathcal{D}(X; V)V$ is the rate of dissipation. For quadratic dissipation potentials, it is easily checked that $D_V\mathcal{D}(X; V)V = 2\mathcal{D}(X; V)$. Therefore, the free energy $\mathcal{F}$ is a Lyapunov function of the dynamics. This also shows that Onsager's principle complies with the second law of thermodynamics by construction, as long as $\mathcal{D}$ satisfies a set of minimal requirements. Finally, we note that this notion of stability is fully nonlinear and does not assume a quadratic form for the dissipation or free energy potentials.

2.6 Diffusion, Low Re Hydrodynamics, and Osmosis in a Fluid with a Solute Interacting with a Semipermeable Membrane

We consider now a simple problem coupling diffusion, hydrodynamics, and mechanics. This problem also exemplifies the treatment of moving interfaces. The physical model is described in Fig. 2. Because of the presence of a semipermeable membrane, that selectively blocks the passage of solute molecules (red dots in the figure) but lets solvent molecules go through (blue background medium), this model will allow us to examine osmotic effects. The semipermeable membrane is rigid, but can

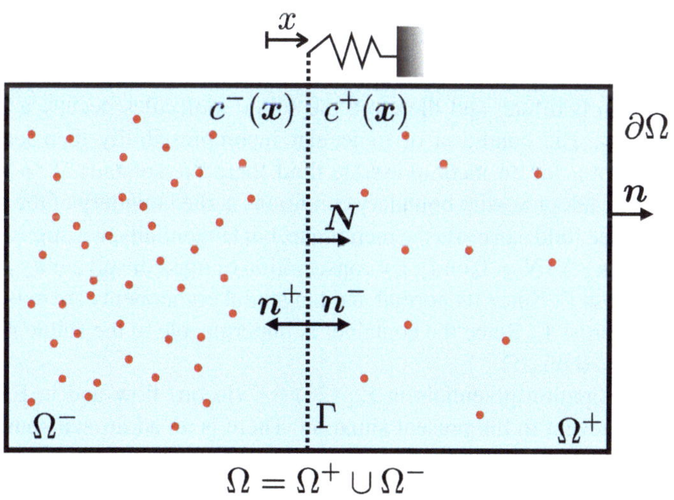

Fig. 2 Impermeable fluid container Ω with a semipermeable membrane dividing the container in two subdomains. The semipermeable membrane is rigid but can move laterally and is connected to a spring. The fluid contains solute molecules

move laterally at the expense of storing elastic energy in a spring. Thus, this model conceptually recapitulates a number the ingredients relevant to membrane physics. Indeed, lipid bilayers are semipermeable membranes embedded in a solution at high osmolar strength, and their deformation stores elastic energy. Because the impermeable barrier does not allow solute molecules to pass, the concentration of these molecules may be discontinuous across Γ. On the other hand, the solvent can cross the semipermeable barrier, but this passage involves some friction characterized by a permeation coefficient (Staykova et al. 2013).

Let us address this problem using Onsager's principle, and let us try to follow the systematic procedure outlined in the previous section. First, we need to identify suitable state variables, which in this case are the concentration field c of solute molecules, which can be discontinuous across Γ (and for this reason we distinguish between c^+ and c^- on the membrane), and the position of the moving semipermeable membrane x. Combining ingredients introduced in the previous examples, we can form the free energy depending on $X = \{c, x\}$ as

$$\mathcal{F}(X) = RT \int_\Omega c\,(\log c - 1)\, dV + \int_\Omega c\mu_0 dV + \frac{k}{2}x^2. \tag{55}$$

Let us discuss the process variables. These include the diffusive velocity $\boldsymbol{w}$ characterizing changes in c and the velocity of the semipermeable membrane $v_m = \dot{x}$. Furthermore, it is clear that the motion of this membrane will displace the fluid, which cannot be assumed to be quiescent as in Sect. 2.4. Therefore, the velocity of the fluid $\boldsymbol{v}$ will also be part of the process variables. Now, since the background fluid is moving, we need to decide whether $\boldsymbol{w}$ describes the absolute velocity of the solutes or their velocity relative to the fluid. We choose the latter, since this relative velocity is the one that is meaningful to describe dissipation during diffusion. Thus, the process variables are $V = \{\boldsymbol{w}, v_m, \boldsymbol{v}\}$.

Let us discuss now the constraints affecting the process variables. We shall assume that the solution is dilute, and therefore the solute molecules occupy a negligible volume fraction. The condition of molecular incompressibility then leads to the common condition for an incompressible fluid (here the solvent) $\nabla \cdot \boldsymbol{v} = 0$ in Ω. For the fluid, we adopt no-slip boundary conditions at the boundary of the container, $\boldsymbol{v} = \boldsymbol{0}$ on $\partial\Omega$. The fluid can cross the membrane, but tangentially, we impose a no-slip condition $\boldsymbol{v} - (\boldsymbol{v} \cdot \boldsymbol{N})\boldsymbol{N} = 0$ on Γ. By conservation of mass of solvent $\boldsymbol{v} \cdot \boldsymbol{N}$ must be continuous across Γ. Since its normal and tangential components are continuous, $\boldsymbol{v}$ is continuous across Γ. Since the container is impermeable to the solute molecules, we have $\boldsymbol{w} \cdot \boldsymbol{n} = 0$ on $\partial\Omega$.

The two dissipation potentials in Eq. (20) for viscous flow and in Eq. (37) for diffusion are relevant to the present situation. There is an additional source of dissipation associated to solvent permeation through the semipermeable membrane. In agreement with commonly used models for permeation, we postulate that the dissipation potential density per unit area is quadratic in the normal component of the velocity of fluid across the interface $\boldsymbol{v} \cdot \boldsymbol{N} - v_m$. Thus, the dissipation potential for this problem can be written as

$$\mathcal{D}(X;V) = \eta \int_\Omega \boldsymbol{d} : \boldsymbol{d}\, dV + \frac{f N_A}{2} \int_\Omega c|\boldsymbol{w}|^2 dV + \frac{\bar{\eta}}{2} \int_\Gamma (\boldsymbol{v} \cdot \boldsymbol{N} - v_m)^2\, dS, \quad (56)$$

where $\bar{\eta}$ is a permeation coefficient.

Following the systematic procedure outlined in the previous section, we now turn to the process operator. This operator relating $\partial_t X$ and V contains the trivial component $\dot{x} = v_m$, and another component stating the conservation of solute molecules, which now takes the form

$$0 = \partial_t c + \nabla \cdot (\boldsymbol{j}_D + c\boldsymbol{v}) = \partial_t c + \nabla \cdot [c(\boldsymbol{w} + \boldsymbol{v})], \quad (57)$$

since the solute molecules can be transported either diffusively or by advection. In addition to these two equations, there is another important process relation at the semipermeable membrane. Because the solutes cannot cross the membrane, their diffusive velocity needs to coincide with the membrane velocity on either side of the domain

$$v_m = (\boldsymbol{w}^\pm + \boldsymbol{v}) \cdot \boldsymbol{N}. \quad (58)$$

Since $\boldsymbol{v} \cdot \boldsymbol{N}$ is continuous across the interface, we conclude that $\boldsymbol{w} \cdot \boldsymbol{N}$ is continuous across the interface. Therefore, the process operator can be summarized by the three relations

$$\dot{x} = v_m, \quad 0 = \partial_t c + \nabla \cdot [c(\boldsymbol{w} + \boldsymbol{v})] \text{ in } \Omega, \quad v_m = (\boldsymbol{w} + \boldsymbol{v}) \cdot \boldsymbol{N} \text{ on } \Gamma. \quad (59)$$

Now, we are in a position to compute the rate of change of the free energy, a key point in the theory. Recalling the definition of the chemical potential in Eq. (36) and applying Reynolds transport theorem in Ω^- and Ω^+ separately to account for the internal moving boundary, we obtain

$$\begin{aligned}
\dot{\mathcal{F}} =& \frac{d}{dt} \int_{\Omega^-} \left[RTc(\log c - 1) + c\mu_0\right] dV + \frac{d}{dt} \int_{\Omega^+} \left[RTc(\log c - 1) + c\mu_0\right] dV + kx\dot{x} \\
=& \int_{\Omega^-} \mu \partial_t c\, dV + v_m \int_\Gamma \left[RTc^-(\log c^- - 1) + c^-\mu_0\right] dS + \\
& \int_{\Omega^+} \mu \partial_t c\, dV - v_m \int_\Gamma \left[RTc^+(\log c^+ - 1) + c^+\mu_0\right] dS + kx\dot{x} \\
=& \int_{\Omega^-} \mu \partial_t c\, dV + \int_{\Omega^+} \mu \partial_t c\, dV - v_m \int_\Gamma \left[\!\left[RTc(\log c - 1) + c\mu_0\right]\!\right] dS + kx\dot{x} \\
=& \int_{\Omega^-} \mu \partial_t c\, dV + \int_{\Omega^+} \mu \partial_t c\, dV - v_m \int_\Gamma \left([\![c\mu]\!] - RT\,[\![c]\!]\right) dS + kx\dot{x},
\end{aligned} \quad (60)$$

where $[\![f]\!]$ denotes the jump of a function f across an interface $f^+ - f^-$. Now, using the first and second process equations in Eq. (59), the divergence theorem, the boundary conditions on $\partial\Omega$, and the fact that $\boldsymbol{n}^- = -\boldsymbol{n}^+ = \boldsymbol{N}$ on Γ, we have

$$\begin{aligned}\dot{\mathcal{F}} =& -\int_{\Omega^-} \mu\nabla\cdot[c(\boldsymbol{w}+\boldsymbol{v})]\,dV - \int_{\Omega^+} \mu\nabla\cdot[c(\boldsymbol{w}+\boldsymbol{v})]\,dV \\ & - v_m \int_\Gamma (⟦c\mu⟧ - RT\,⟦c⟧)\,dS + kxv_m \\ =& \int_\Omega c\nabla\mu\cdot(\boldsymbol{w}+\boldsymbol{v})dV + \int_\Gamma ⟦c\mu⟧\,(\boldsymbol{w}+\boldsymbol{v})\cdot\boldsymbol{N}dS \\ & - v_m \int_\Gamma (⟦c\mu⟧ - RT\,⟦c⟧)\,dS + kxv_m \end{aligned} \tag{61}$$

Finally, using the third process equation in Eq. (59), we obtain

$$\dot{\mathcal{F}} = RT\int_\Omega \nabla c\cdot(\boldsymbol{w}+\boldsymbol{v})dV + v_m\left(RT\int_\Gamma ⟦c⟧\,dS + kx\right). \tag{62}$$

In the abstract formalism of the previous section, the equation above is a workable expression of $D\mathcal{F}(X)\cdot P(X)V$. The second term already shows that, in addition to the elastic force kx, the semipermeable membrane experiences an osmotic force that agrees with the classical van't Hoff formula, which naturally follows from the present formalism.

We can now form the constrained Rayleighian (accounting for solvent incompressibility), which takes the form

$$\begin{aligned}\mathcal{L}[c,x;\boldsymbol{w},v_m,\boldsymbol{v},p] =& RT\int_\Omega \nabla c\cdot(\boldsymbol{w}+\boldsymbol{v})dV + v_m\left(RT\int_\Gamma ⟦c⟧\,dS + kx\right) \\ & + \eta\int_\Omega \boldsymbol{d}:\boldsymbol{d}\,dV + \frac{fN_A}{2}\int_\Omega c|\boldsymbol{w}|^2dV \\ & + \frac{\bar{\eta}}{2}\int_\Gamma (\boldsymbol{v}\cdot\boldsymbol{N} - v_m)^2\,dS - \int_\Omega p\,\nabla\cdot\boldsymbol{v}\,dV. \end{aligned} \tag{63}$$

Making this functional stationary with respect to $\boldsymbol{w}$ leads, as in the pure diffusion example, to Fick's law

$$c\boldsymbol{w} = -\frac{k_BT}{f}\nabla c \quad \text{in } \Omega. \tag{64}$$

The variation with respect to v_m leads to balance of forces acting on the semipermeable membrane

$$0 = RT\int_\Gamma ⟦c⟧\,dS + kx + \bar{\eta}\int_\Gamma (v_m - \boldsymbol{v}\cdot\boldsymbol{N})dS. \tag{65}$$

Variation with respect to p recovers the incompressibility condition $0 = \nabla\cdot\boldsymbol{v}$. Finally, variation with respect to $\boldsymbol{v}$ leads to

$$
\begin{aligned}
0 = & RT \int_{\Omega} \nabla c \cdot \delta \boldsymbol{v} dV + 2\eta \int_{\Omega} \boldsymbol{d} : \nabla \delta \boldsymbol{v} \, dV - \int_{\Omega} p \nabla \cdot \delta \boldsymbol{v} \, dV \\
& + \bar{\eta} \int_{\Gamma} (\boldsymbol{v} \cdot \boldsymbol{N} - v_m) \delta \boldsymbol{v} \cdot \boldsymbol{N} dS. \qquad (66)
\end{aligned}
$$

Performing integration by parts carefully over the two subdomains, and recalling the homogeneous boundary conditions on $\partial\Omega$ and the fact that $\boldsymbol{v}$ (and hence $\delta\boldsymbol{v}$) is continuous across Γ, we find

$$
\begin{aligned}
0 = & \int_{\Omega} (RT \nabla c - 2\eta \nabla \cdot \boldsymbol{d} + \nabla p) \cdot \delta \boldsymbol{v} \, dV - \int_{\Gamma} \boldsymbol{N} \cdot [\![2\eta \boldsymbol{d} - p \boldsymbol{I}]\!] \cdot \delta \boldsymbol{v} \, dS \\
& + \bar{\eta} \int_{\Gamma} (\boldsymbol{v} \cdot \boldsymbol{N} - v_m) \delta \boldsymbol{v} \cdot \boldsymbol{N} dS. \qquad (67)
\end{aligned}
$$

Thus, identifying the stress tensor as $\boldsymbol{\sigma} = 2\eta \boldsymbol{d} - p\boldsymbol{I}$, the above equation leads to

$$
0 = \nabla \cdot \boldsymbol{\sigma} - RT \nabla c \quad \text{in } \Omega \qquad (68)
$$

and to

$$
\boldsymbol{N} \cdot [\![\boldsymbol{\sigma}]\!] = \bar{\eta} (\boldsymbol{v} \cdot \boldsymbol{N} - v_m) \boldsymbol{N} \quad \text{on } \Gamma. \qquad (69)
$$

Finally, we can eliminate $\boldsymbol{w}$ from the formulation by plugging Fick's law in Eq. (64) into the two process relations in Eq. (59) encoding mass conservation to obtain

$$
\partial_t c - \frac{k_B T}{f} \Delta c + \boldsymbol{v} \cdot \nabla c = 0 \quad \text{in } \Omega, \qquad (70)
$$

and the two equations

$$
\frac{k_B T}{f} \frac{\partial c}{\partial N}^{\pm} = c^{\pm} (\boldsymbol{v} \cdot \boldsymbol{N} - v_m) \quad \text{on } \Gamma. \qquad (71)
$$

In summary, we have deduced using Onsager's principle the governing equations for the system depicted in Fig. 2. These equations are a *Stokes/advection-diffusion system* in the bulk reflecting conservation of mass of solutes and solvent and balance of linear momentum in the fluid, together with Fick's law and the constitutive relation for a Newtonian fluid:

$$
\left.
\begin{aligned}
0 &= \partial_t c - \frac{k_B T}{f} \Delta c + \boldsymbol{v} \cdot \nabla c \\
0 &= \nabla \cdot \boldsymbol{v} \\
0 &= \nabla \cdot \boldsymbol{\sigma} - RT \nabla c
\end{aligned}
\right\} \quad \text{in } \Omega,
$$

where $\boldsymbol{\sigma} = 2\eta \boldsymbol{d} - p\boldsymbol{I}$ with boundary conditions

$$\boldsymbol{v} = \boldsymbol{0} \quad \text{and} \quad \frac{\partial c}{\partial n} = 0 \quad \text{on } \partial\Omega. \tag{72}$$

These equations are supplemented with conditions at the moving semipermeable membrane. These conditions are a no-slip condition in the tangential direction

$$\boldsymbol{v} - (\boldsymbol{v} \cdot \boldsymbol{N})\boldsymbol{N} = 0 \quad \text{on } \Gamma, \tag{73}$$

a *global force balance on the membrane* involving permeation, osmotic, and elastic forces

$$\bar{\eta} \int_\Gamma (\boldsymbol{v} \cdot \boldsymbol{N} - v_m) dS = RT \int_\Gamma [\![c]\!] \, dS + kx. \tag{74}$$

a *local force balance in the fluid at the interface* involving the jump of fluid tractions and the permeation tractions

$$\bar{\eta}(\boldsymbol{v} \cdot \boldsymbol{N} - v_m)\boldsymbol{N} = \boldsymbol{N} \cdot [\![\boldsymbol{\sigma}]\!] \quad \text{on } \Gamma, \tag{75}$$

and *interface conditions resulting from conservation of solute and solvent*

$$[\![\boldsymbol{v}]\!] = \boldsymbol{0}, \quad [\![\boldsymbol{w}]\!] \cdot \boldsymbol{N} = 0, \quad \text{and} \quad \frac{k_B T}{f} \frac{\partial c}{\partial N}^{\pm} = c^{\pm}(\boldsymbol{v} \cdot \boldsymbol{N} - v_m) \quad \text{on } \Gamma. \tag{76}$$

We note that, at the interface, we impose simultaneously Dirichlet and Neumann-like jump conditions for $\boldsymbol{v}$, and Dirichlet and Robin-like jump conditions for c. This is possible because the interface is moving.

We could have directly arrived at this set of equations with sufficient physical insight and invoking constitutive relations such as Fick's law, van't Hoff's relation, or that of a Newtonian fluid. It is also clear that one can easily make errors in such a direct derivation. Instead, all these relations have followed systematically from the rather simple modeling assumptions behind $\mathcal{F}$ and $\mathcal{D}$, the use of conservation of mass for solvent and solute to define the process operator, and Reynolds transport theorem. Furthermore, just by looking at the final equations, it is not easy to see how is $\mathcal{F}$ driving this system and decreasing during the dynamics or how is energy being dissipated. Finally, Onsager's principle allows us to systematically construct more complex models by adding additional building blocks. For instance, the membrane could be made flexible and endowed with tension or curvature elasticity. Or, using the elementary models presented in the next sections, the solute molecules could be chemically active and react with other species, adsorb to surfaces, or preferentially react while adsorbed on a catalyzer.

We would like to make a final point regarding this example. We discussed earlier that the diffusion equation can be formally derived from Onsager's principle starting from the energy and dissipation potentials in Eqs. (31) and (32), not founded on the microscopic physics of diffusion. The reader can easily become convinced that these functionals, however, dramatically fail when diffusion acts in concert with other physics. Indeed, they cannot be meaningfully combined in a Rayleighian with the functionals encoding additional ingredients such as hydrodynamics or permeation through a semipermeable membrane. Instead, the approach described above naturally produces, for instance, entropic effects such as osmotic forces.

2.7 *Reaction–Diffusion of Two Species in a Quiescent Fluid*

We introduce next a new item into our catalog of phenomena amenable to Onsager's principle: chemical reactions. As a minimal model system, we want to identify the variational structure of a system of two coupled linear reaction–diffusion equations for two chemical species. We consider a domain Ω, whose boundary is assumed for simplicity to be impermeable to both substances.

Let us first describe this simple model. The state of the system is described by two molar concentration fields c_1 and c_2, one for each one of the species X_1 and X_2, which transform through the simple reaction

$$X_1 \underset{k_b}{\overset{k_f}{\rightleftharpoons}} X_2. \tag{77}$$

We assume that this reaction follows the law of mass action, by which in this simple example the molar rate per unit volume of transformation of species X_1 to species X_2, the forward rate r_f, is proportional to the concentration of the reactant, $r_f = k_f c_1$. Conversely, the backward rate is given by $r_b = k_b c_2$, and thus the net forward rate is $r = k_f c_1 - k_b c_2$. Then, the dynamics of this system can be modeled through the linear system of reaction–diffusion equations

$$\left.\begin{aligned} \partial_t c_1 &= D_1 \Delta c_1 - k_f c_1 + k_b c_2 \\ \partial_t c_2 &= D_2 \Delta c_2 + k_f c_1 - k_b c_2 \end{aligned}\right\} \quad \text{in } \Omega, \tag{78}$$

where D_1 and D_2 are the diffusion coefficient of each chemical species, supplemented by initial and boundary conditions. In equilibrium, the concentrations will be uniform and $r = 0$, and thus $c_1^{eq}/c_2^{eq} = k_b/k_f = K$ is a constant called equilibrium constant of the reaction.

In the previous sections, we showed that molecular diffusion can be understood as a process of entropic free energy minimization, dragged by the resistance exerted by the solvent. Can we integrate this phenomenology with that of chemical reactions between the diffusing species? In other words, can we find the appropriate dissipation and free energy potentials so that the diffusion–reaction dynamics emerge from

Onsager's principle? The answer is yes and due to Mielke (2012). Let us develop such a model.

We first model the free energy of the system. Assuming that the concentrations are dilute, we write the chemical free energy of the system building on that of an ideal gas in Eq. (34) as a function of $X = \{c_1, c_2\}$

$$\begin{aligned}\mathcal{F}(X) =& RT\int_\Omega c_1\,(\log c_1 - 1)\,dV + \int_\Omega c_1\mu_{0,1}dV\\ &+RT\int_\Omega c_2\,(\log c_2 - 1)\,dV + \int_\Omega c_2\mu_{0,2}dV,\end{aligned} \tag{79}$$

where $\mu_{0,i}$ are the standard chemical potentials for each species. In the single-species diffusion case, this was an irrelevant constant. We shall see that for two reacting species, the difference between these two constants determines the equilibrium constant of the reaction.

Now, in addition to the diffusive velocities $\boldsymbol{w}_1$ and $\boldsymbol{w}_2$ for each substance, the concentrations can evolve as a result of chemical reactions quantified for instance by the net forward reaction rate r. Thus, the process variables are now $V = \{\boldsymbol{w}_1, \boldsymbol{w}_2, r\}$. Note that, in 3D, we need only two scalar fields to describe the state of the system and as many as 7 (two vector fields and a scalar field) to describe the rate of change of the system. As we show later, we do need that many degrees of freedom in V to properly model dissipation.

Accounting for chemical reactions, the process operator is then given by the equations

$$\begin{aligned}\partial_t c_1 + \nabla\cdot(c_1\boldsymbol{w}_1) + r = 0,\\ \partial_t c_2 + \nabla\cdot(c_2\boldsymbol{w}_2) - r = 0,\end{aligned} \tag{80}$$

encoding balance of mass for the dissolved species. The conditions $0 = \boldsymbol{w}_i\cdot\boldsymbol{n}$ in $\partial\Omega$, reflecting the fact that $\partial\Omega$ is impermeable, can also be viewed as part of the process operator. With the free energy and the process operator at hand, and following a similar calculation as in Sect. 2.4, we can write the rate of change of the energy as

$$\begin{aligned}\dot{\mathcal{F}}(X;V) =& -\int_\Omega \mu_1\nabla\cdot(c_1\boldsymbol{w}_1)\,dV - \int_\Omega \mu_2\nabla\cdot(c_2\boldsymbol{w}_2)\,dV\\ &+\int_\Omega(\mu_2-\mu_1)r\,dV\end{aligned} \tag{81}$$

with the chemical potentials given by

$$\mu_i(c) = \mu_{0,i} + RT\log c_i. \tag{82}$$

After integration by parts using the fact that $\partial\Omega$ is impermeable, we obtain

$$\begin{aligned}\dot{\mathcal{F}}(X;V) =& RT\int_{\Omega}\nabla c_1\cdot \boldsymbol{w}_1\,dV + RT\int_{\Omega}\nabla c_2\cdot \boldsymbol{w}_2\,dV\\ &+\int_{\Omega}(\mu_2-\mu_1)\,r\,dV.\end{aligned} \tag{83}$$

This expression already contains interesting information about equilibrium. Indeed, the equilibrium state should minimize the free energy, and from the expression above three stationary conditions can be extracted. Stationarity with respect to diffusive velocities $\boldsymbol{w}_i$ implies that in equilibrium both concentrations are uniform. Stationarity with respect to the reaction rate r implies that $\mu_2=\mu_1$, and thus

$$K=\frac{c_1^{eq}}{c_2^{eq}}=\exp\frac{\Delta\mu_0}{RT}, \tag{84}$$

where $\Delta\mu_0=\mu_{0,2}-\mu_{0,1}$ is the difference of reference chemical potentials. Thus, $K=k_b/k_f$ is a purely thermodynamic quantity (although both k_b and k_f contain kinetic information).

Having examined equilibrium, we introduce the dissipation potential, which accounts for the dissipation during diffusion and reaction. Recalling Eq. (37), we consider

$$\mathcal{D}(X;V)=\frac{f_1N_A}{2}\int_{\Omega}c_1|\boldsymbol{w}_1|^2dV+\frac{f_2N_A}{2}\int_{\Omega}c_2|\boldsymbol{w}_2|^2dV+\frac{1}{2}\int_{\Omega}\frac{1}{\bar{k}}r^2dV, \tag{85}$$

where f_i are the molecular drag coefficients of the two species. We postulate that the dissipation potential is quadratic in the rate r (all the dissipation potentials examined so far have been quadratic), and leave the coefficient $\bar{k}$ unspecified for the moment. This parameter should be nonnegative for consistency with the second law of thermodynamics, as discussed in Sect. 2.5.

Forming the Rayleighian $\mathcal{R}=\dot{\mathcal{F}}+\mathcal{D}$ and minimizing it with respect to $\boldsymbol{w}_i$, we recover Fick's law for each species

$$c_i\boldsymbol{w}_i=-\frac{k_BT}{f_i}\nabla c_i \quad \text{in } \Omega. \tag{86}$$

Minimization with respect to r leads to

$$r=\bar{k}(\mu_1-\mu_2). \tag{87}$$

Now, let us remember that our goal here was to identify Onsager's variational structure for the reaction–diffusion system in Eq. (78). We can directly established the diffusion part by introducing Eq. (86) into the process equations in (80). To establish the reaction part, we need to express the reaction rate in Eq. (87) in the form

$r = k_f c_1 - k_b c_2$. Examining the expression of the chemical potentials in Eq. (82), it is clear that $\bar{k}$ will need to be a complicated function of the concentrations.

Consider now the following choice for the concentration-dependent kinetic coefficient

$$\begin{aligned}\bar{k}(c_1, c_2) &= k\frac{c_1 - e^{\Delta\mu_0/(RT)}c_2}{\mu_1 - \mu_2} \\ &= \frac{ke^{-\mu_{0,1}/(RT)}}{RT}\frac{e^{\mu_1/(RT)} - e^{\mu_2/(RT)}}{\mu_1/(RT) - \mu_2/(RT)},\end{aligned} \tag{88}$$

where $k > 0$ is a kinetic constant not depending on the concentrations. Using Eq. (82) it is easily shown that these two expressions for $\bar{k}$ are equivalent. The first form is useful because when plugged into Eq. (87), we immediately find the sought after expression

$$r = \underbrace{k}_{k_f} c_1 - \underbrace{ke^{\Delta\mu_0/(RT)}}_{k_b} c_2, \tag{89}$$

which furthermore shows that we recover Arrhenius equation. The second form of the kinetic coefficient in Eq. (88) is important because, since the exponential function is monotonically increasing, it clearly shows that $\bar{k} \geq 0$ for any choice of concentrations.

Thus, we have recovered the reaction–diffusion system in Eq. (78) using Onsager's principle. This derivation has showed that both reaction and diffusion are driven by the same chemical energy in Eq. (79), which decreases during the dynamics. This free energy contains an entropic component, but also an enthalpic one given by the difference of reference chemical potentials between the reacting species $\Delta\mu_0$. The newest and maybe surprising ingredient in this model has been the form of the coefficient encoding dissipation during the chemical reaction in Eq. (88). In the numerator, we have "Δc" measuring the deviation from the equilibrium condition in Eq. (84), and in the denominator we have $\Delta\mu$. This expression can be generalized to more complex chemical reactions obeying the law of mass action (Mielke 2012). By understanding the basic structure of the reaction–dissipation potential, we have a new building block for modeling that can be easily adapted to different settings and combined with different physics, as shown in subsequent sctions. We are not aware of a compelling microscopic interpretation of Eq. (88).

During the derivation of the equations, we have identified the diffusion constants as $D_i = k_B T/f_i$. Furthermore, we have understood that the forward and backward rates contain not only kinetic information, but also thermodynamic information in that their ratio depends on $\Delta\mu_0$. Onsager's principle has allowed us to untangle the kinetic and thermodynamic components of the reaction dynamics. Thus, this example further exemplifies two benefits of Onsager's principle: (1) it provides a systematic method to derive models for dissipative systems from a library of building blocks, and (2) it highlights the energetic-dissipative structure of such systems, providing physical insight into the model parameters.

3 Surface Sorption and Diffusion

Having considered reaction–diffusion systems in the bulk, we are in a position to examine chemical adsorption. We consider diffusing solutes in a bulk fluid adsorbing or desorbing on the surface enclosing it and diffusing on it. For simplicity, we assume that the surface has a fixed shape and the fluid is quiescent. The bulk fluid is represented by Ω and the surface by $\Gamma = \partial\Omega$ as shown in Fig. 3. The bulk fluid can exchange solutes with the surrounding surface through sorption—the process encompassing adsorption and desorption.

In the context of biological membranes, adsorption from the bulk is a possible mechanism for protein incorporation. Another important mechanism involves fusion with vesicles loaded with membrane proteins. Proteins may adsorb by weakly scaffolding the membrane or by inserting amphiphilic domains into one leaflet or the entire bilayer. Irrespective of whether the process of adsorption induces a conformational change or not, we consider that the chemical reaction is the transformation from a molecule in solution to one bound to the surface. Thus, we treat the solute molecules in the bulk X_B and those on the surface X_S–the adsorbate–as two different species transforming through the elementary reaction

$$X_B \rightleftharpoons X_S. \tag{90}$$

We will denote by c the molar concentration of solute molecules in the bulk. To adhere with the literature, we will express the concentration of adsorbates on the surface through the area fraction of surface covered by adsorbed molecules ϕ. Thus, the state variables are the bulk and surface fields $X = \{c, \phi\}$. The molar surface concentration can then be recovered as ϕ/a_0, where a_0 is the molar area of the adsorbate.

In this section, we will assume that c is small, which allows us to safely consider the ideal gas mixing entropy introduced earlier. However, we will consider the possibility that the area fraction is finite, and even large. Large area coverage of membrane proteins is common in synthetic systems (Sorre et al. 2012; Zhu et al. 2012) and in

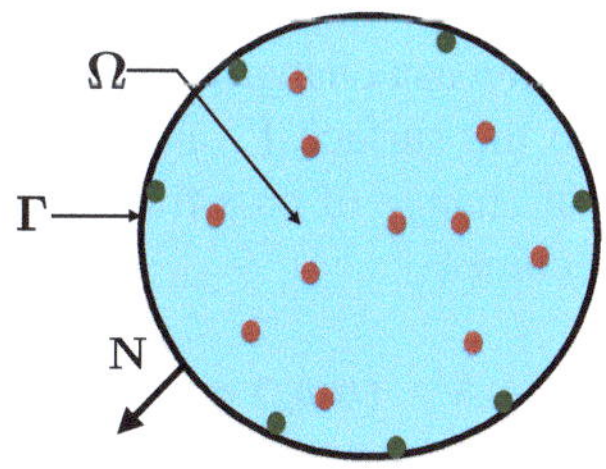

Fig. 3 Elementary model for sorption–diffusion. The solutes and adsorbates are labeled with *red* and *green dots*, respectively. The bulk domain representing quiescent fluid is Ω and the surface of fixed geometry is $\Gamma = \partial\Omega$

cell organelles (Shibata et al. 2009; Terasaki et al. 2013). Molecular crowding of proteins on the membrane can then lead to nonlinear chemical effects as discussed in the introduction. If the adsorbates modify the preferred curvature of the bilayer by any of the proposed mechanisms such as scaffolding, wedging, or crowding (Shibata et al. 2009; Stachowiak et al. 2012), then adsorption may induce significant shape transformations. In this section, however, we ignore such a coupling between chemistry and mechanics, which we only examine in a simple model in Sect. 4, and focus here on the sorption/diffusion system.

3.1 Onsager's Principle for Linear Sorption–Diffusion

First, we assume that adsorbates are very dilute. This situation is very similar to that in Sect. 2.7, and therefore, we will provide a concise presentation highlighting the main differences. In close analogy with that section, we write the chemical energy in the surface and the bulk as

$$\begin{aligned} \mathcal{F} = {} & \frac{RT}{a_0}\int_\Gamma \phi(\log\phi - 1)\,dS + \frac{1}{a_0}\int_\Gamma \mu_{0,a}\phi\,dS \\ & + RT\int_\Omega c\left(\log\frac{c}{c_0} - 1\right)dV + \int_\Omega \mu_{0,s} c\,dV, \end{aligned} \tag{91}$$

where the molar area of adsorbate a_0 is introduced in the first line because the standard form of the chemical energy is for a concentration, not an area fraction. To maintain dimensional consistency of the formulation, we have explicitly introduced a reference concentration c_0 at which the chemical potential of the solute is precisely $\mu_{0,s}$. This does not involve a real additional parameter in the model because $\mu_{0,s}$ is defined relative to c_0.

The state of the system in the bulk can change due to the solute diffusive velocity $\boldsymbol{w}_s$ according to the continuity equation

$$\partial_t c + \nabla\cdot(c\boldsymbol{w}_s) = 0 \quad \text{in } \Omega. \tag{92}$$

In the surface, the surface fraction can change due to the adsorbate diffusive velocity $\boldsymbol{w}_a$, a vector field *tangent* to the surface Γ, and due to the rate of adsorption r, which we express as a rate of change of area fraction. Therefore, the statement of conservation of adsorbate becomes

$$\partial_t\phi + \nabla_s\cdot(\phi\boldsymbol{w}_a) = r \quad \text{on } \Gamma, \tag{93}$$

where ∇_s denotes here the covariant derivative on the surface. Besides these two equations, we need an additional equation expressing the fact that the rate of adsorbed molecules is balanced by a flux of molecules in solution exiting Ω:

$$c\,\boldsymbol{w}_s \cdot \boldsymbol{N} = \frac{r}{a_0} \quad \text{on } \Gamma, \tag{94}$$

where a_0 is required to convert r into rate of change of molar areal concentration. Equations (92–94) are the process equations relating the state variables $X = \{c, \phi\}$ and the process variables $V = \{\boldsymbol{w}_s, \boldsymbol{w}_a, r\}$.

Since the surface does not have boundary and the fluid is quiescent, the rate of change of the free energy takes the form

$$\dot{\mathcal{F}}(X; V) = \frac{1}{a_0}\int_\Gamma \mu_a \partial_t \phi \, dS + \int_\Omega \mu_s \partial_t c \, dV, \tag{95}$$

where the chemical potentials for adsorbate and solute resulting from this calculation are

$$\begin{aligned} \mu_a(\phi) &= \mu_{0,a} + RT \log \phi, \\ \mu_s(c) &= \mu_{0,s} + RT \log \frac{c}{c_0}. \end{aligned} \tag{96}$$

Using the process equations and the divergence theorem, we obtain

$$\begin{aligned} \dot{\mathcal{F}}(X; V) = &\frac{RT}{a_0}\int_\Gamma \nabla_s \phi \cdot \boldsymbol{w}_a \, dS + RT \int_\Omega \nabla c \cdot \boldsymbol{w}_s \, dV \\ &+ \frac{1}{a_0}\int_\Gamma r(\mu_a - \mu_s)\, dS \end{aligned} \tag{97}$$

We note from this expression that the chemical potential of the solute plays a role only at the interface, where it undergoes a reaction. Making the free energy stationary, we conclude that in equilibrium $\mu_a = \mu_s$, and therefore, the equilibrium constant of the reaction is

$$\frac{c^{eq}}{\phi^{eq} c_0} = \exp \frac{\Delta \mu_0}{RT}, \tag{98}$$

where $\Delta\mu_0 = \mu_{0,a} - \mu_{0,s}$.

Similar to the previous section, the dissipation potential accounting for diffusion of solute, of adsorbate and the sorption reaction is given by

$$\mathcal{D}(X; V) = \frac{f_s N_A}{2}\int_\Omega c|\boldsymbol{w}_s|^2 \, dV + \frac{f_a N_A}{2a_0}\int_\Gamma \phi |\boldsymbol{w}_a|^2 \, dS + \frac{1}{2a_0}\int_\Gamma \frac{1}{\bar{k}} r^2 \, dS, \tag{99}$$

where f_a, the drag coefficient of an adsorbed molecule on the membrane, will depend strongly on the membrane interfacial viscosity and weakly on the molecule size according to the theory by Saffman and Delbruck (1975). We consider now a concentration-dependent kinetic coefficient with the same structure as in the previous section and taking the form

$$\bar{k} = k\frac{c/c_0 - e^{\Delta\mu_0/(RT)}\phi}{\mu_s(c) - \mu_a(\phi)} \tag{100}$$

To prove that, as thermodynamically required, $\bar{k} \geq 0$, we recall Eqs. (96) and (98). Then, a direct calculation shows that

$$\bar{k} = \frac{ke^{-\mu_{0,s}/(RT)}}{RT} \frac{e^{\mu_s/(RT)} - e^{\mu_a/(RT)}}{\mu_s/(RT) - \mu_a/(RT)}, \tag{101}$$

which is manifestly nonnegative because the exponential function is strictly increasing.

Forming the Rayleighian combining Eqs. (97) and (99), and making it stationary with respect to r, Onsager's principle leads to

$$r = \bar{k}(\mu_s - \mu_a). \tag{102}$$

Recalling our choice for kinetic coefficient in Eq. (100), we immediately conclude that

$$r = \underbrace{\frac{k}{c_0}}_{k_A} c - \underbrace{ke^{\Delta\mu_0/(RT)}}_{k_D} \phi, \tag{103}$$

which allows us to identify the adsorption and desorption rates k_A and k_D for a model obeying the law of mass action. As in the previous example, we recognize that their ratio is a thermodynamic quantity, while the purely kinetic information about the reaction is given by the rate constant k.

Stationarity of the Rayleighian with respect to the surface and bulk diffusive velocities leads to Fick's law in the bulk and the surface. Finally, replacing these relations in the process equations we obtain the diffusion–sorption equation on the surface

$$\partial_t \phi = \frac{k_B T}{f_a} \Delta_s \phi + k_A c - k_D \phi \quad \text{on } \Gamma, \tag{104}$$

where Δ_s is the surface Laplacian, the diffusion equation in the bulk

$$\partial_t c = \frac{k_B T}{f_s} \Delta c \quad \text{in } \Omega, \tag{105}$$

and a condition on the surface matching bulk flux and surface reaction

$$\frac{k_B T}{f_s} \frac{\partial c}{\partial N} = \frac{1}{a_0} (k_D \phi - k_A c) \quad \text{on } \Gamma. \tag{106}$$

These are the equations that we could have postulated a priori, but now we have a clear understanding of the free energy driving the system and of the dissipative mechanisms dragging the dynamics. This example shows that Onsager's principle naturally deals with interfacial phenomena coupled to bulk phenomena.

While this linear sorption–diffusion model is reasonable in a dilute limit, that is for small values of c and ϕ, an obvious conceptual drawback apparent from Eq. (98) is that ϕ^{eq} can be made arbitrarily large by either increasing c^{eq} at fixed $\Delta\mu_0$, or by considering a negative $\Delta\mu_0$ of increasing the magnitude at fixed c^{eq}. However, the area fraction of adsorbates cannot be larger than 1.

3.2 *Onsager's Principle for Langmuir Sorption–Diffusion*

The above limitations of the linear sorption model can be overcome with the classical Langmuir model (Masel 1996). In a nutshell, this model introduces the notion that, for a molecule in solution X_B to become adsorbed, X_S, it needs to react with a free site on the surface X_F, which is thus viewed as an additional reactant/product in the adsorption/desorption reaction

$$X_B + X_F \rightleftharpoons X_S. \tag{107}$$

In this way, as the area coverage of adsorbate increases, fewer free sites become available, which slows down the adsorption reaction and fixes the issue of unbounded area coverage in the linear sorption model. In a continuum model, if ϕ is the area fraction of adsorbates, then the free area fraction is $1-\phi$. Because in some systems the maximum area fraction of adsorbates ϕ_m saturates before reaching unity, we can slightly generalize the area fraction of free sites as $\phi_m - \phi$. Then, from the reaction above and the law of mass action, we can postulate the following form of the adsorption reaction rate

$$r = k_A c(\phi_m - \phi) - k_D \phi, \tag{108}$$

where k_A and k_D are adsorption/desorption rate coefficients. It is clear that in a dilute limit $\phi_m - \phi \approx \phi_m$ and we essentially recover the reaction rate in Eq. (103). In equilibrium, $r = 0$, which leads to the following expression for the equilibrium area fraction of adsorbates

$$\phi^{eq} = \frac{k_A \phi_m c^{eq}}{k_A c^{eq} + k_D}. \tag{109}$$

It is now clear that as the bulk concentration becomes larger, $c^{eq} \to +\infty$, the area fraction tends to the saturation value $\phi^{eq} \to \phi_m$ as expected. To couple this adsorption model with diffusion in the bulk and the surface, it seems reasonable to replace the reaction rate in Eq. (103) by that in Eq. (108), which results in simply replacing $k_A c$ by $k_A c(\phi_m - \phi)$ in Eqs. (104) and (106). The reaction–diffusion system that follows from this reasonable modeling approach is nonlinear in the reaction terms.

Our question now is whether Langmuir's sorption model can be derived from Onsager's principle, and if so, what is the appropriate notion of free energy and dissipation potential. If this is the case, Onsager's principle provides a natural way to couple the sorption reaction with diffusion. Then, a second question is whether the resulting sorption–diffusion system is indeed that discussed in the previous paragraph.

Let us focus on the mixing entropy on the surface. In the ideal gas model in Eq. (91), we accounted for the entropy of the adsorbates with a term of the form $\phi \log \phi$ (up to normalization factors). Now, since we view empty sites as a new reacting species, it makes sense to consider also their entropic contribution, which will be of the form $(\phi_m - \phi)\log(\phi_m - \phi)$. Since the empty sites are immaterial, it does not make sense to include an enthalpic term analogous to $\mu_{0,a}\phi$. This argument leads to following expression for the free energy of the system:

$$\begin{aligned}\mathcal{F}[c,\phi] = &\frac{RT}{a_0}\int_\Gamma \left[\phi\log\phi + (\phi_m - \phi)\log(\phi_m - \phi)\right] dS + \frac{1}{a_0}\int_\Gamma \mu_{0,a}\phi\, dS \\ &+RT\int_\Omega c\left(\log\frac{c}{c_0} - 1\right) dV + \int_\Omega \mu_{0,s} c\, dV.\end{aligned} \tag{110}$$

The first term is in fact the well-known Flory–Huggins expression for the entropy of mixing (Huggins 1941; Flory 1942), introduced originally in the context of polymer blends. In the Flory–Huggins theory, an additional enthalpic term is added to the free energy density to account for the interaction between the mixing species of the form $(RT/a_0)\chi\phi(\phi_m - \phi)$, where χ is a dimensionless parameter. In this context, it makes more sense to interpret this term as a self-interaction term of adsorbate molecules of the form $-(RT/a_0)\chi\phi^2$ plus a term proportional to ϕ that can be included in $\mu_{0,a}$. Such a free energy has been invoked to examine equilibrium in the context of adsorption of curving proteins on lipid membranes (Sorre et al. 2012; Singh et al. 2012).

Let us examine next the consequences of considering this free energy in the framework of Onsager's principle. The process equations of the previous section remain unchanged. Likewise, a direct calculation shows that the rate of change of the free energy adopts the form

$$\begin{aligned}\dot{\mathcal{F}}(X;V) = &\frac{1}{a_0}\int_\Gamma \nabla_s\mu_a \cdot (\phi \boldsymbol{w}_a)\, dS + \int_\Omega \nabla\mu_s \cdot (c\boldsymbol{w}_s)\, dV \\ &+ \frac{1}{a_0}\int_\Gamma r(\mu_a - \mu_s)\, dS,\end{aligned} \tag{111}$$

where now the chemical potentials of the adsorbate in the surface and of the solute in the bulk are

$$\mu_a(\phi) = \mu_{0,a} + RT \log \frac{\phi}{\phi_m - \phi},$$
$$\mu_s(c) = \mu_{0,s} + RT \log \frac{c}{c_0}. \tag{112}$$

From the first expression, we recognize that $\mu_{0,a}$ is the chemical potential of the adsorbate when the area fractions of adsorbate and free sites are equal. Noting that now

$$\nabla_s \mu_a = RT \frac{\phi_m}{\phi_m - \phi} \frac{\nabla_s \phi}{\phi} \tag{113}$$

we find that

$$\begin{aligned}\dot{\mathcal{F}}(X; V) = &\frac{RT}{a_0} \int_\Gamma \frac{\phi_m}{\phi_m - \phi} \nabla_s \phi \cdot \boldsymbol{w}_a \, dS + RT \int_\Omega \nabla c \cdot \boldsymbol{w}_s \, dV \\ &+ \frac{1}{a_0} \int_\Gamma r(\mu_a - \mu_s) \, dS\end{aligned} \tag{114}$$

In equilibrium, $\mu_a = \mu_s$ and therefore, similarly to earlier, we find that

$$\frac{\phi_m - \phi^{eq}}{\phi^{eq}} \frac{c^{eq}}{c_0} = e^{\Delta\mu_0/(RT)}, \tag{115}$$

where $\Delta\mu_0 = \mu_{0,a} - \mu_{0,s}$.

We adopt the same structure of dissipation potential as in Eq. (99) with the following natural choice for the kinetic coefficient

$$\bar{k} = k \frac{\frac{1}{c_0} c(\phi_m - \phi) - e^{\Delta\mu_0/(RT)} \phi}{\mu_s(c) - \mu_a(\phi)}, \tag{116}$$

which can be shown to be nonnegative with analogous arguments to those leading to Eq. (101). Invoking Onsager's principle and making the Rayleighian stationary with respect to r, we recover Langmuir's adsorption model and identify the thermodynamic/kinetic components behind the reaction rates

$$r = \underbrace{\frac{k}{c_0}}_{k_A} c(\phi_m - \phi) - \underbrace{k e^{\Delta\mu_0/(RT)}}_{k_D} \phi. \tag{117}$$

Therefore, this derivation establishes that the Langmuir's adsorption model can be viewed as a consequence of the Flory–Huggins form of the mixing entropy. Now, making the Rayleighian stationary with respect to $\boldsymbol{w}_a$, we find that the adsorbates undergo non-Fickian transport in that

$$\boldsymbol{j}_a = \phi \boldsymbol{w}_a = -\frac{k_B T}{f_a} \frac{\phi_m}{\phi_m - \phi} \nabla_s \phi. \tag{118}$$

The term multiplying $\nabla_s \phi$ can be interpreted as a diffusion coefficient dependent on area fraction. When plugged into the corresponding process operator, this relation leads to the nonlinear diffusion equation on the surface

$$\partial_t \phi = \frac{k_B T}{f_a} \nabla_s \cdot \left(\frac{\phi_m}{\phi_m - \phi} \nabla_s \phi \right) + k_A c (\phi_m - \phi) - k_D \phi \quad \text{on } \Gamma. \tag{119}$$

Thus, according to this derivation, the nonlinear Langmuir adsorption model would be paired with a nonlinear diffusion of adsorbates, both following from the Flory–Huggins entropy of mixing.

It is instructive to note that we can recover Fickian diffusion on the surface by defining the surface contribution to the dissipation potential as

$$\frac{f_a N_A \phi_m}{2a_0} \int_\Gamma \frac{\phi}{\phi_m - \phi} |\boldsymbol{w}_a|^2 \, dS. \tag{120}$$

Without a compelling physical interpretation, however, this remains nothing but a mathematical trick. As we argue next, removing the nonlinearity with such a trick is artificial and does not seem justified from a physical point of view.

Indeed, something unsettling about the nonlinear diffusion equation in (119) is that it has been obtained from a free energy that tries to account for the finite area coverage of the adsorbates to deal with inconsistencies in the dilute limit. However, the dissipation contribution due to adsorbates has the structure

$$\frac{N_a \phi}{a_0} \cdot \frac{f_a}{2} |\boldsymbol{w}_a|^2 \tag{121}$$

where the first factor represents the number of molecules per unit area and the second factor is the dissipation potential for a single molecule. Thus, it is the superposition of the effect of an isolated molecule, which can be expected to be valid only in a dilute limit. A more pertinent modeling approach would be to couple the Flory–Huggins entropy to a better approximation of dissipation in a crowded solution of adsorbates. It is natural to expect that accounting for crowding will introduce an additional source of nonlinearity in the dissipation, which will not be in general of the form of that in Eq. (120). For instance, in a bulk solution and accounting for first-order interaction effects, the hydrodynamic drag coefficient of spherical solutes of radius a can be approximated as

$$f_s(c) = 6\pi\eta a \left(1 + B v_0^{1/3} c^{1/3}\right), \tag{122}$$

where v_0 is the molar volume of solute molecules and B is a nondimensional positive constant (Happel and Brenner 2012). Such a concentration-dependent drag will introduce additional nonlinear effects and lead to non-Fickian diffusion. We are not aware of similar approximations capturing the influence of area coverage and applicable to molecules moving on a two-dimensional fluid, that is an expression for $f_a(\phi)$ extending the theory by Saffman and Delbruck (1975) to crowded membranes.

Thus, Onsager's approach vividly shows that a concentration-dependent diffusion coefficient (Ramadurai et al. 2009) can have its origin in both the free energy driving diffusion and in the dissipation dragging it. Furthermore, *Onsager's approach provides a framework to model systems at multiple scales*, that it connects effective coefficients such as diffusivity or reaction rates to microscopic thermodynamic and kinetic quantities, which can in principle be estimated with microscopic theories or experiments.

4 A Minimal Model for Curvature Sensing and Generation in a Membrane Tube

Having established the Onsager variational structure behind the Langmuir adsorption model, in this section, we study the coupling between adsorption and mechanics. Rather than considering a general model with concentration gradients and general shapes, we focus on a uniform tubular lipid membrane to highlight how this coupling allows us to understand curvature sensing and generation by membrane proteins.

We consider a membrane tube of length ℓ and radius ρ, subjected to a longitudinal force F. We will assume that the membrane is inextensible, and therefore, the product $\ell\rho$ is constant. We will consider two different ensembles, one in which the length (and therefore the radius) is fixed and the force can adjust, and another in which the force is fixed and the length and radius are allowed to adjust. In neither of these situations the membrane tension is constant. A constant membrane tension ensemble would require changes in membrane area and therefore lipid flows from a reservoir, which is at odds with a simple uniform dynamical model. Zhu et al. (2012); Prévost et al. (2015) have developed models that allow for exchange of lipids and proteins with a reservoir, albeit in equilibrium. We further assume that both the osmolarity and the hydraulic pressure are the same inside and outside of the tube. Therefore, there is no pressure difference across the membrane.

When proteins are dissolved in the bulk fluid, only those closest to the membrane will adsorb, thereby creating a gradient in concentration followed by diffusion of dissolved proteins toward the membrane. Thus, if we assume that the concentration of proteins in the bulk is uniform, we are assuming that the timescale associated with equilibration of bulk protein gradients by diffusion is much smaller than the timescale of the adsorption reaction. Let us check if this is reasonable. The typical diffusion coefficient for proteins in the bulk is in the range of $D \approx 1$–$10\,\mu\text{m}^2/\text{s}$, (Elowitz et al. 1999). Therefore, the timescale for radial diffusion can be estimated as $\tau_d = \rho^2/D$. The timescale associated with protein adsorption on membranes is typically of a few tens of minutes, $\tau_a \approx 600$ s (Sorre et al. 2012). Requiring that $\tau_d \ll \tau_a$ is then tantamount to requiring that $\rho \ll 25$–$80\,\mu\text{m}$, which is the case in most situations of interest where tubules have radii of tens of nanometers. Therefore, it is reasonable to assume a uniform concentration in the bulk.

The state of the system is hence characterized by the area fraction of proteins on the surface, the radius of the cylinder and the length of the cylinder, $X = \{\phi, \rho, \ell\}$. We will model the bending elasticity of the membrane using Helfrich's curvature model (Helfrich 1973; Lipowsky 1991). According to this classical model, the free energy density takes the form $\frac{\kappa}{2}(H - C_0)^2$ where κ is the bending rigidity, H is the mean curvature (here $H = 1/\rho$), and C_0 is the spontaneous curvature. Thus, it penalizes deviations of the mean curvature from the spontaneous curvature. As discussed earlier, membrane proteins can change the preferred curvature of the bilayer in various ways (McMahon and Gallop 2005; Zimmerberg and Kozlov 2006; Sens et al. 2008; Shibata et al. 2009; Antonny 2011). This effect is generally modeled by considering that C_0 is a function of the area coverage of proteins, the simplest model being that it is proportional to ϕ (see Breidenich et al. (2000) for a microscopic justification). Thus, the free energy of the system considering the elastic and chemical contributions and the potential of the longitudinal force is

$$\mathcal{F}(X) = 2\pi\ell\rho\left\{\frac{\kappa}{2}\left(\frac{1}{\rho} - C_0\phi\right)^2 + \frac{RT}{a_0}\left[\phi\log\phi + (\phi_m - \phi)\log(\phi_m - \phi)\right] + \frac{\mu_{0,a}\phi}{a_0}\right\} - F\ell. \tag{123}$$

Note that in this simple uniform example, $\mathcal{F}$ is an algebraic function of the state variables. We chose to impose the inextensibility constraint $\ell\rho$ =constant, or equivalently $\dot{\ell}\rho + \ell\dot{\rho} = 0$, later using a Lagrange multiplier, which will allow us to identify the membrane tension. The nontrivial process equation is the balance of adsorbed molecules analogous to Eq. (93). Because the system is uniform, this equation does not involve diffusive fluxes. However, because we are not imposing yet the inextensibility condition of the surface, this equation includes a second term involving the rate of change of area (Rahimi et al. 2013):

$$\dot{\phi} = r - \phi\frac{\dot{\ell}\rho + \ell\dot{\rho}}{\ell\rho}. \tag{124}$$

Following a lengthy but direct calculation that uses the process equation above, the rate of change of the free energy can be computed as

$$\begin{aligned}\dot{\mathcal{F}} = &\frac{2\pi\rho\ell}{a_0}\left\{\mu_{0,a} + RT\log\frac{\phi}{\phi_m - \phi} - a_0\kappa\left(\frac{1}{\rho} - C_0\phi\right)C_0\right\}r \\ &+ 2\pi\ell\left\{\frac{RT}{a_0}\phi_m\log(\phi_m - \phi) - \frac{\kappa}{2}\left(\frac{1}{\rho} - C_0\phi\right)^2\right\}\dot{\rho} \\ &+ 2\pi\rho\left\{\frac{RT}{a_0}\phi_m\log(\phi_m - \phi) + \frac{\kappa}{2}\left(\frac{1}{\rho^2} - C_0^2\phi^2\right)\right\}\dot{\ell} - F\dot{\ell}.\end{aligned} \tag{125}$$

During this calculation, we identify the chemical potential of adsorbates as

$$\mu_a = \underbrace{\mu_{0,a} + RT \log \frac{\phi}{\phi_m - \phi}}_{\mu_a^{chem}} \underbrace{- a_0 \kappa \left(\frac{1}{\rho} - C_0 \phi \right) C_0}_{\mu_a^{mech}}, \tag{126}$$

involving a chemical component, but also a mechanical component.

To model the effect of the fixed bulk concentration of proteins $\bar{c}$, which are thus at a fixed chemical potential $\bar{\mu}_s = \mu_{0,s} + RT \log(\bar{c}/c_0)$, we introduce the external chemical power

$$\mathcal{P} = -\frac{2\pi \ell \rho \bar{\mu}_s}{a_0} r, \tag{127}$$

where we recall that a_0 allows us to dimensionally reconcile a chemical potential per unit mole with a reaction rate measuring the rate of change of adsorbate area fraction. Finally, the only dissipative mechanism operative here is the sorption reaction, and therefore,

$$\mathcal{D} = \frac{\pi \ell \rho}{a_0 \bar{k}} r^2. \tag{128}$$

With these ingredients, we can form the constrained Rayleighian as

$$\mathcal{L} = \dot{\mathcal{F}} + \mathcal{D} + \mathcal{P} + 2\pi\sigma \left(\dot{\ell} \rho + \ell \dot{\rho} \right). \tag{129}$$

Let us focus first on adsorption. The optimality condition $\partial \mathcal{L} / \partial r = 0$ resulting from Onsager's principle leads to

$$r = \bar{k}(\bar{\mu}_s - \mu_a). \tag{130}$$

Now, we are confronted with the modeling choice of defining the coefficient $\bar{k}$ characterizing dissipation during reaction. A direct analogy with the previous section would suggest

$$\bar{k} = k \frac{\frac{\bar{c}}{c_0}(\phi_m - \phi) - e^{\Delta\mu_0/(RT)} \phi}{\bar{\mu}_s - \mu_a}, \tag{131}$$

where as before $\Delta\mu_0 = \mu_{0,a} - \mu_{0,s}$. Combining the two equations above, a direct calculation shows that this leads to an adsorption rate of the Langmuir form $r = k_A \bar{c}(\phi_m - \phi) - k_D \phi$. However, the choice in Eq. (131) has at least two important drawbacks. First, such a model does not capture the phenomenology by which curving proteins bind at a higher rate to curved membranes–*curvature sensing*. Second, because μ_a depends also on the curvature of the membrane, see Eq. (126), it is not possible to express $\bar{k}$ as defined in Eq. (131) as a positive coefficient times a term of the form $(e^a - e^b)/(a - b)$, and therefore, in general, we will not be able to guarantee that $\bar{k} \geq 0$ as thermodynamically required. Therefore, a standard Langmuir adsorption model insensitive to curvature $r = k_A \bar{c}(\phi_m - \phi) - k_D \phi$ is thermodynamically inconsistent with our form of free energy in Eq. (123). Therefore, we discard Eq. (131).

One way around thermodynamic inconsistency is to define the reaction–dissipation coefficient as

$$\bar{k} = k\frac{\frac{\bar{c}}{c_0}(\phi_m - \phi) - e^{\Delta\mu_0/(RT)}\phi}{\bar{\mu}_s - \mu_a^{chem}} \tag{132}$$

$$= \frac{ke^{-\mu_{0,s}/(RT)}(\phi_m - \phi)}{RT} \frac{e^{\bar{\mu}_s/(RT)} - e^{\mu_a^{chem}/(RT)}}{\bar{\mu}_s/(RT) - \mu_a^{chem}/(RT)}, \tag{133}$$

which has the right structure to guarantee its nonnegativity. Importantly, with Eq. (132) we are postulating that dissipation due to the adsorption reaction does not depend on membrane curvature. This choice for $\bar{k}$ does lead to a reaction rate sensitive to curvature because, when combined with Eq. (130), we find

$$r = [k_A\bar{c}(\phi_m - \phi) - k_D\phi]\left[1 + \frac{a_0\kappa\left(\frac{1}{\rho} - C_0\phi\right)C_0}{\bar{\mu}_s - \mu_a^{chem}(\phi)}\right]$$

$$= [k_A\bar{c}(\phi_m - \phi) - k_D\phi]\left[1 + \frac{a_0\kappa\left(\frac{1}{\rho} - C_0\phi\right)C_0}{RT\log\frac{k_A\bar{c}(\phi_m - \phi)}{k_D\phi}}\right] \tag{134}$$

with k_A and k_D defined as in Eq. (117). We recover the Langmuir equation when $C = 0$. In the dilute limit, the second term between brackets becomes 1, $\phi_m - \phi \approx \phi_m$, and we recover the linear adsorption model.

Therefore, we conclude that the model emanating from Eq. (132) *is thermodynamically acceptable and physically meaningful.* Of course, there are other possible choices. One could be replacing the two instances of μ_a^{chem} in Eq. (133) by μ_a, which leads to positive dissipation, introduces an explicit curvature dependence in $\bar{k}$, but strangely, this dependence has a structure dictated by thermodynamics. Therefore, it would be a rather artificial choice. A more realistic model accounting for the curvature sensitivity in the dissipation could consider k to be a function of curvature. To our knowledge, none of these issues have been examined before. Again, we see how Onsager's principle provides a systematic framework to think about modeling.

Variations of the constrained Rayleighian with respect to $\dot{\rho}$ and $\dot{\ell}$ lead to the radial and longitudinal equilibrium equations

$$0 = \frac{RT}{a_0}\phi_m \log(\phi_m - \phi) - \frac{\kappa}{2}\left(\frac{1}{\rho} - C_0\phi\right)^2 + \sigma, \tag{135}$$

$$0 = \frac{RT}{a_0}\phi_m \log(\phi_m - \phi) + \frac{\kappa}{2}\left(\frac{1}{\rho^2} - C_0^2\phi^2\right) - \frac{F}{2\pi\rho} + \sigma. \tag{136}$$

Not being dragged by any dissipative mechanism, these equations are instantaneously satisfied by the system at any value of ϕ by adjusting σ and F in the first ensemble, or σ, ℓ and ρ in the second ensemble. Subtracting these two equations, we can relate the applied force to the shape and area fraction of protein as

$$F = 2\pi\kappa\left(\frac{1}{\rho} - C_0\phi\right). \tag{137}$$

4.1 Protein Sorption at Fixed Shape of the Tube

Suppose that the length of the tube is held fixed. As a result of the inextensibility of the membrane, the radius is also fixed, However, the force required to maintain the fixed length will depend on protein area coverage. Recalling the expression for the sorption rate, we have

$$\dot{\phi} = [k_A\bar{c}(\phi_m - \phi) - k_D\phi]\left[1 + \frac{a_0\kappa\left(\frac{1}{\rho} - C_0\phi\right)C_0}{RT\log\frac{k_A\bar{c}(\phi_m-\phi)}{k_D\phi}}\right]. \tag{138}$$

We consider the parameters reported in Sorre et al. (2012) for amphiphysins interacting with a model lipid membrane: $C_0 = 10\ \text{nm}^{-1}$, $\bar{c} = 100$ nMolar, $k_D/k_A = 35$ nMolar. This is a first-order nonlinear differential equation, which can be solved numerically.

Figure 4 (top) shows the results for $\phi(t)$ assuming no initial area coverage of protein, $\phi(0) = 0$ for tubes of varying curvature between 0.001 C_0 and 1.5 C_0. The figure shows that the adsorption dynamics strongly depends on curvature. Both the initial adsorption rate and the saturation area fraction in equilibrium increase with increasing curvature. This curvature-dependent adsorption of proteins is generally referred to as *curvature sensing*, and generally quantified by ratios of the equilibrium area coverage for various curvatures. Curvature sensing may provide a mechanism for the chemical organization of membrane organelles.

Even though shape is fixed in this section, the adsorbing proteins have a mechanical manifestation in the force required to maintain the tubule shape, see Fig. 4 (bottom). Force is computed using Eq. (137). The decrease of required force needed to maintain a tube as adsorption proceeds is an expected consequence of the scaffolding effect of the curving proteins and is consistent with experiments in tethers pulled out of vesicles. Along with this, our simple model provides an explanation for observations of no retraction of membrane tubules covered with proteins upon removal of the applied force (Sorre et al. 2012). Beyond equilibrium, the model can also predict the time and area fraction of protein at which we can observe no retraction upon force removal or even compressive buckling.

Figure 5 represents the adsorption isotherms for tubes of various curvatures. These curves represent the equilibrium area coverage–the horizontal asymptotes

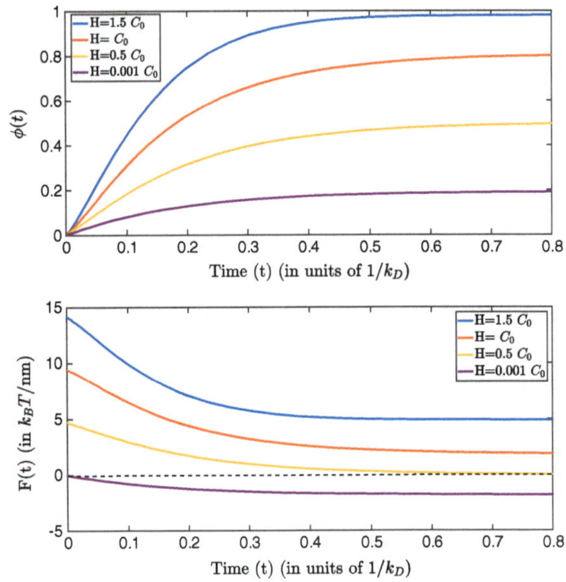

Fig. 4 (*Top*) Area fraction of adsorbed proteins as a function of time for various fixed curvatures of the membrane. We observe a curvature-dependent adsorption response: both the initial rate of change of ϕ (slope of the curve at $t \sim 0$) and the equilibrium area coverage increase with increasing curvature of tube $(1/\rho)$. (*Bottom*) Tube longitudinal force F as a function of time. The force decreases with time due to increased area fraction of the protein. The unit of force is in $k_BT/\text{nm} \approx 4.1$ pN

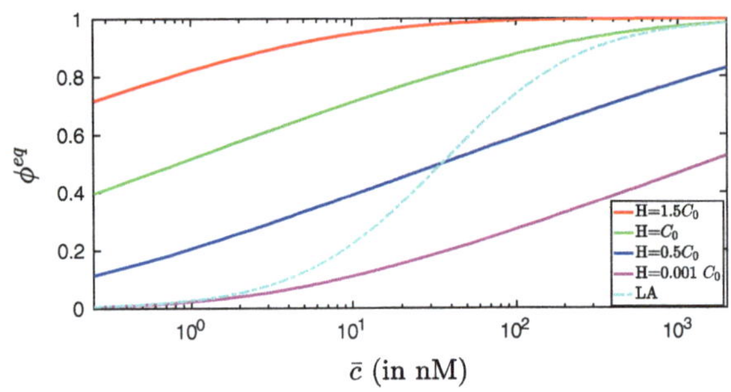

Fig. 5 Adsorption isotherms in tubules of different curvature, i.e., equilibrium values ϕ^{eq} as a function of concentration of proteins in the bulk in a semi-log scale. The legend 'LA' stands for the Langmuir adsorption model

in Fig. 4 (top)–as a function of the bulk concentration. In equilibrium, $\dot{\phi} = 0$, and therefore, the right-hand side in Eq. (138) is equal to zero. In the Langmuir model, recovered when $C_0 = 0$, the first factor is zero and the isotherm is given by $\phi^{eq}(\bar{c}) = \phi_m/[1 + (k_D/k_A\bar{c})]$. If the adsorbed molecules are mechanically active, however, the isotherm $\phi^{eq}(\bar{c})$ is implicitly defined by the relation

$$0 = RT \log \frac{k_A\bar{c}(\phi_m - \phi^{eq})}{k_D\phi^{eq}} + a_0\kappa \left(\frac{1}{\rho} - C_0\phi^{eq}\right) C_0. \tag{139}$$

The figure shows how, for proteins with preferred curvature ($C_0 \neq 0$), the isotherms significantly deviate from the Langmuir model. Notably, when the membrane is nearly planar (a similar model can be derived for spherical shapes more pertinent to tense vesicles, yielding similar results), the behavior closely follows the Langmuir model at very small concentrations, but strongly deviates from it at large concentrations. Yet, Langmuir's model is still used to interpret adsorption experiments onto large vesicles of proteins with preferred curvature (Sorre et al. 2012). The present model suggests a methodology to interpret similar experiments, where in addition to the parameters k_D/k_A and ϕ_m present in the Langmuir isotherm, the additional parameters a_0, κ, and C_0 could be fit from equilibrium or dynamical observations at different bulk concentration and curvature.

4.2 Protein Sorption at Fixed Force

While the fixed shape ensemble of the previous section allowed us to examine the curvature sensing capability of proteins with preferred curvature, it did not provide insight about their ability to generate curvature. Toward an elementary model for shape generation, we consider now a membrane tube at fixed force, which can adapt its radius and length to meet the mechanical equilibrium equations and the inextensibility constraint during adsorption.

Now, we integrate in time the differential-algebraic system for ϕ and ρ given by Eqs. (137) and (138). Equation (135) allows us to compute the tension σ. By plugging Eq. (137) into (138), we find that

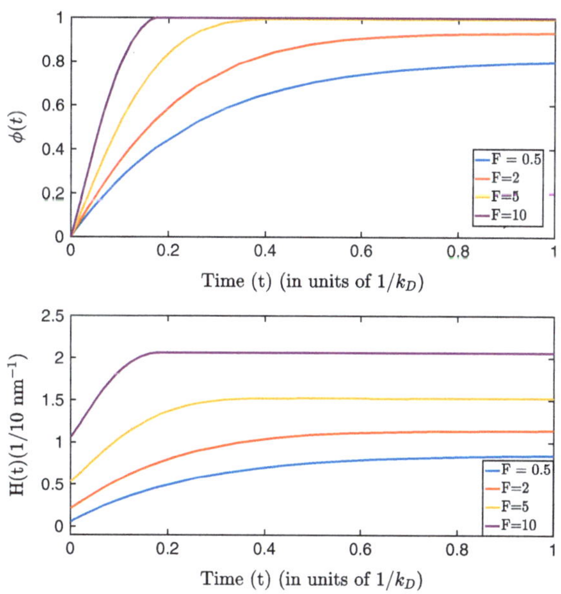

Fig. 6 Area fraction of proteins on the membrane (*top*) and mean curvature $1/\rho$ (*bottom*) as a function of time. The bulk concentration of proteins is $\bar{c} = 100$ nM and the unit of force is $k_B T/nm \approx 4.11$ pN

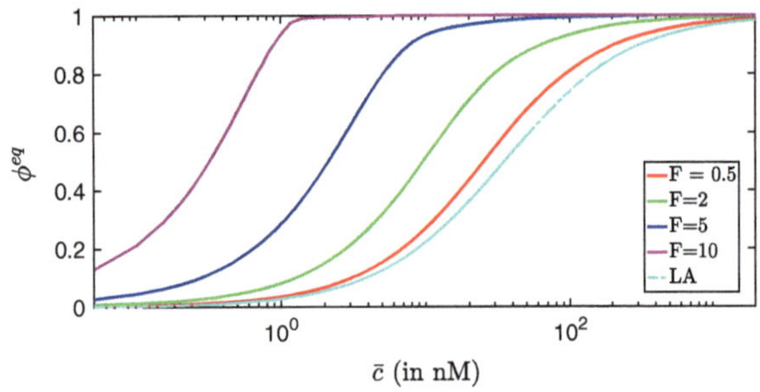

Fig. 7 Adsorption isotherms for fixed force adsorption of proteins. The isotherms deviate from the Langmuir isotherm with increase in area fraction as a function of concentration. The unit of force is $k_B T/nm \approx 4.11$ pN

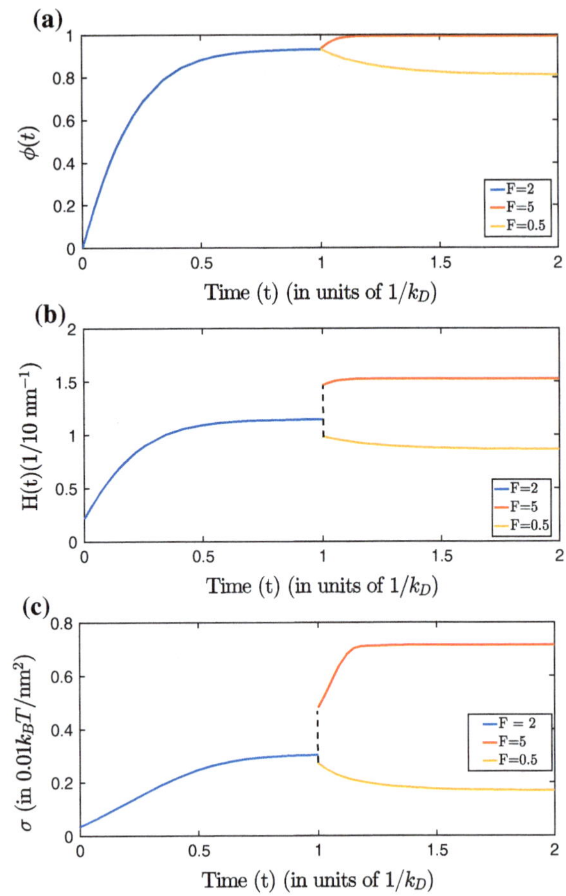

Fig. 8 Dynamics of a tubule during adsorption in response to a sudden increase (decrease) of the applied force. **a** Area fraction, **b** mean curvature and **c** surface tension as a function time. The bulk concentration of proteins is $\bar{c} = 100$ nM and the unit of force is $k_B T/nm \approx 4.11$ pN

$$\dot{\phi} = [k_A\bar{c}(\phi_m - \phi) - k_D\phi]\left[1 + \frac{a_0 C_0 F/(2\pi)}{RT \log \frac{k_A\bar{c}(\phi_m-\phi)}{k_D\phi}}\right], \tag{140}$$

and therefore the Langmuir model is recovered for $F = 0$.

Figure 6 shows the dynamics of area coverage and of tube curvature. We observe that both quantities increase in time up to an equilibrium value. Thus, as molecules adsorb, they modify the shape of the membrane. The rate of increase and magnitude of area coverage and curvature increase with the magnitude of the applied load, consistent with experiments. The adsorption isotherms are shown in Fig. 7. As predicted theoretically, the low force limit approaches the Langmuir model. The figure also shows a very large sensitivity of the equilibrium area coverage to the applied force, particularly at intermediate bulk concentrations.

To highlight the chemo-mechanical coupling captured by our simple model, we consider that during adsorption, and close to the equilibrium plateau, the force F is suddenly increased (decreased), see Fig. 8. It can be observed how, to adapt to such a disturbance, the protein coverage, tube curvature and tension increase (decrease). Interestingly, the mechanical quantities F and σ adjust discontinuously but the area fraction adjusts continuously, since the rate of adsorption/desorption is penalized by the dissipation potential.

References

S. Aimon, A. Callan-Jones, A. Berthaud, M. Pinot, G.E.S. Toombes, P. Bassereau, Membrane shape modulates transmembrane protein distribution. Dev. Cell **28**(2), 212–218 (2014)

B. Antonny, Mechanisms of membrane curvature sensing. Annu. Rev. Biochem. **80**, 101–123 (2011)

M. Arroyo, A. DeSimone, Relaxation dynamics of fluid membranes. Phy. Rev. E **79**, 031915 (2009)

M. Arroyo, A. DeSimone, L. Heltai, The role of membrane viscosity in the dynamics of fluid membranes. arXiv **2007**, 1–21 (2010)

M. Arroyo, L. Heltai, D. Millán, A. DeSimone, Reverse engineering the euglenoid movement. Proc. Natl. Acad. Sci. U. S. A. **44**, 17874–17879 (2012)

J.W. Barrett, H. Garcke, R. Nürnberg, *A Stable Numerical Method for the Dynamics of Fluidic Membranes*, vol. 134 (Springer, Berlin, 2016)

M. Breidenich, R.R. Netz, R. Lipowsky, The shape of polymer-decorated membranes. **49**, 431–437 (2000)

A. Callan-Jones, M. Durand, J.-B. Fournier, Hydrodynamics of bilayer membranes with diffusing transmembrane proteins. Soft Matter **12**(6), 1791–1800 (2016)

R. Capovilla, J. Guven, Stresses in lipid membranes. J. Phy. A: Math. Gen. **35**(30), 6233–6247 (2002)

M. Doi, Onsager's variational principle in soft matter. J. Phys.: Condens. Matter **23**(28), 284118 (2011)

C.M. Elliott, B. Stinner, Computation of two-phase biomembranes with phase dependent material parameters using surface finite elements. Commun. Comput. Phys. **13**(2), 325–360 (2013)

M.B. Elowitz, M.G. Surette, P.E. Wolf, J.B. Stock, S. Leibler, Protein mobility in the cytoplasm of Escherichia coli. J. Bacteriol. **181**(1), 197–203 (1999)

A. Embar, J. Dolbow, E. Fried, Microdomain evolution on giant unilamellar vesicles. Biomech. Model. Mechanobiol. **12**(3), 597–615 (2013)

E. Evans, A. Yeung, Hidden dynamics in rapid changes of bilayer shape. Chem. Phy. Lipids **73**(1–2), 39–56 (1994)

F. Feng, W.S. Klug, Finite element modeling of lipid bilayer membranes. J. Comput. Phys. **220**(1), 394–408 (2006)

P.J. Flory, Thermodynamics of high polymer solutions. J. Chem. Phys. **10**(1), 51–61 (1942)

J.-B. Fournier, N. Khalifat, N. Puff, M.I. Angelova, Chemically triggered ejection of membrane tubules controlled by intermonolayer friction. Phys. Rev. Lett. **102**(1), 018102 (2009)

J.B. Fournier, On the hydrodynamics of bilayer membranes. Int. J. Non-Linear Mech. **75**, 67–76 (2015)

H. Goldstein, *Classical Mechanics*, World student series (Addison-Wesley, Reading, 1980)

W.T. Góźdź, Shape transformation of lipid vesicles induced by diffusing macromolecules. J. Chem. Phys. **134**(2) (2011)

J. Happel, H. Brenner, *Low Reynolds Number Hydrodynamics: with Special Applications to Particulate Media*, vol. 1 (Springer Science & Business Media, 2012)

M. Heinrich, A. Tian, C. Esposito, T. Baumgart, Dynamic sorting of lipids and proteins in membrane tubes with a moving phase boundary. Proc. Natl. Acad. Sci. U. S. A. **107**(16), 7208–7213 (2010a)

M.C. Heinrich, B.R. Capraro, A. Tian, J.M. Isas, R. Langen, T. Baumgart, Quantifying membrane curvature generation of drosophila amphiphysin N-BAR domains. J. Phys. Chem. Lett. **1**(23), 3401–3406 (2010b)

W. Helfrich, Elastic properties of lipid bilayers: theory and possible experiments. Z. Naturforsch. **28c**, 693–703 (1973)

M.L. Huggins, Solutions of long chain compounds. J. Chem. Phys. **9**(5), 440–440 (1941)

R. Jordan, D. Kinderlehrer, F. Otto, The variational formulation of the Fokker-Planck equation. SIAM J. Math. Anal. **29**(1), 1–17 (1998)

F. Jülicher, R. Lipowsky, Domain-induced budding of vesicles. Phys. Rev. Lett. **70**(19), 2964–2967 (1993)

N. Khalifat, N. Puff, S. Bonneau, J.-B. Fournier, M.I. Angelova, Membrane deformation under local pH gradient: mimicking mitochondrial cristae dynamics. Biophys. J. **95**(10), 4924–4933 (2008)

N. Khalifat, M. Rahimi, A.-F. Bitbol, M. Seigneuret, J.-B. Fournier, N. Puff, M. Arroyo, M.I. Angelova, Interplay of packing and flip-flop in local bilayer deformation. How phosphatidylglycerol could rescue mitochondrial function in a cardiolipin-deficient yeast mutant. Biophys. J. **107**(4), 879–890 (2014)

A.J. Kosmalska, L. Casares, A. Elosegui-Artola, J.J. Thottacherry, R. Moreno-Vicente, V. González-Tarragó, M.Á. del Pozo, S. Mayor, M. Arroyo, D. Navajas, X. Trepat, N.C. Gauthier, P. Roca-Cusachs, Physical principles of membrane remodelling during cell mechanoadaptation. Nat. Commun. **6**, 7292 (2015)

L.D. Landau, E.M. Lifshitz, *Fluid Mechanics: Landau and Lifshitz: Course of Theoretical Physics*, vol. 6 (Elsevier, Amsterdam, 2013)

A. Lew, J.E. Marsden, M. Ortiz, M. West, Variational time integrators. Int. J. Numer. Meth. Eng. **60**(1), 153–212 (2004)

R. Lipowsky, The conformation of membranes. Nature **349**(6309), 475–481 (1991)

R. Lipowsky, Spontaneous tubulation of membranes and vesicles reveals membrane tension generated by spontaneous curvature. Faraday Discuss. **161**, 305–331 (2013)

J. Liu, Y. Sun, D.G. Drubin, G.F. Oster, The mechanochemistry of endocytosis. PLoS Biol. **7**(9), e1000204 (2009)

R.I. Masel, *Principles of Adsorption and Reaction on Solid Surfaces*, vol. 3 (Wiley, New York, 1996)

H.T. McMahon, J.L. Gallop, Membrane curvature and mechanisms of dynamic cell membrane remodelling. Nature **438**(7068), 590–596 (2005)

A. Mielke, A gradient structure for reactiondiffusion systems and for energy-drift-diffusion systems. Nonlinearity **24**(4), 1329–1346 (2011a). doi:10.1088/0951-7715/24/4/016. ISSN 0951-7715

A. Mielke, On thermodynamically consistent models and gradient structures for thermoplasticity. GAMM Mitt. **34**(1), 51–58 (2011b)

A. Mielke, Thermomechanical modeling of energy-reaction-diffusion systems, including bulk-interface interactions. Discrete Continuous Dyn. Syst. - Ser. S **6**(2), 479–499 (2012)

L. Onsager, Irreversible processes. Phys. Rev. **37**, 237–241 (1931a)

L. Onsager, Reciprocal relations in irreversible processes I. Phys. Rev. **37**(4), 405–426 (1931b)

M. Ortiz, E.A. Repetto, Nonconvex energy minimization and dislocation structures in ductile single crystals. J. Mech. Phys. Solids **47**(2), 397–462 (1999)

H.C. Öttinger, *Beyond Equilibrium Thermodynamics* (Wiley, New York, 2005)

F. Otto, The geometry of dissipative evolution equations: The porous medium equation. Commun. Partial Differ. Equ. **26**(1–2), 101–174 (2001)

W. Pauli, C.P. Enz, *Thermodynamics and the kinetic theory of gases*, vol. 3 (Courier Corporation, 2000)

C. Peco, A. Rosolen, M. Arroyo, An adaptive meshfree method for phase-field models of biomembranes. Part II: A Lagrangian approach for membranes in viscous fluids. J. Comput. Phys. **249**, 320–336 (2013)

M.A. Peletier, Variational modelling: energies, gradient flows, and large deviations, February 2014

C. Prévost, H. Zhao, J. Manzi, E. Lemichez, P. Lappalainen, A. Callan-Jones, P. Bassereau, IRSp53 senses negative membrane curvature and phase separates along membrane tubules. Nat. Commun. **6**, 8529 (2015)

I. Prigogine, *Introduction to Thermodynamics of Irreversible Processes* (Interscience Publishers, 1967)

M. Rahimi, M. Arroyo, Shape dynamics, lipid hydrodynamics, and the complex viscoelasticity of bilayer membranes. Phys. Rev. E **86**(1), 011932 (2012)

M. Rahimi, A. DeSimone, M. Arroyo, Curved fluid membranes behave laterally as effective viscoelastic media. Soft Matter **9**(46), 11033 (2013)

S. Ramadurai, A. Holt, V. Krasnikov, G. van den Bogaart, J.A. Killian, B. Poolman, Lateral diffusion of membrane proteins. J. Am. Chem. Soc. **13135**, 12650–12656 (2009)

P. Rangamani, A. Agrawal, K.K. Mandadapu, G. Oster, D.J. Steigmann, Interaction between surface shape and intra-surface viscous flow on lipid membranes. Biomech. Model. Mechanobiol. **12**(4), 833–845 (2013)

R. Rangarajan, H. Gao, A finite element method to compute three-dimensional equilibrium configurations of fluid membranes: Optimal parameterization, variational formulation and applications. J. Comput. Phys. **297**, 266–294 (2015)

Rayleigh, Proc. Math. Soc. London **363**, 357 (1873)

D.S. Rodrigues, R.F. Ausas, F. Mut, G.C. Buscaglia, Numerical modeling of tether formation in viscous. **XXXII**, 19–22 (2013)

A. Rustom, R. Saffrich, I. Markovic, P. Walther, H.-H. Gerdes, Nanotubular highways for intercellular organelle transport. Sci. (New York, N.Y.) **303**(5660), 1007–1010 (2004)

P.G. Saffman, M. Delbruck, Brownian motion in biological membranes. Proc. Natl. Acad. Sci. U. S. A. **72**(8), 3111–3113 (1975)

R.A. Sauer, T.X. Duong, K. Mandadapu, D. Steigmann, A stabilized finite element formulation for liquid shells and its application to lipid bilayers. J. Comput. Phys. **330**, 1–19 (2017)

U. Seifert, S.A. Langer, Viscous modes of fluid bilayer membranes. Europhys. Lett. (EPL) **23**(1), 71–76 (1993)

U. Seifert, Configurations of fluid membranes and vesicles. Adv. Phys. (1997). July 2011

P. Sens, L. Johannes, P. Bassereau, Biophysical approaches to protein-induced membrane deformations in trafficking. Curr. Opin. Cell Biol. **20**(4), 476–482 (2008)

Z. Shi, T. Baumgart, Membrane tension and peripheral protein density mediate membrane shape transitions. Nat. Commun. **6**, 5974 (2015). May 2014

Y. Shibata, H. Junjie, M.M. Kozlov, T.A. Rapoport, Mechanisms shaping the membranes of cellular organelles. Annu. Rev. Cell Dev. Biol. **25**, 329–354 (2009)

P. Singh, P. Mahata, T. Baumgart, S.L. Das, Curvature sorting of proteins on a cylindrical lipid membrane tether connected to a reservoir. Phys. Rev. E - Stat. Nonlinear, Soft Matter Phys. **85**(5), 1–10 (2012)

B. Sinha, D. Köster, R. Ruez, P. Gonnord, M. Bastiani, D. Abankwa, R.V. Stan, G. Butler-Browne, B. Vedie, L. Johannes, N. Morone, R.G. Parton, G. Raposo, P. Sens, C. Lamaze, P. Nassoy, Cells respond to mechanical stress by rapid disassembly of caveolae. Cell **144**(3), 402–413 (2011)

B. Sorre, A. Callan-Jones, J.-B. Manneville, P. Nassoy, J.-F. Joanny, J. Prost, B. Goud, P. Bassereau, Curvature-driven lipid sorting needs proximity to a demixing point and is aided by proteins. Proc. Natl. Acad. Sci. U. S. A. **106**(14), 5622–5626 (2009)

B. Sorre, A. Callan-Jones, J. Manzi, B. Goud, J. Prost, P. Bassereau, A. Roux, Nature of curvature coupling of amphiphysin with membranes depends on its bound density. Proc. Natl. Acad. Sci. U. S. A. **109**(1), 173–178 (2012)

H. Sprong, P. van der Sluijs, G. van Meer, How proteins move lipids and lipids move proteins. Nat. Rev. Mol. Cell Biol. **2**(7), 504–513 (2001)

J.C. Stachowiak, E.M. Schmid, C.J. Ryan, H.S. Ann, D.Y. Sasaki, M.B. Sherman, P.L. Geissler, D.A. Fletcher, C.C. Hayden, Membrane bending by proteinprotein crowding. Nat. Cell Biol. **14**(9), 944–949 (2012)

M. Staykova, M. Arroyo, M. Rahimi, H.A. Stone, Confined bilayers passively regulate shape and stress. Phys. Rev. Lett. **110**, 028101 (2013)

D.J. Steigmann, On the relationship between the Cosserat and Kirchhoff-Love theories of elastic shells. Math. Mech. Solids **4**, 275–288 (1999)

M. Terasaki, T. Shemesh, N. Kasthuri, R.W. Klemm, R. Schalek, K.J. Hayworth, A.R. Hand, M. Yankova, G. Huber, J.W. Lichtman, T.A. Rapoport, M.M. Kozlov, Stacked endoplasmic reticulum sheets are connected by helicoidal membrane motifs. Cell **154**(2), 285–296 (2013)

Z.C. Tu, Z.C. Ou-Yang, A geometric theory on the elasticity of bio-membranes. J. Phys. A: Math. Gen. **37**(47), 11407–11429 (2004)

N. Yamaguchi, T. Mizutani, K. Kawabata, H. Haga, Leader cells regulate collective cell migration via Rac activation in the downstream signaling of integrin $\beta 1$ and PI3K. Sci. Rep. **5**, 7656 (2015)

C. Zhu, S.L. Das, T. Baumgart, Nonlinear sorting, curvature generation, and crowding of endophilin N-BAR on tubular membranes. Biophys. J. **102**(8), 1837–1845 (2012)

J. Zimmerberg, M.M. Kozlov, How proteins produce cellular membrane curvature. Nat. Rev. Mol. Cell Biol. **7**(1), 9–19 (2006)

MIX
Papier aus verantwortungsvollen Quellen
Paper from responsible sources
FSC® C105338
FSC
www.fsc.org

If you have any concerns about our products,
you can contact us on
ProductSafety@springernature.com

In case Publisher is established outside the EU,
the EU authorized representative is:
Springer Nature Customer Service Center GmbH
Europaplatz 3, 69115 Heidelberg, Germany

Printed by Libri Plureos GmbH
in Hamburg, Germany